Landscaping PATAGONIA

The David J. Weber Series in the New Borderlands History

Andrew R. Graybill and Benjamin H. Johnson, editors

EDITORIAL BOARD
Juliana Barr
Sarah Carter
Maurice Crandall
Kelly Lytle Hernández
Cynthia Radding
Samuel Truett

The study of borderlands—places where different peoples meet and no one polity reigns supreme—is undergoing a renaissance. The David J. Weber Series in the New Borderlands History publishes works from both established and emerging scholars that examine borderlands from the precontact era to the present. The series explores contested boundaries and the intercultural dynamics surrounding them and includes projects covering a wide range of time and space within North America and beyond, including both Atlantic and Pacific worlds.

Published with support provided by the William P. Clements Center for Southwest Studies at Southern Methodist University in Dallas, Texas.

A complete list of books published in the David J. Weber Series in the New Borderlands History is available at https://uncpress.org/series/david-j-weber-series-in-the-new-borderlands-history.

Landscaping PATAGONIA

SPATIAL HISTORY AND NATION-MAKING IN CHILE AND ARGENTINA

María de los Ángeles Picone

THE UNIVERSITY OF NORTH CAROLINA PRESS

Chapel Hill

This book was published with the assistance of the Clough Center for the Study of Constitutional Democracy and the Morrissey College of Arts and Sciences at Boston College.

© 2025 The University of North Carolina Press
All rights reserved

Designed by Jamison Cockerham
Set in Arno, Scala Sans, Archive Roundface Script, and Irby
by Jamie McKee, MacKey Composition

Portions of chapters 2 and 3 originally appeared as María de los Ángeles Picone, "Legitimizing and Resisting Spatial Violence in Southern Chile (1890s–1910s)," *Historia Crítica*, no. 82 (2021): 55–78.

Cover art: Map of the Pérez Rosales Pass, 1904, in *Chile y Arjentina*; inside back cover of *Guía del Veraneante*, 1942, courtesy of Memoria Chilena; Casa Pangue, 1913, courtesy of Tulane University Digital Library; photo of Primo Capraro, Fondo Exequiel Bustillo, box 3354, Departamento de Documentos Escritos, Archivo General de la Nación, Buenos Aires.

Library of Congress Cataloging-in-Publication Data
Names: Picone, María de los Ángeles, author.
Title: Landscaping Patagonia : spatial history and nation-making in Chile and Argentina / María de los Ángeles Picone.
Other titles: David J. Weber series in the new borderlands history.
Description: Chapel Hill : University of North Carolina Press, [2025] | Series: The David J. Weber series in the new borderlands history | Includes bibliographical references and index.
Identifiers: LCCN 2024041047 | ISBN 9781469686134 (cloth ; alk. paper) | ISBN 9781469686141 (paperback ; alk. paper) | ISBN 9781469686158 (epub) | ISBN 9781469686165 (pdf)
Subjects: LCSH: Settler colonialism—Patagonia (Argentina and Chile)—History. | National characteristics, Argentine. | National characteristics, Chilean. | Cultural landscapes—Political aspects—Patagonia (Argentina and Chile) | Patagonia (Argentina and Chile)—History, Local—20th century. | Patagonia (Argentina and Chile)—History, Local—19th century. | Patagonia (Argentina and Chile)—Geography—Social aspects.
Classification: LCC F2936 .P497 2025 | DDC 982/.704—dc23/eng/20241101
LC record available at https://lccn.loc.gov/2024041047

This book will be made open access within three years of publication thanks to Path to Open, a program developed in partnership between JSTOR, the American Council of Learned Societies (ACLS), the University of Michigan Press, and the University of North Carolina Press to bring about equitable access and impact for the entire scholarly community, including authors, researchers, libraries, and university presses around the world. Learn more at https://about.jstor.org/path-to-open/.

For the four loves of my life,

TOTU, VITU, MANU, and GINO

Contents

List of Illustrations ix

Acknowledgments xiii

Introduction Taming the Monster 1

Chapter 1 Science for the Nation 19

Chapter 2 Settling Patagonia 45

Chapter 3 The Materiality of Space 79

Chapter 4 Spatial Discourses for a Healthy Nation 111

Chapter 5 National Aesthetics in the Argentine Locality 141

Chapter 6 The Outdoor Destination 175

Conclusion The Opportunity of Spatial History 203

Notes 209

Bibliography 251

Index 291

Illustrations

FIGURES

0.1. Detail of Gutiérrez and Cock's map, 1562   *xxiv*

2.1. Map of land concessions in southern Chile, 1908   *58*

2.2. *Plano demostrativo del estado de la tierra pública en los Territorios Nacionales del Sud*, 1900   *74*

3.1. Old and new roads on Bernardino Barría's property, 1912   *87*

3.2. Settlements and railroads proposed by the Hydrologic Studies Commission for the environs of Lake Nahuel Huapi, 1914   *108*

4.1. Overlay of "Mapa geográfico-comercial," 1923   *126*

5.1. Partial view of the Bariloche Civic Center, ca. 1950   *142*

5.2. Map of the Pérez Rosales Pass, 1904   *146*

5.3. Carlos Wiederhold's store, La Alemana, ca. 1902   *148*

5.4. Main building of the Chile-Argentina Trading and Cattle-Breeding Company in Bariloche, ca. 1910   *149*

5.5. A visual account of the Argentine National Parks Bureau's impact on travel, 1938   *157*

5.6. Bariloche Civic Center, ca. 1955   *164*

5.7. View of Llao Llao Grand Hotel, 1942   *171*

6.1. Cover of *En Viaje*, February 1938   *180*

6.2. Cover of *En Viaje*, January 1944   *181*

6.3. Back cover of *Guía del Veraneante*, 1942   *182*

6.4. Inside back cover of *Guía del Veraneante*, 1942   *183*

6.5. Cover of *En Viaje*, November 1941   *186*

6.6. Cover of *En Viaje*, March 1942   *186*

6.7. Cover of *En Viaje*, January 1937   *187*

6.8. Aggregate frequency of photographic features in *En Viaje*'s centerfold by region, 1933–1945   *190*

6.9. Page from Club Andino Bariloche's excursion logbook, 1931   *193*

6.10. Annotated photograph of an excursion, 1931   *198*

6.11. Map of Pérez Rosales Pass, 1937   *200*

7.1. President Roosevelt in Casa Pangue, 1913   *204*

7.2. The author in Casa Pangue, 2017   *205*

MAPS

0.1. Chile and Argentina   *xviii*

0.2. Provinces in southern Chile and national territories and the province of Buenos Aires in Argentina, 1910   *xix*

0.3. Localities in southern Chile and northern Patagonia   *xx*

0.4. The northern Patagonian Andes, 1910   *xxi*

0.5. The environs of Lake Llanquihue, 1920s   *xxii*

0.6. The environs of Lake Nahuel Huapi, 1940s   *xxiii*

1.1. The Palena and Futaleufú River valleys   *35*

1.2. The Calén/Baker Sound   *39*

3.1. Simplified sketch of northwestern Neuquén Territory   97

3.2. Proposed division of the national territories in Patagonia, 1914   105

6.1. Mapping of aggregate frequency of photographic features in *En Viaje*'s centerfold, 1933–1945   189

TABLES

4.1. Statistics for analyzing the number of police officers in each department of Neuquén relative to its population and area, 1910s   125

6.1. Number of occurrences of each visual component on the available covers of *En Viaje*, 1933–1945   185

Acknowledgments

This book was born over a cup of hot cocoa on a snowy afternoon in Bariloche in 2010. I was trying to write a thesis proposal, but my then adviser, the late Pedro Navarro Floria, was trying to convince me to write an application for a PhD program. Good advisers see beyond the immediate into the horizon of possibilities, and Pedro was no exception. That day, when Pedro was helping me imagine the after-graduation world, I wrote down the following in my thesis notebook: *How does the relationship between people and environment shape a region with a shared past?* In this vague wording in mid-2010, I was very clumsily asking about how our ideas about geographical space underpin how we see the world. To Pedro I owe his conviction that my questions were interesting questions and that finding the answers is one of the coolest things we do as academics.

At Emory, this project gained shape and substance. I am deeply thankful to Jeff Lesser and Tom Rogers for their generosity, counsel, and never-ending patience, including after graduation. They nurtured my curiosity, taught me to write more clearly, and helped me find my voice as a scholar. They especially encouraged my creativity, and much of that mentorship certainly appears in these pages. Yanna Yannakakis's questions helped me think about broader theoretical implications that I had not foreseen. She asks the type of questions that will force you to rethink your approach. I thank her for being a role model as a woman in academia, a creative scholar, and a great mentor. Jeff, Tom, Yanna, thank you. My deepest gratitude goes to other Latin Americanists at Emory that shaped the questions of this book: Karen Stolley, Ana Teixeira, Hernán Feldman, Phil MacLeod, and Katherine Ostrom. Fellow graduate students of Latin America helped me think through questions about the environment, space, and the nation: Ariel Svarch, Ben Nobbs-Thiessen, Andrew Britt, Jonathan Coulis, Audrey Henderson, Alexandra Lemos Zagonel, Shari Wejsa, Marissa Nichols, Alexander Cors, Anthony Tipping, and Hannah Abrahamson. I thank Andrew Zondermann, Julia López Fuentes, Danielle Wiggins, and Jennifer Morgan for innumerable conversations around research, academia, and writing. I cherish Ashley Parcells, Anastasiia Strakhova, Emma Meyer, Ashleigh Dean, and Shatam Ray for helping me think in other scales. I especially thank Abigail Meert and Jennifer Schaefer for their constant, unapologetic encouragement, comments, and advice.

At Boston College, this book found the sharp questions of Stacie Kent, the spontaneous conversations on spatial history with Franziska Seraphim, and the numerous discussions on environmental history with Conevery Bolton Valencius in her backyard. Sylvia Sellers-García offered invaluable advice on writing, historiography, and structure, as well as constant support. Lynn Johnson offered vital conversations that grounded this work. I thank Ginny Reinburg and Robin Fleming for their advice over wine and cheese, and Eddie Bonilla and Mike Glass for providing a soccer group chat that works as a supportive space for new faculty. Yajun Mo offered so much knowledge on the writing process; I am forever grateful. Prasannan Parthasarathi helped me protect my time and locate funding for the publication of this book. In the context of the Bodies and Places group, early drafts found engaging questions from Martin Summers, Nicole Eaton, Laura Clerx, Alexander D'Alisera, Meghan McCoy, Robin Radner, and Eric Grube. This would not have been the same without the strenuous work of Bee Lehman and Erin Scheopner in locating published works. The life of a department rests on the amazing work of administrators. Kim DeMeo and Stacy Moulis were critical in setting up assistance with scanning, submitting expense reports, and making sure there was always coffee available. I hope readers find in these pages the interdisciplinary conversations that nurtured my writing. For providing such spaces, I thank the Fox Center for Humanistic Inquiry at Emory University, the Clough Center for the Study of Constitutional Democracy at Boston College, and the Weatherhead Center for International Affairs at Harvard University. We write in a community. I appreciate the communities created by Erin Goodman and Christopher Jones that allowed me to finish the manuscript. For their comments, I thank the anonymous reviewers. I also thank the Clements Center for Southwest Studies at Southern Methodist University for subsidizing publication for this book. I particularly thank Andrew Graybill and Benjamin Johnson, the series editors, for keeping me accountable and offering guidance. My gratitude also goes to Jim Keenan, convener of the

Tenure-Track Faculty Group at Boston College's Jesuit Institute, for creating a space for junior faculty to support one another and share experiences.

A number of scholars and archivists in Chile and Argentina have helped me enormously in locating files, sharing documents, providing advice, and giving support. The generosity of Giuletta Piantoni, Melina Piglia, Laura Méndez, Federico Silin, Maia Gattás, Eduardo Bessera, Ricardo "Chip" Vallmitjana, and Jorge Muñoz Sougarret transpired in this book in the form of document-sharing and advice. Two mentors, one in Buenos Aires and one in Bariloche, animated the intellectual questions of this book. My thanks go to Elena Piñeiro and Paula Núñez. In the Argentine National Archives, I am profoundly grateful to Elizabet Cipoletta and Graciela Swiderski, who helped locate documents I had no idea I needed. Even when a strike hit the city, the staff in the National Archives opted for being flexible and accommodated me in a short visit. This speaks to their call for service, even in hard times. In Carmen de Patagones, Alejandro Zangrá coordinated access to digitized copies of local newspapers at the Emma Nozzi Museum. Finally, I thank my friend Mariano Cuevillas, from the Servicio Histórico del Ejército in Buenos Aires, for his help in finding documents that had not been catalogued yet. In Chile, I thank Macarena Acuña and Carmen Duhart, from the Archive of the Ministry of Foreign Affairs in Santiago; research librarian Jimena Rosenkranz, from the National Library; and the staff at the Periodicals Collection in this library, the National Archives in Santiago, and the Regional Archive in Temuco. In the Archivo Histórico de Osorno, I received invaluable help from Raciel Gallardo, Gabriel Peralta Vidal, and Patricio López Cárdenas, my fellow researcher. In the Archivo Histórico de Puerto Montt, I am in debt to César Sánchez. We all suffer setbacks during research, including a pandemic. I had the fortune of enlisting the incredible work of research assistants. For their invaluable work in Argentina, I thank Brenda Froschauer, Maryluna Santos, José Jesús Lara Yagia, Naiara Gnes, and Nicholas Jones. In Boston, Asa Ackerly, Shruthi Sriram, John Kooken, James Pritchett, and Alexandra Lermond assisted with scans, mapping, and research.

Numerous colleagues read early versions of chapters over the years. I am indebted to Kyle Harvey for his sharp suggestions, generous advice, and constant support as we both finished our manuscripts. Kyle and I put together the New Borderlands History seminar, whose generous feedback resulted in the New Borderlands History Group. This group has commented on chapter drafts and provided a space to discuss literature in the context of border regions. For their comments I thank Javier Cikota, Christine Mathias, Geraldine Davies Lenoble, Sarah Foss, Sarah Sarzynsky, Jesse Zarley, Alberto

Harambour, Ximena Sevilla, and Hannah Greenwald. I thank Josh Savala, Ryan Edwards, Javier Puente, Sarah Hines, Mateo Carrillo, James Mestaz, Dave Glovsky, Corinna Zeltsmann, Emily Wakild, John Soluri, Jeffrey Erbig Jr., and Julia Sarreal for their comments, their suggestions, and their encouragement. My special gratitude goes to Mark Healey, Heather Roller, and Fred Freitas, who read the entire manuscript from beginning to end and spent a whole day in Boston offering generous, engaging feedback. The book is better because of you.

The community surrounding this book kept me grounded and sane in the revision phase. Trinidad Rico, my academic big sister, never saw a word of this book. However, her advice is imprinted in these pages. She is the person from another discipline that filled conversations on book proposals, on approaching presses, on seeing the big picture, and on remaining passionate about our work. ¡Gracias Trin! It is hard to be away from loved ones, but my friends in Argentina made it feel like they were right here, asking how the book was going. I thank all of them, especially Gala, Caro, Dani, Sol, Dolo, and Chof, and their families. I thank my group of Argentine friends in Boston for their unapologetic encouragement and many *juntadas*. A special thanks goes to Chechu, who has no idea what I'm doing but supports me no matter what. I'm thankful to my field hockey team in Boston for our weekly games and my rowing friends and coaches at Community Rowing Inc. for providing a space on and out of the water to figure out the next move. Rowing and writing are very much alike; one stroke after the other, much like words, covers great distances.

It takes a village. I cherish the constant, unconditional support of my family. The Picones and the Clericis offered asados, mates, homestays, and outings that kept me going during archival research. I especially thank my godparents, Rita and Juan Carlos, who celebrated every little milestone of this book and who can't wait for it to be published in Spanish. My thanks go to Patricia and Guillermo's support especially during fieldwork. I am grateful to my cousins and their families, Caro, Juan, Luli, Maru, Andy, and Santi, for their wonderments about writing, about doing research, and about why it takes so long. I thank my siblings and their spouses, Esteban, Agustina, Inés, and Diego, for constantly keeping me accountable. I thank my four niblings, who reminded me of the important things in life, like monster trucks, glittery unicorns, and tickle monsters. Finally, I thank my father, Alejandro, and my late mother, Mónica, who have taught me, above all, to carve my way in life staying true to myself. They were supportive of this project from day one, even if it meant moving 5,000 miles away. I also thank them for moving the family to my hometown, Bariloche, the most beautiful place on earth.

Landscaping
PATAGONIA

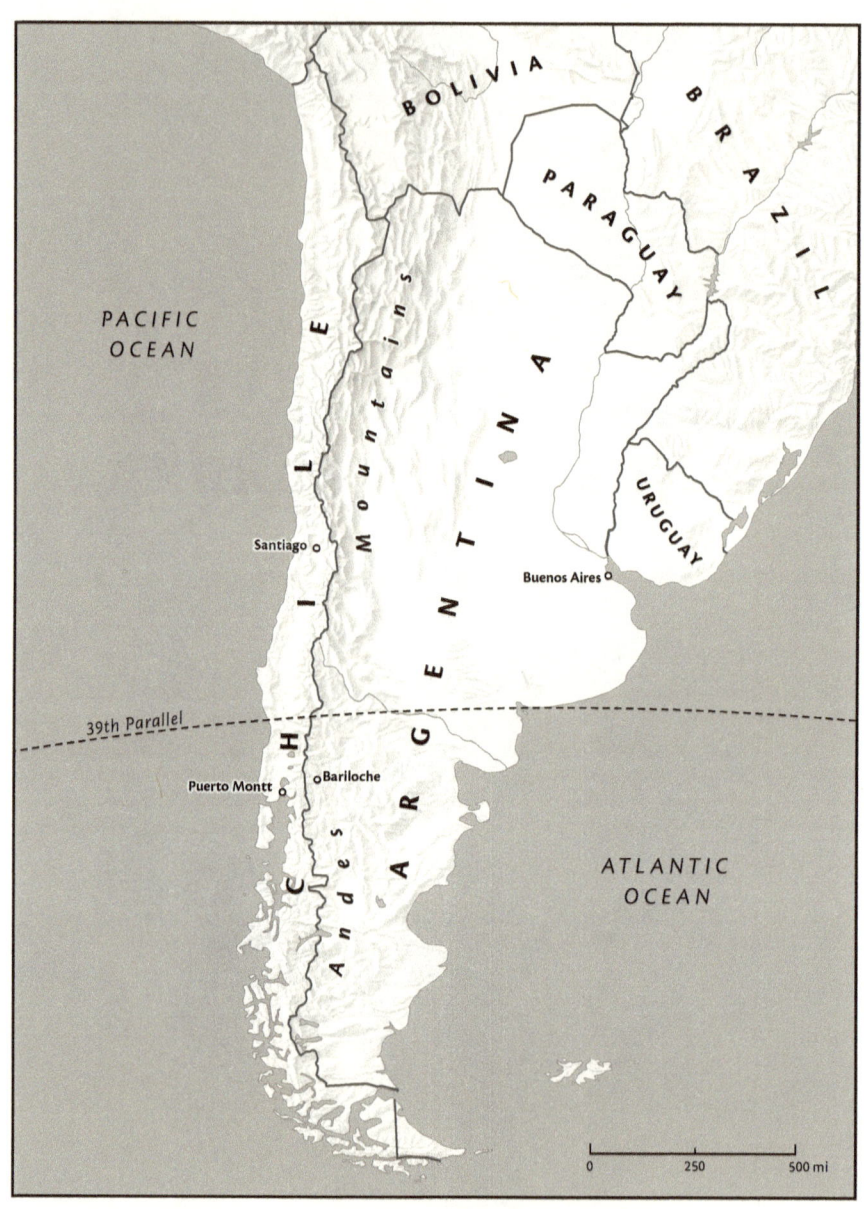

Map 0.1. Chile and Argentina.
Map by Erin Greb.

Map 0.2. Provinces in southern Chile and national territories and the province of Buenos Aires in Argentina, 1910. Map by Erin Greb.

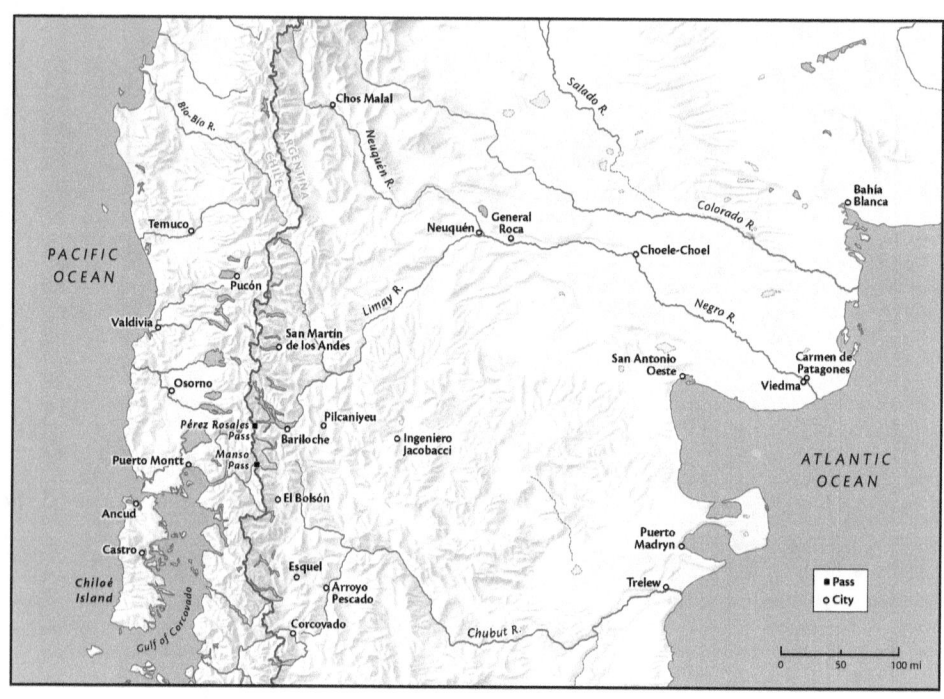

Map 0.3. Localities in southern Chile and northern Patagonia. Map by Erin Greb.

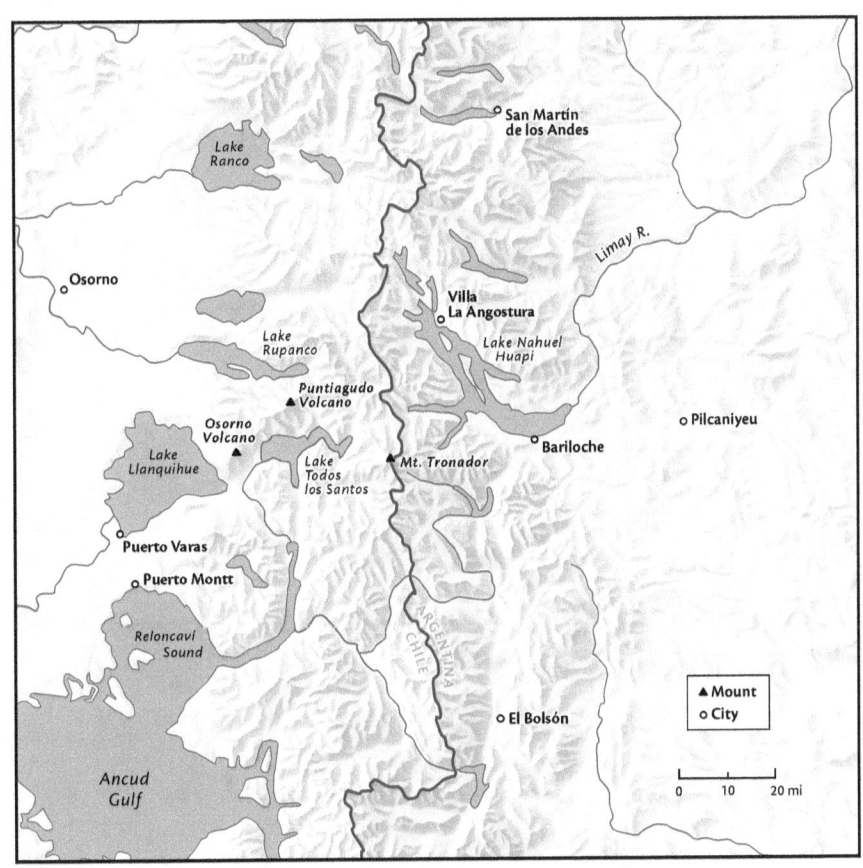

Map 0.4. The northern Patagonian Andes, 1910.
Map by Erin Greb.

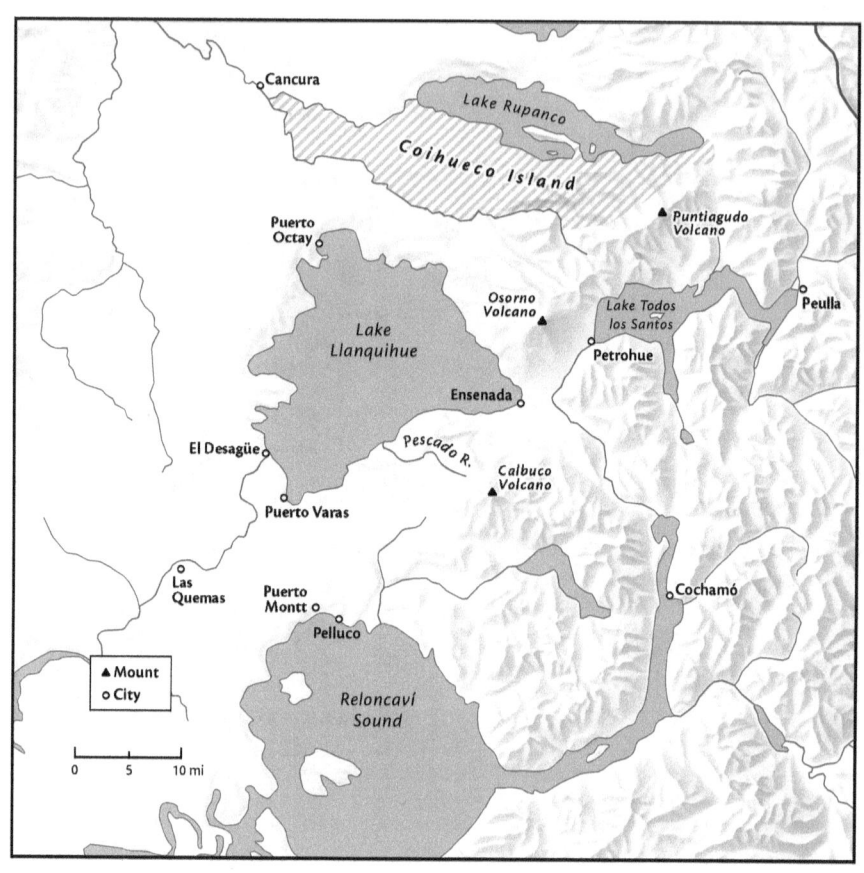

Map 0.5. The environs of Lake Llanquihue, 1920s.
Map by Erin Greb.

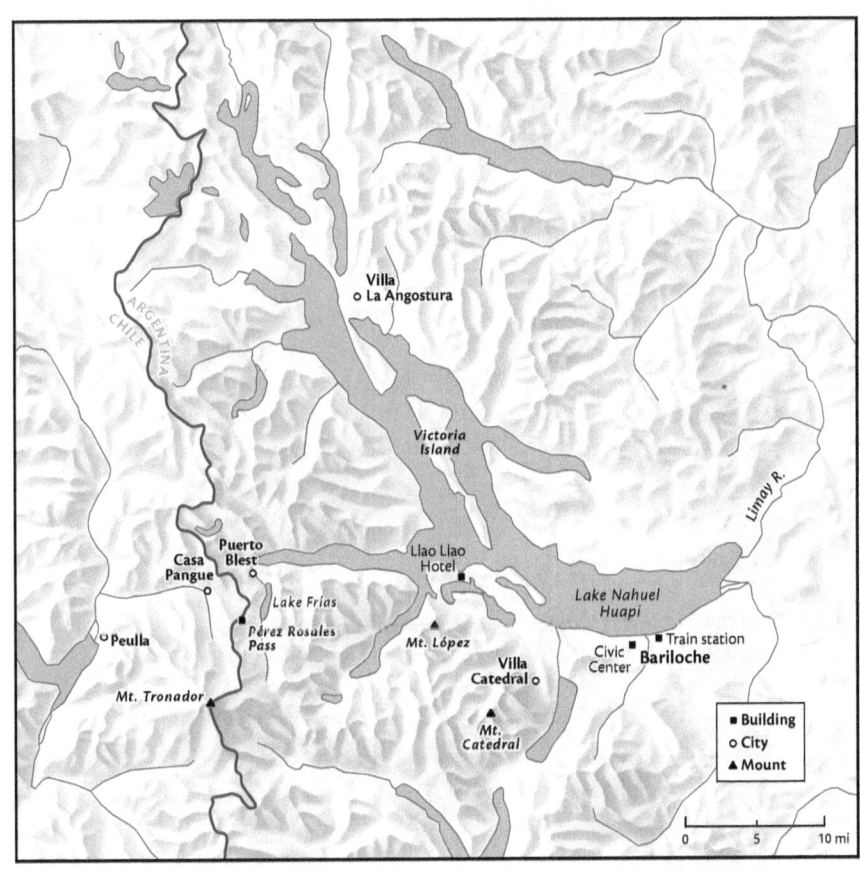

Map 0.6. The environs of Lake Nahuel Huapi, 1940s.
Map by Erin Greb.

Figure 0.1. Detail of Gutiérrez and Cock's map illustrating Patagonia as a "Kingdom of Giants," 1562. Courtesy of David Rumsey Map Collection.

Introduction

TAMING THE MONSTER

Since Ferdinand Magellan's expedition around the globe (1519–22), explorers have mapped Patagonia as a space inhabited, quite literally, by monsters. Spanish cartographer Diego Gutiérrez and Dutch artist Hieronymus Cock, for instance, produced a map of the Americas in 1562 that illustrated how monsters of all sorts tormented sailors across the Pacific and Atlantic Oceans. On dry land, the only space that depicted such images was Patagonia, where the mapmakers drew giants and labeled it "Kingdom of Giants" (fig. 0.1).[1] This was a European reference to the Indigenous Aónikenk, who wrapped their feet in animal skins to keep them warm, leaving behind footprints that Europeans misinterpreted as unusually large.[2] Challenging classifications and categorizations of natural science, monsters represented the unknown, difference, that which was beyond the boundaries of the known natural world. In geography, monsters evoked a violation of civil and moral law by escaping colonial power.[3] In the nineteenth century, once most Latin American countries had gained independence, authorities and explorers described those areas as Deserts.[4] Sometimes, this label coincided with areas of low humidity, soil aridity, and scarce vegetation, such as the Patagonian plateau or the Atacama Desert. Other times, however, Deserts evoked areas beyond state control but that were not necessarily arid.[5] The independence of Chile (1818) and Argentina (1816) did not erase colonial understandings of Patagonia as a space marked by otherness. On the contrary, imperial views of Patagonia persisted through the rest of the nineteenth century in the image of the Desert. For new Chilean and Argentine authorities, this region represented the absence of the rule of law and the emptiness of history. It was a non-place: a site with no relational identity, a monster.[6] How could the new states of Chile and Argentina make it their own?

Landscaping Patagonia tells the story of how people living in, governing, and traveling through northern Patagonia sought to construct versions of "the nation" based on their ideas about and experiences in geographical space. Chilean and Argentine authorities of the second half of the nineteenth and early twentieth centuries attempted to "eliminate," "occupy," "pacify," and "conquer" the Desert. To achieve this transformation, they deployed myriad nationalizing policies, from military campaigns to hotels, all in the name of a stronger nation. But the "nation" meant different things to the many people who came and went through the northern Patagonian Andes. Beyond the governing halls in Santiago and Buenos Aires, explorers, migrants, local authorities, bandits, and visitors made sense of the nation through their everyday lives. They surveyed passes, opened roads, claimed land titles or leases, traveled miles to the nearest police station, rode miles on horseback escaping the police, and participated in commemorative ceremonies. In various capacities and to different extents, they contributed to local imaginings of "Chile" and "Argentina," giving distinctive (and sometimes conflicting) meanings to geographical space. The consequences of these meanings differed depending on people's standing in society. For example, state authorities in southern Chile preferred for settlers to clear a path from their lot to the nearest public road. Paths and roads signaled settlement. However, when someone's road went through somebody else's lot and the lot owner complained to law enforcement, authorities tended to side with the person they knew best, often through family or business ties. These gradations of power, therefore, also shaped how people gave meaning to space and the consequences of these meanings. Geographical space was a fundamental aspect of people's ideas of the nation.

This book relies on a wide variety of primary documents from sixteen archives, coupled with a myriad of digital materials made available especially during the COVID-19 pandemic. I examined government documents, including published reports from different departments, congressional debates, daily bulletins, and census data. Unpublished materials include copybooks, where officials wrote on tracing paper to have a copy of dispatches they sent;

military reports; and correspondence. People made their concerns known by appealing to their authorities, either by sending letters or filing complaints; I found many of these at the Ministry of Foreign Affairs in Chile, where the Office of Colonization functioned, and the Ministry of the Interior in Argentina, where the Office of Lands and Colonies worked. Sometimes, we see in the case files the route a case took between offices, but sometimes it is hard to know whether a case was even acknowledged. In Puerto Montt, Chile, I found boxes of letters to the governor from all over Llanquihue with varied requests. In Osorno, Chile, municipal records shed light on local politics. In Bariloche, Argentina, a repository exists in excellent condition in the Patagonia Museum, whose collection includes personal papers of Francisco Moreno, the Argentine expert in the border negotiations with Chile, and Emilio Frey, a multifaceted local resident. All other remaining local materials are housed in the home of a local historian, Ricardo Vallmitjana, who has devoted his life to collecting historical documents and writing snippets of local histories. His work is invaluable. This book also relies on newspapers, magazines, and travel publications in addition to photographs, sketches, and a wide array of cartographic materials found in books, files, and map collections. Finally, it was invaluable to carry out fieldwork on-site, walking the streets, navigating the lakes, looking at the same buildings that many of the people in these pages did. Perhaps the pandemic has accentuated my longing for archival work, but it has been enlightening to travel across the northern Patagonian Andes envisioning how 100 years ago others understood that same space around them.

THE DESERT AND THE CORDILLERA

In the following pages, I examine the tension between the Desert, a space imagined without history, and the promise of the cordillera, which evoked a possible future. To paraphrase John Dixon Hunt, we give meanings to places based on our ideas and our experiences in other sites.[7] In late nineteenth-century Patagonia, earlier ideas about the region shaped how authorities, settlers, and travelers imagined it. Legend has it that in the 1520s, a Spanish ship wrecked in the Strait of Magellan and its survivors interned the Patagonian steppe, where they built a city of fantastic riches, the City of Caesars, on the shores of a great lake.[8] In southern Chile, the rumor of the mythical city spread among adventurers and permeated the ranks of the Jesuits stationed on the island of Chiloé, who launched expeditions to find it.[9] Imagining a city of riches in territories beyond colonial control allowed Spaniards to make sense of an unknown space. By piecing together reports and stories, they symbolically conquered that space, filled empty maps with references, and fueled expeditions to find the mythical sites. Specifically, Jesuit missionaries insisted on the existence of such rich site on the shores of a large lake, which they believed could be Lake Nahuel Huapi.[10] In the nineteenth century, the allure of the City of Caesars survived in the certainty that regions beyond state control, those Deserts, hid resources only waiting to be exploited—resources that would bring wealth to the nation. Even though the City of Caesars did not survive the scrutiny of nineteenth-century exploration, its symbolism lingered as an elusive future. If the greatness of the nation (may that be Chile or Argentina) depended on the productive use of natural resources, the absence of such enterprise, the Desert, reminded authorities that they were not there quite yet. Mythical placemaking, of course, was not unique to Patagonia, but it did permeate the role Patagonia played in nation-making in the late nineteenth century.[11]

Explorers and authorities used the image of the Desert to describe Patagonia as an "empty" region that needed to be "filled." The constitution of the Desert as "backward," "barbaric," and "uncultivated" symbolized across the Americas areas that escaped the reach of emergent nation-states.[12] Nineteenth-century Latin American cities represented "small oases of civilization" that were surrounded "by an untilled plain... rarely interrupted by any settlement of consequence."[13] The emptiness of the Desert symbolized the absence of the rule of law and capitalist production; it reminded new societies of the material boundaries of their perceived freedom.[14] For Chilean and Argentine governing elites of the nineteenth century, Indigenous presence attested to the continuous failure of nascent states to control national territories. Consequently, governments recognized an external frontier (with other nation-states) and an internal frontier, an area where state and others-beyond-the-state met. Beyond the frontier was the Desert, as space marked by otherness.[15] In this context, political and intellectual elites of the Americas viewed Indigenous peoples as constitutive of the Desert. Before the region's first contact with Europeans in the sixteenth century, several ethnic and cultural groups lived in Patagonia.[16] In the mid-nineteenth century, northern Patagonia, broadly construed, was mostly inhabited by Indigenous groups. The Mapuche herded cattle back and forth across the Andes looking for pastures.[17] Cattle ranching underpinned social differentiation, with some *lonkos* (leaders) increasing their wealth, their *conas* (foot soldiers), and their influence, especially across the Andes. There, the Aónikenk of northern Patagonia (or Tehuelche, as the Mapuche called them) appropriated the Mapuche's language and beliefs.[18] Academics and authorities of the late nineteenth century labeled this process the "araucanization of the Pampas" and used it to inscribe the Mapuche in "ethnic foreignness." In doing so, argues Diana Lenton, intellectual elites created a category of local Indigenous barbarism, which was redeemable, and of foreign groups, "invaders," demarcating the limits of the nation-state. The ramifications of such essentialist discourses reverberate across this book and into the present day.[19]

National authorities in Chile and Argentina called for the "suppression of the Desert."[20] This consisted first in portraying the Desert as empty and available, and then conquering it both physically and symbolically.[21] Imagining regions beyond state control as available land comprised the violent erasure of Indigenous presence. In Chile, authorities removed the Mapuche from their territories in two ways: through a series of violent attacks over two decades and through the simultaneous introduction of European immigrants to replace them. In 1843, Chileans built Fort Bulnes on the shores of the Strait of Magellan at the southernmost tip of the American continent, which created the illusion of continuous control over the Chilean territory, from roughly the twenty-fifth parallel southward. Within a decade, they initiated the colonization of Llanquihue, where they founded Puerto Montt on the Reloncaví Sound (1852) and Puerto Varas on Lake Llanquihue (1853). From that moment through the end of the century, the Chilean government actively recruited European immigrants to settle in the southern provinces of Valdivia and Llanquihue. Jeffrey Lesser reminds us that in Latin America, governing elites saw immigrants as agents for improving an imperfect nation.[22] Population growth in Europe outpaced work opportunities, pushing young couples and families out of their countries and across the Atlantic. In the Americas, governments offered European immigrants "empty" land at low cost and travel subsidies.[23] In the North, Chile defeated Bolivia and Peru in the War of the Pacific (1879–83), consolidating its control over the nitrate-rich regions. The Chilean territory, however, remained dislocated as Araucanía, the Mapuche heartland, prevented communication from north to south by land. Hence, the government initiated a series of military campaigns in 1861 that sought to subdue the Mapuche and make them conform to state laws and pushed them off their land. Military operations ended in 1883, followed by forced migration so that new settlers could gain access to Indigenous land.[24]

In Argentina, where political stability was only achieved in the 1860s, authorities attempted to first resolve the problem of the Desert with a military campaign to the Negro River in 1879–81. While in previous decades there had been some incursions into Indigenous territory, "private interests became a national project" only in 1879, resulting in the military operations that followed.[25] Additionally, demographic pressure from Chile and the need to define Argentine sovereignty propelled the government to lead a military raid against the Mapuche/Tehuelche of northern Patagonia in 1879. The government forced Natives to dissolve their tribal structures, accept Argentine culture, learn Spanish and abandon their languages, and serve in the army or navy.[26] Their land became property of the state.

The fertile valleys of the cordillera were a counterpoint to the barren Desert. *Landscaping Patagonia* shows that portrayals of the Patagonian Andes often oscillated between versions of the Desert, as a political program to annihilate unwanted inhabitants and behaviors, and the image of fertile valleys, as a trope for the potential greatness of the nation. Indeed, for explorers, travelers, residents, and authorities, the riches of the mountain valleys symbolized a promise of a prosperous future for both countries. This promise mobilized a myriad of nationalizing policies following the military campaigns in southern Chile and Argentina. Such an array of patriotic work, from surveying the Andes to building roads, police stations, and hotels, would "stop those deserts from being deserts." They would "open the road" for "civilization" to reach the Andes and replace the "lazy Indian by the hard-working man in the plains, the woodlands, and the mountains, [which are] all fertile and rich."[27] The solution for the problem of the Desert was the postcolonial colonization of Patagonia.[28] National authorities hoped that, once developed, "the opulent Andean region" would become a beacon of "civilized life and security to all the southern continent."[29] The oscillation between Patagonia as Desert and Patagonia as promised land drove much of the anxieties, conflicts, and experiences in the decades during and after the border negotiations. The prospect of the Desert activated policy, and the promise of fruitful valleys attracted migrants, local authorities, and other observers to the Andes.

THE NORTHERN PATAGONIAN ANDES

Googling the term "Patagonia," like many other regional labels, gives different results in Argentina, Chile, or the United States. This is because Patagonia can evoke different meanings for people in different places. For most people in the United States, it represents an outdoor clothing brand. In Chile, Patagonia roughly refers to the area south of Puerto Montt, covering the present-day districts of Magallanes, Aysén, and southern Los Lagos. In Argentina, Patagonia has often brought to mind the area south of the Colorado River. But meanings are not static. In the period covered in this book, "Patagonia" stood for an undefined region south of the thirty-ninth parallel in South America, approximately. For clarity, I use the term "Patagonia" when I mean the continental landmass and the channels south of the thirty-ninth parallel. At that latitude, the main chain of the Andes changes significantly. The Patagonian Andes are lower and more broken up than the chains farther north. They also turn westward and go into the Pacific Ocean. Such topographic variations allow for more water-dense clouds that form in the Pacific to cross to the east and precipitate in the wide Patagonian valleys. Andean lakes, the remnants of Ice Age glaciers, also begin appearing at approximately the thirty-ninth parallel, sprinkling the cordillera with several west–east bodies of water. While it might look like these lakes are located east of the Andes, several drain into the Pacific Ocean.

Landscaping Patagonia focuses on the people governing, living, and traveling in the northern Patagonian Andes. The northern Patagonian Andes consist of the cordillera and valleys from the thirty-ninth to forty-sixth parallels, which marks the northern tip of the Northern Ice Field. The Northern and Southern Ice Fields are the remnants of the Patagonian ice mass that covered the Andes and Chilean channels in the global Last Glacial Maximum (about 21,000 years ago). The ice fields prevent trans-Andean crossing. The lakes in the northern Patagonian Andes that appear in this book (Llanquihue, Rupanco, and Nahuel Huapi, to name a few) are glacial lakes. They formed when glaciers retreated and meltwater was trapped by terminal moraines in Andean valleys, where lush forests blossomed.[30] Culturally, the northern Patagonian Andes sometimes extended beyond the valleys into the steppe in the east and the Pacific ports in the west. "Northern Patagonian Andes" is not a term that appears in the sources, but it is a useful shorthand to refer to the Andes on both sides of the international border. When talking about the Chilean side, I use the administrative divisions as reference, mostly the province of Llanquihue in its largest extension, between the Bueno River and the forty-seventh parallel.[31] When I use "southern Chile," I mostly refer to the provinces of Valdivia and Llanquihue, where governing elites deployed similar policies, especially regarding land tenure. In Argentina, I use "northern Patagonia" to reference the southern part of the National Territory of Neuquén, the western third of the National Territory of Río Negro, and the northwestern part of Chubut. Again, this is only based on how the sources interacted with one another; it certainly does not offer a definitive contour of northern Patagonia. If anything, all these labels are equivocal.

The low altitude, the wide valleys, the weather patterns, and the river basins all influenced settlement, trans-Andean crossings, and border negotiations. Between Temuco and Puerto Montt, the Chilean territory is characterized by a wide fertile valley sandwiched between a rugged coastal range on the Pacific coast in the west and the Andes mountains in the east. At this latitude, the highest peaks range between 3,600 and 4,100 meters (between 11,000 and 13,451 feet), significantly lower than in any other section of the Chilean-Argentine cordillera. The warm, humid air from the South Pacific brings regular rain and temperate weather, which has allowed the dense Valdivian Forest, an evergreen ecoregion, to thrive roughly south of the thirty-ninth parallel and north of the forty-eighth. This area coincides with the northern Patagonian Andes. As clouds travel east and dry up, they precipitate less. On the Argentine side of the Andes, this translates into a rapid shift from temperate forest in the mountains to arid steppe within fifty miles. Multiple glacial lakes flanked by sedimental moraines decorate the Andean valleys, providing freshwater, fertile soils, and fish supply. Deep rivers flow from these lakes to the Atlantic Ocean through the Patagonian plateau, a series of bushy plains at different altitudes, consistently brushed by cold, dry winds.

LANDSCAPING PATAGONIA

Landscaping Patagonia asks how people viewed and altered the space around them in advancing ideas of the nation. To answer this question, I draw from the field of spatial history as an analytical framework to examine historical understandings and transformations of space. Rather than viewing geographical space as a white canvas on which events happen, spatial history considers geographical space part of history and contingent on it.[32] The different ways in which people instilled meanings and experienced the geography of northern Patagonia created "spatial discourses" about the nation, a notion I have borrowed from Paul Carter's analysis of James Cook's naming of Australian sites.[33] In what follows, I look at the creation of these spatial discourses through six prisms, one per chapter, in an attempt to bring together not only state, top-down ideas of the nation but also the ways in which midlevel officials, local businesspeople, laborers, and travelers, for example, adapted and contested those ideas. The lenses of creation of knowledge, settlement, movement, transgression, built environment, and the outdoors help disentangle the experiences of multiple actors, with varying degrees of political, economic, and social power. The spatial history of the northern Patagonian Andes destabilizes monolithic narratives of nation-making by unearthing kaleidoscopic, subjective ideas (or a "multiplicity of trajectories," as Doreen Massey puts it) of the nation that lie at the center of historical processes.[34]

Even though geographical space is a reality we all inhabit, the meanings it carried for historical actors should not be taken for granted.[35]

Recent scholarship of Latin America has examined nation-making through the lens of spatial history, delving especially into how political centers drew and made sense of the national territory.[36] Scholars have also analyzed specific practices that shaped the way different people made sense of the space around them. These practices included collecting data to build corpora of knowledge, surveying territories and interviewing Indigenous people to draw maps, and creating national parks to bolster a national project.[37] Scholars of Chile and Argentina made significant contributions to the history of nation-making in the late nineteenth and early twentieth centuries. *Landscaping Patagonia* unpacks the underlying assumption about the homogenizing force of nation-making into a set of changing, sometimes contradictory, ideas about "Chile" and "Argentina." Shifting the focus from centers of political power to a border region shows that the significations governing elites tried to impose on space were quite unstable. The focus on a border region allows me to examine national authorities' spatial discourses in addition to ways in which other actors viewed the space they inhabited, how this environment affected their lives, how they transformed it, how they received decisions made by others, and how, in sum, this underpinned, if at all, a local idea of Chile and of Argentina.

By repositioning the analytical focus from the nation-state to the transnational region, *Landscaping Patagonia* challenges the center-periphery paradigm that typifies scholarship of border regions, of Latin America, and of Patagonia. Different, even contradictory, spatial discourses in border spaces contributed to the same nationalizing goals spurred in the capital cities. The international boundary was one of many ways in which the northern Patagonian Andes were perceived and transformed, together with cross-border trade, transnational land tenure, and the criminalization of foreigners, to name a few. Separating nation-making from state formation provides an opportunity to unveil understudied spatial interpretations of the nation, which advances our understanding of the relationship between nature and culture. Hence, *Landscaping Patagonia* tells a story of nation-making independent from, but of course connected to, national politics.

At heart, this book shows we can write a history of border regions without depending on a history of the nation-state. I foreground alternative histories of nation-making in the border regions of Latin America that are often seen through the lens of state formation. Shifting the analytical focus away from the nation-state and toward people's cross-border, regional experiences reveals that the border is not necessarily constitutive of the region, as some scholars have argued.[38] Border regions unveil an intriguing paradox: they constitute an undefined area that generates a distinct way of life around a marker of difference, the borderline. Such geographical imagining of a national territory allowed people to differentiate between "here" and "there." It is through this geographical awareness, writes Susan Schulten, that people made sense of the world around them. Simultaneously, overriding forces, such as trade or kinship bonds, bring these areas together. Simply put, borderlands are places of encounter and differentiation.[39]

Landscaping Patagonia asks how people constructed ideas about the nation through their everyday experiences in a transnational region, the northern Patagonian Andes, where the mountains and lakes constitute epicenters of human settlement and movement.[40] I do not cast border regions as contours of an analytical unit, the nation-state, but as a lens to investigate historical questions about the movement of people and goods, state- and nation-making, conflict, and territoriality. By critically engaging with geographical space, how people represented it and experienced it, I join a cohort of scholars that have stressed the centrality of border regions in the history of nation-making.[41]

In the next six chapters, I show how different people invested distinct meanings in the environments around them. Drawing from these symbolic environments provides what Thomas Greider and Lorraine Garkovich call landscapes: concrete sites imbued with specific cultural meanings. As a result, landscapes reflect "definitions of ourselves."[42] The physical space in northern Patagonia underpinned how different people—authorities, travelers, or settlers—both imagined the nation and sought to create it. At the beginning of the period covered in this book, in the 1890s and 1900s, diplomats viewed the Andes as a clear border between two countries. For local ranchers, its valleys opened trans-Andean herding routes for summer grazing. In the 1920s, local elites were on alert because it seemed the valleys harbored fugitives. A decade later, tourists recognized in the mountains of the northern Patagonian Andes the essence of the nation, as if the whole notion of "Chile" or "Argentina" could fit on one postcard. I join a growing number of environmental scholars who have examined the making of interconnected cultural and material environments, from Mexico to Brazil, highlighting change and continuity in how and why people landscaped Patagonia.[43]

THE BACKDROP

Our story begins in the 1890s. By then, Chilean and Argentine armies had invaded Mapuche territory and diplomats had agreed to draw a borderline. States used geographical knowledge not only to buttress military operations but also to consistently reinforce historical narratives as "transhistorical truths."[44] In essence, authorities sought to define the physical contours of a latent nation that would give its citizens a national identity. Among governing elites existed an imagined national territory; now it was time to complete its contours. Crews of explorers were compiling data up and down the Andes and sending their results to Santiago and Buenos Aires. In Chile, where nitrate dividends flowed into the state coffers, this was the moment when a civil war put an end to the presidentialist republic established after independence and installed a parliamentary regime, which lasted until 1925. Argentines were also enjoying the export boom of wheat, corn, linseed, and chilled and canned beef. The expansion of agriculture in both countries created pressure to populate the regions south of the thirty-ninth parallel for market-oriented production of foodstuffs. Hence, in the 1890s, scientists sought to construct knowledge about the area to better negotiate with the other country and to design legislation for distributing lands among non-Indigenous settlers. While explorers surveyed the Andes in the 1890s, a Chilean man of German descent opened a general store on the southeastern shore of Lake Nahuel Huapi, in the northern Patagonian Andes, with significant historical ramifications. Around the store grew a town, Bariloche, and the ranchers and farmers from around the town used the store to export goods to Chile. This snippet from the 1890s illustrates two conflicting understandings of the Andes: authorities viewed the cordillera as a natural boundary; locals viewed it as part of a shared environment.

Diplomats finalized Chilean-Argentine border negotiations in 1901 (except for fifty kilometers, or about thirty-one miles, along the Southern Ice Field). By then, governments on both sides of the Andes had introduced legislation to organize settlement in Patagonia. Bringing people to the fertile northern Patagonian Andes was crucial to developing the mountains' agrarian potential. Productivity for capitalist markets would drown the Desert. However, the legislation suffered from contradictions and in some cases ended up sustaining the Desert it was trying to erase. In Chile, some immigrants refused to accept the lands they were assigned, and more importantly, Indigenous and mestizo Chileans who were already living in distributed lots protested removals. In Argentina, scarce infrastructure deterred newcomers. Settlement efforts introduced by both governments from the 1880s through the late 1910s resonate with what scholars in the English-speaking world have identified as settler colonialism. National governments relied on settler colonial practices in their efforts to eliminate Indigenous culture in southern Chile and Argentine northern Patagonia. In reference to domestic colonization, scholars of settler colonialism have challenged the assumption that Indigenous populations were "internal" to nation-states. The Spanish *pueblos originarios*, which could be translated as "first nations," illustrate the chronological precedence of Native peoples before the Chilean and Argentine states. The governments' urgency to bring settlers to Patagonia, farm the land, graze the fields, and export goods was animated by what Barbara Arneil identifies as a "colonial impulse." In the liberal republics of late nineteenth-century South America, she argues, it was John Locke's political thought that "justified both dispossession and assimilation of Indigenous peoples" and framed the colonization of Patagonia.[45]

So in the first two decades of the twentieth century, the northern Patagonian Andes were bustling. Immigrants settled, merchants traveled back and forth, businesses grew, and people (both Indigenous and not) applied for land leases. National states deployed police officers, judges, land surveyors, and land inspectors and made plans to open roads and build railroads, which arrived in 1912 at Puerto Montt and in 1934 at Bariloche. Displaced Chileans, often of mestizo background, migrated to Neuquén and Río Negro to breed cattle for the Chilean market. People moved a lot. During these first two decades, Buenos Aires especially received more immigration than any other Latin American country. Increasingly, the ruling classes in both countries watched with concern migrants' conglomeration in closed quarters and viewed sharing leftist ideas and agitating their fellow workers as particularly dangerous. In the early decades of the twentieth century, both governments did not hesitate to violently repress strikes and demonstrations. Governments in both countries introduced legislation that could refuse entry to and expel foreigners who were deemed dangerous to the established order.[46] This made it easier for national authorities to replicate a sense of danger in parts of their countries with larger immigrant populations, such as the northern Patagonian Andes. The onset of World War I dislocated international trading circuits, and Patagonia was no exception. The customs office in Puerto Montt, which gravitated toward trans-Andean trade, collected twenty times less revenue in October 1914 than in the previous month. When the war ended, European countries dropped their demand for South American products. In light of falling commercial balances, Santiago and Buenos Aires increased tariffs for imports in the cordillera. As a result, trans-Andean traffic of cattle and wool waned. By now, Bariloche and its environs had 2,000 inhabitants, a sharp contrast with the population on the Chilean side (18,484 people).[47]

The first decades of the twentieth century also brought profound changes in Chile and Argentina. In Argentina, pressure from workers, the professional middle class, and college students resulted in the granting of voting rights to all male citizens (excluding those in the national territories). In the presidential election of 1916, Hipólito Yrigoyen defeated the conservative oligarchy as part of the Unión Cívica Radical (Radical Civic Union), a political party of the urban, educated, growing middle class. Unión Cívica Radical ruled until 1930, when it was overthrown by the first military coup in Argentine history. The Unión administrations, argues Luis Alberto Romero, had to navigate establishing democratic institutions while negotiating the demands of social reform.[48] These demands were ubiquitous across Latin America, and perhaps the most dramatic and with most lasting effects were the Mexican Revolution and its Constitution of 1917. In Chile, the Parliamentary Republic inaugurated in 1891 launched a period of intense political negotiations and very little governing. The pressure for social reform was evident in the emergence of leftist political parties. In 1920, the growing middle class, workers, and college graduates elected to the highest office Arturo Alessandri Palma, a candidate who rode to the presidency by critiquing the parliamentary system. But frustration ensued. Alessandri found it difficult to get new legislation passed in the conservative Congress. While World War I had pushed Chilean nitrate to record-breaking exports, the war's aftermath doomed the industry, as demand fell drastically.[49] The economic crisis coupled with Alessandri's inability to introduce labor laws underscored a political crisis that gave the final blow to an already declining system centered in Congress. In 1924, the lower-rank military, also in deep discontent with Congress, forced legislators to pass laws that would address their salaries. The newly formed junta did not introduce as many changes as others in the force expected. As a result, in 1925 a second faction carried out a second coup, forcing the junta to resign and requesting Alessandri to come back to finish his term. Under the tutelage of the mastermind officer Carlos Ibáñez del Campo, the president sponsored a constitutional reform that was in effect until 1981.

The 1930s in Chile and Argentina were tumultuous times, like in much of the world. The economic depression that followed the Wall Street crash of 1929 destabilized political regimes around the world. In Chile and Argentina, political elites allied with the military to keep a strong grip on the economy. Unlike during previous financial crises, the global Depression resulted in unprecedented state involvement in economic planning in Chile and

Argentina and throughout the Americas.[50] State intervention in the economy led to multiple development plans that affected all aspects of life, including the environment. Chilean and Argentine governments tightened control over the economy, as well as politics and society. In Argentina, Gen. José Félix Uriburu led a military coup against President Hipólito Yrigoyen (1928–30). With the support of some factions of the armed forces and the paramilitary far-right Patriotic League—which had risen to prominence during the 1920s—Uriburu inaugurated a long period in Argentine history in which the armed forces controlled politics to, in their view, protect democracy from its own dangers.[51] The administration called this moment the Conservative Restoration, alluding to its roots in pre-1916 Argentina. The opposition called this period the Infamous Decade, because conservatives maintained power through fraud. Uriburu called for elections, and in 1932 Gen. Agustín P. Justo rose to the presidency (1932–38), with Julio A. Roca Jr. as his vice president. Justo dominated the political sphere for the following ten years through an alliance with conservatives, Radicals, and independent socialists.[52] But as fascism was rising elsewhere, national politics realigned, with a "national front" against an emerging "popular front" led by the Unión Cívica Radical. While the government remained in power thanks to fraud, it eroded its alliances. When several political leaders passed away in the span of a couple of months in 1943, leaving no clear candidates for the upcoming election of 1944, the weakened conservative administration courted the military to intervene. On June 4, 1944, a second military coup interrupted democracy and—though nobody knew this then—eventually helped propel the political career of a young lieutenant colonel, Juan Domingo Perón.[53]

Chile followed a similar path. In 1925, the country began experimenting with a new constitution that centralized authority in the executive branch. Stability, however, rested on the rise to power of Ibáñez del Campo, elected in 1927. Ibáñez del Campo introduced a reform agenda that set parameters on how the state would participate in the revitalization of the economy in the 1930s and 1940s by his successors. Part of his centralization efforts comprised increasing control of the national government over state agencies, such as the General Controlling Office; the National Police (*Carabineros*); the Chilean Air Force, founded in 1930; and a national airline, established under his watch.[54] Despite this, Ibáñez del Campo lost support when he could not weather the Depression and resigned in 1931. New elections reinstalled Alessandri, who stabilized the economy and restored trust in the political system. He left the country in much better shape than when he had received it, with a robust party system but still plagued by widespread fraud. In 1938, a fragile alliance of center-left parties, known as the Popular Front, won the presidency and ruled for fourteen years (they were succeeded by Ibáñez del Campo's second presidency in 1952).[55] Ultimately, Chile's political trajectory toward the left diverged from Argentina's inclinations to the right.

FIFTY YEARS IN SIX CHAPTERS

Over the course of six chapters and through the analysis of written, visual, and geospatial sources, *Landscaping Patagonia* shows how residents, authorities, and passersby constructed their versions of Chile and Argentina through their everyday spatial experiences, which often crossed the Andes. The book begins in the 1890s, when the Chilean and Argentine governments were negotiating the Patagonian boundary. While crews of explorers surveyed the Andes, new trans-Andean trading networks replaced Indigenous routes. Local experiences collided with the ways in which diplomats envisioned the nation. For the next fifty years, the Chilean and Argentine governments introduced nationalizing efforts to expand their often scrawny control over the Patagonian Andes. *Landscaping Patagonia* examines policy and everyday life in the environment to offer an experiential understanding of the nation in both countries through six lenses, developed in each chapter: creation of knowledge, settlement, movement, transgression, the built environment, and the outdoors.

After the military campaigns against the Mapuche in northern Patagonia (1879–84), the governments of Chile and Argentina set out to finalize the international boundary. For political elites, the Andes Cordillera worked as a "natural" wall. However, scientific surveys of the Andes from both countries revealed that far from being a tidy succession of peaks, the Patagonian Andes were broken links of a chain, clusters of peaks separated from one another by wide valleys. Crews of explorers surveyed the terrain, taking copious notes about flora, fauna, weather, soil, and people, and transferred that data to extensive reports and detailed maps.[56] In doing so, they gave space "a stable signification" to be "more effectively appropriated, transformed, and regulated."[57] Drawing the nation literally and symbolically represented the cornerstone of state-building in peripheral regions. Topography challenged imagined versions of the boundary as a line that ran along the Andean peaks. Chapter 1, "Science for the Nation," goes behind the border negotiations of the late nineteenth century into scholarly debates, illustrating how the creation of knowledge emerged both from the tension between universalist aspirations of scientific inquiry and the historically contingent territorial anxieties of the late nineteenth century. Scientist-explorers put forth a rhetoric of emptiness to violently displace Indigenous communities and their epistemologies to make room for the colonization of the Patagonian space. The circulation of new knowledge was vital for reconciling existing ideas about nature and the nation as a means to draw frontiers and govern them. In the case of the northern Patagonian Andes, the borderline remained an idea at times and a strict reality at others. The flexibility of the border elucidated how cross-border networks and shared ideas about space challenged top-down versions of the nation that authorities in Santiago and Buenos Aires repeatedly—and to this day—attempted to impose. While authorities and scientists invented the Desert, they made it their mission to civilize it and create a metaphorical garden. At heart, this chapter highlights the imperial origins of nation-making in Patagonia.

For national authorities, border-making represented the final constitution of the national territories of Chile and Argentina. However, it did not mean that regions peripheral to political centers were automatically incorporated in the nations. Chapter 2, "Settling Patagonia," discusses how settler colonial practices animated land legislation and structured state policy, but it also analyzes how conflicts among neighbors and resistance to legislation diluted the optimistic vision of colonization. For authorities in Santiago and Buenos Aires, creating landed property was central to the settlement of foreigners in Patagonia and, ultimately, the elimination of the Desert. However, the policies they introduced, especially new legislation, rested on assumptions about land availability and motivation to settle, which resulted in a redefined Desert. In the last decade of the nineteenth century, during intense border negotiations with Argentina, the Chilean government granted vast extensions of land to holding companies, colliding with existing inhabitants. The Rupanco Colonizing, Farming, and Cattle-Breeding Company removed several mostly Indigenous families from their land on Coihueco Island, a fertile plain south of Lake Rupanco. These expropriations resonated with a history of violence against the Indigenous peoples of southern Chile and sparked vigorous resistance in the courts, to local authorities, and ultimately in protests.

Distribution of landed property was only one aspect of settlement. Additionally, people's ability to herd cattle, access towns, carry cargo, and travel safely constituted a crucial expectation among settlers. Chapter 3, "The Materiality of Space," examines roads and railroads as visible expressions of how different actors envisioned the settlement of a modern nation. Roads made space legible for authorities, but they also represented a crucial means for rural settlers to access markets and carry out personal business. When new fences went up, cutting off improvised trails, people appealed to the courts to do good by them. Hence, judges and land inspectors found themselves between the letter of the law and the pragmatism of everyday use of space. This chapter also examines roads as necessary routes of trans-Andean trade, a constitutive aspect of rural northern Patagonia. Finally, it delves into how railroad extensions reoriented the organization of space, effectively disarticulating trading networks in the long term.

The governments of Chile and Argentina favored the settlement of European immigrants in their southern territories. Chile began bringing farmers, woodworkers, ranchers, and manufacturers to Llanquihue Province as early as the 1850s, while the first scattered non-Indigenous farmers came to the environs of Lake Nahuel Huapi as late as the early 1890s. The settlement of non-Indigenous people in the northern Patagonian Andes, however, did contribute to social conflict. In Chapter 4, "Spatial Discourses for a Healthy Nation," I show how earlier markers of difference persisted within the new society. Ranchers, professionals, businesspeople, and local authorities of the northern Patagonian Andes portrayed the region as a dangerous landscape in order to argue for a more substantial presence of the state in the 1920s. Using police reports, census data, news articles, and government reports, I delve into local elites' and authorities' everyday efforts to appropriate a vocabulary on public health to diagnose why Patagonia was, in their eyes, a moral desert. This rhetoric situated undesired behaviors as being out of place, perpetuating earlier ideas of Patagonia as a space with no order that lent itself to the crimes of bandits. Conversely, "progress" captured what local economic and political elites envisioned as the best interest for the advancement of civilization: owning land, growing crops, breeding animals, producing a surplus for trade, erecting buildings, and settling with a family. Bandits' criminal behaviors opposed the elites' attempts to bring economic prosperity to the region. Thus, in the desert of moral values and civilization, the businesspeople, officers, farmers, and immigrants brought "progress" to northern Patagonia.

So far, governing elites had attempted to nationalize Patagonia through military violence, scientific generation of knowledge, legislation, and policing behaviors. The northern Patagonian Andes thus far had been a Desert and a site of hope; a place marked by otherness and where otherness—the boundary—showed itself; a home for migrants but not for Natives; and a dangerous region whose people threatened the core of the nation. But that was not it. In a stark contrast from the previous decade, in the 1930s and through the mid-1940s, the governments of Chile and Argentina developed what John Urry calls the tourist gaze, depicting northern Patagonia as extraordinary, where visitors could recognize their belonging to the nation by experiencing the beauty of Andean landscapes.⁵⁸ The last two chapters use different lenses to make a similar argument. They explore how governments created a national landscape to evoke a sense of patriotism among Chilean and Argentine citizens. To do so, they deliberately re-signified agrarian tropes and hopes for the northern Patagonian Andes as picturesque, attractive views for tourists. Chapter 5, "National Aesthetics in the Argentine Locality," studies the aesthetics of the built environment in the environs of Lake Nahuel Huapi. It analyzes the transition from a trans-Andean architectural style in the first decades of the twentieth century to an aesthetic that came to symbolize an Argentine landscape. Chapter 6, "The Outdoor Destination," examines how national authorities framed the re-significations of the northern Patagonian Andes around tourism and local enjoyment of the outdoors. Building on earlier ideas of Patagonia as an empty space, authorities in both countries increasingly used the powerful imagery of the "she-land," a feminized version of space. Overall, guidebooks and newspaper articles portrayed the Patagonian Andes as passive, subject to the power of political and economic centers of power, but also as seductive "virgin" spaces that needed protection (hence the creation of national parks) and penetration (hence the expansion of tourist facilities).⁵⁹ The built environment and the outdoors were both constructed as metaphoric locations of the nation.⁶⁰

By focusing on geographical space, *Landscaping Patagonia* shows how different people, including explorers, settlers, authorities, visitors, and bandits, sought to make Patagonia their own by transforming it from a non-place to a national landscape, a collection of sites that could evoke a shared past and a common future. Or, I should say, they created a place that could evoke imagined versions of the past to buttress ideas about what the future of Patagonia (and, by extension, of Chile and Argentina) should look like. All in all, the multiple ways of making sense of the nation through spatial discourses emerged from seeing Patagonia as a monstrous emptiness, a metaphoric Desert whose elimination underpinned the making of the nation in a border region.

Chapter One

SCIENCE FOR THE NATION

1890s

Overlooking the Patagonian plateau that he had been surveying for several weeks in late 1897, Argentine explorer Francisco Moreno concluded that "Mount Palique . . . is not a mountain: it is a simple hill."[1] Moreno's assertion responded directly to Chilean claims that Mount Palique was indeed a mountain and therefore part of the Andes mountains. Amid fierce border negotiations between Chile and Argentina in December 1897, the scientific inclusion or exclusion of Mount Palique in the Andes would have affected where the international boundary was to be drawn. This debate encapsulates the distance between diplomatic border demarcation and scientific border negotiations. Political authorities imagined the Andes as a "natural" wall between Chile and Argentina, and they believed science would support this idea. Far from relying on "kings, advisors, or lawyers," Argentine and Chilean diplomats prided themselves on trusting the work of "geometricians and geographers" to delineate a borderline.[2] Scientific surveys, however, fell short. Rather than unveiling an absolute truth regarding the Andes, their work exposed the internal contradictions within academic interpretations of Patagonia. Explorers found themselves between the truth-finding mission that typified scientific endeavors and the Chilean-Argentine border negotiations that sculpted the immediate goals of the expeditions.

This chapter shifts the attention from the diplomatic talks that often typify the scholarship of border negotiations toward the people gathering data about border regions.[3] Scientist-explorers constructed the borderline more as a result of scholarly debates based on scientific explorations than the revelation of an objective truth, as authorities hoped. This offers a window into the ideological framework that underscored Chilean and Argentine nationalizing efforts for the next fifty years. The chapter begins with an account of the border negotiations between Chile and Argentina that framed the production of knowledge. I then outline the major institutions that shaped scientific traditions in Chile and Argentina. Museums of natural history, cartographic offices within the military, and learned societies in both countries sponsored the generation of knowledge about Patagonia, broadly construed. Perhaps more important, they also cemented the emptiness of the Desert as an image to represent the region. Finally, I illustrate the scholarly debate about the relationship between border and nature by delving into the topographic and toponymic debate between two of the leading scientist-explorers, Hans Steffen, who worked for the Chilean government, and Argentine Francisco Moreno. The generation and circulation of scientific knowledge was historically contingent, locally based, and globally circulated.[4]

Scientist-explorers navigated the narrow threshold between international recognition and national sponsorship incorporating the mandates of global (that is, Western) science and a geopolitical agenda.[5] "Science" was a shorthand for a kaleidoscopic combination of geography, anthropology, archaeology, geology, botany, zoology, and even medicine. Latin American states of the nineteenth century used science as an avenue for imposing a liberal order, expanding an export-oriented economy, and consolidating a national narrative.[6] For geographers, geologists, and naturalists, the production of knowledge relied on observing phenomena, collecting data, and making connections. Nineteenth-century scholarly circles in Europe and the Americas recognized in Alexander von Humboldt the best example (or at least the best-known example) of using that methodology to know the natural world.[7] The work of scientist-explorers, however, was first and foremost a service to the state. Indeed, as Álvaro Bello puts it, explorations of "unknown territories" and the geographical knowledge they produced were crucial for "the understanding of space" and for "putting together fragments of a national territory."[8] The Chilean and Argentine governments paid scientist-explorers' honoraria, provided equipment, and facilitated transportation to the field. Scientist-explorers also had other jobs, usually within the state apparatus. For instance, in Chile, members of subcommissions frequently taught at the University of Chile, the Instituto Pedagógico (the national teachers college), and the Instituto Nacional (Chile's flagship secondary school). In Argentina, many worked at the Museum of Natural History, funded by the province of Buenos Aires.[9] Scientist-explorers, then, found themselves torn between their call to discover a universal truth and their service to the nation.

This nationalizing policy did not begin or end on the border, as this and the following chapters demonstrate. The urgency for a well-bounded territory animated not only the border negotiations but also military operations in Indigenous communities in Araucanía and northern Patagonia. In Araucanía, the Mapuche heartland, the Chilean government began a military campaign in the 1860s that disarticulated Indigenous polities and their trans-Andean trading networks. After military operations concluded in 1883, Indigenous peoples were forcibly relocated to allow new settlers to occupy their lands.[10] In Argentina, the government waged war against the Mapuche, Tehuelche, and Puelche from what later became the National Territories of La Pampa, Río Negro, and Neuquén. Authorities compelled Indigenous peoples to dismantle their tribal systems, assimilate into Argentine culture, forsake their Native languages, and enlist in the military. Their land was appropriated and became state property.[11]

Behind the military columns came the explorers, theorizing about the geological origins of rivers, sketching flora and fauna, and measuring temperature, atmospheric pressure, and mountain height. Their scientific gaze provided an "optic consistency" for the symbolic Deserts that were southern Chile and northern Patagonia.[12] They filled maps with names and museums with artifacts. Two of the most prominent experts in the context of border negotiations were Hans Steffen, a German-born geographer working for the Chilean commission, and Francisco Moreno, who led the Argentines in border negotiations.[13] Steffen and Moreno found in each other scholarly interlocutors. Their explorations of Patagonia directly informed their governments' boundary proposals. They both led expeditions in the Patagonian Andes and presented their results at scholarly meetings. They also published extensively in national and international journals, weaving a network of conversations that exemplified the multinodal and multilevel generation of knowledge.[14]

All in all, the generation of scientific knowledge collaborated with, relied on, and amplified the violence of the military raids against Indigenous groups.

DIPLOMATS DRAW THE BORDER

Latin American nation-states began negotiating their international boundaries as soon as they gained independence. In part, this was because of the confusing nature of independence. Were new countries stand-alone in their own right, or were they the residue of imperial dismemberment? Claiming a territory, drawing it on a map, and having other states acknowledge it represented a crucial step in the making of these young nations. Delineating the contours of the territory constituted a very real and relatively modern way of establishing the nation. In Latin America, as John Charles Chasteen and Sara Castro-Klarén remind us, the state preceded the nation. In other words, new states needed to forge a sense of nationality that had not previously existed.[15]

The negotiations of the Chilean-Argentine border went through three stages. First, the new governments compiled their ideas about their border. Based on these, they agreed on a somewhat ambiguous borderline. Ideas about space did not always coincide with the reality on the ground. The final stage consisted of resolving the bulk of these contradictions in the 1890s. After they gained independence in 1816 and 1818, respectively, Argentina and Chile adhered to the Spanish American principle of *uti possidetis iuris*. It meant that each state would govern what it had legally possessed before independence. In other words, it did not matter if Buenos Aires had no effective control over Patagonia; colonial legislation had granted Buenos Aires authority over Patagonia, and so it shall pass to the new government. *Uti possidetis* effectively repelled territorial encroachment of other imperial powers (with some exceptions, of course). But the legal principle did not prevent Latin American countries from competing for territories with one another. In the first decades after independence, debates swirled around what each country had possessed in 1810. Scholars in each country dove into the colonial archive to trace royal decrees that defined and redefined Spanish jurisdictions in Patagonia. At this early stage, both Chileans and Argentines used royal decrees to support claims over all of Patagonia, the ones missing some documents, the others misinterpreting their context. The Treaty of Friendship, Peace, and Navigation (1856) put some of these issues to rest. It recognized the possessions of each country in 1810 with the Andes as the accepted boundary between the two. The treaty implied what diplomats explicitly agreed on decades later: that

Chile would have an orientation toward the Pacific Ocean and Argentina toward the Atlantic.[16]

In 1881, Chilean and Argentine authorities signed the cornerstone treaty to their border negotiations. Both governments were in a relatively strong position to negotiate. Chileans finalized a twenty-year military campaign in Araucanía, the Mapuche heartland, which represented a major triumph over a space Spaniards could never subdue. They were also waging a war against their northern neighbors, Peru and Bolivia, over the mining districts of the Atacama Desert. In 1881, Chilean forces occupied Lima but could not broker peace for another two years. Argentines took advantage of Chileans' focus in Peru. They rode on the victory of the military campaign against the Mapuche of Río Negro and Neuquén to push Chileans to talk about Patagonia. The Treaty of 1881 stated that the borderline between the two countries would "run along the highest peaks of this range that divide watersheds and between slopes that run to one or the other side."[17] The treaty introduced two geographic landmarks, the Andes and the continental watershed, as defining features of the international boundary. For diplomats, the "physical drawing of lines" that followed would be a simple task because it was "entrusted to science and experts' acumen."[18] If science dictated borders, authorities believed there would be little room for discussion. Science proved them wrong.

In the early 1890s, each country's Congress appointed a chief expert (*perito*) to coordinate the drawing of the borderline, Diego Barros Arana for Chile and Francisco Moreno for Argentina.[19] In turn, they each appointed subcommissions of scientists to validate each border point on the ground, mostly from north to south. Barros Arana and Moreno, however, understood the urgency to agree on a boundary and did not wait until work was done in the North to begin surveying the Patagonian Andes. The drawing of the border went as follows. In the early spring (September), experts met to plan the summer work. They each delivered their instructions to their crews about where a specific point of the border should be located. Each crew met at the designated spot, where they pinpointed a border landmark. If the locations coincided, another dot was added to the official boundary. If they did not, Moreno and Barros Arana compiled these disagreements to revise the reasons and propose solutions to a mediating party, the English Crown.[20]

Progress was slow in Patagonia. The topographic reality of the Patagonian Andes challenged authorities' ideas about border regions. The main chain of the Patagonian cordillera, unlike the Andes farther north, decreases in altitude, breaks into different ranges, tilts west, and sinks into the South Pacific Ocean.

In other words, far from being a tidy succession of peaks, the Patagonian Andes resemble a collection of fractured links in a chain that slant westward and submerge into the ocean.[21] As a result, some streams begin flowing eastward but mouth into the Pacific Ocean, signaling the presence of wide trans-Andean valleys. This geography could have given Chile claims over the eastern valleys of the Andes while Argentina could have petitioned for access to the Pacific coastline. The ambiguous wording combined with a general sentiment of mistrust resulted in frequent talks among diplomatic envoys to agree on a border peacefully. A protocol signed in 1893 attempted to solve the dilemma by clarifying that the borderline would follow the highest chain of the Andes that parted the watershed. This prevented Argentina from gaining access to the Pacific Ocean in southern Patagonia, and it helped demarcate the border in Tierra del Fuego. It was a clear victory for Chilean diplomacy. After finishing their visits to the Andes, experts Diego Barros Arana and Francisco Moreno met to match their notes about the terrain to the documents and maps their countries had collected over the years.[22]

Despite the ongoing diplomatic talks, border negotiations strained Chilean-Argentine relations. Both countries had been victorious in wars against their neighbors, Chile in the War of the Pacific (1879–84) and Argentina in the War of the Triple Alliance (1864–70). Both increasingly benefited from the exports of primary goods, and as long as the world needed to use the Strait of Magellan for interhemispheric trade, both had geopolitical interests in Patagonia. A naval arms race ensued, especially in the context of tightening global tensions.[23] Between 1881 and 1902, the Argentine and Chilean navies significantly increased their fleets, incorporating cruisers, destroyers, frigates, corvettes, and torpedo boats. Historian Pablo Lacoste calculates that by 1900, each fleet had 100,000 tons of warships (eighth in the world), a disproportionate size relative to their countries' populations. Chile boasted about 33.33 kilograms of war fleet per capita, the largest in the world, and Argentina about 22.22 kilograms, third globally.[24] In 1891, an opportunity arose to mobilize the navies. While at port in Valparaíso, Chile, USS *Baltimore* captain Winfield Schley conceded shore leave to 120 of his sailors. At the True Blue Saloon, some brawled with Chilean navy men, resulting in between one and five US deaths and several injuries.[25] In the face of Chilean refusal to provide reparations and a public apology, US-Chilean relations reached a new low, opening the path for a possible armed conflict. Estanislao Zeballos, the confrontational Argentine minister of foreign affairs, offered the US government help in the form of supplies and support.[26] But war was averted. The Chilean attackers were arrested in 1892 and sentenced to prison, and Chile paid the United

States a US$75,000 compensation, which alleviated the strained relationship.[27] Chilean-Argentine tensions also eased as European powers renewed their commitment to peace. In this climate, and with Zeballos away from the Ministry of Foreign Affairs, the president of Chile, Federico Errázuriz, and his Argentine counterpart, Julio Roca, called for a peace summit. The Embrace of the Strait (Abrazo del Estrecho) in 1899 appeared contradictory to the increasing friction between Argentina and Chile. On the southernmost tip of the continent, in a hostile climate, the leaders of Chile and Argentina met for the first time to seal the most salient ambiguity of all: a constant oscillation between suspicion of each other and the idea of a "natural" partnership.

While the naval arms race developed in the 1890s, Diego Barros Arana and Francisco Moreno continued to negotiate the borderline. However, the lack of a clear watershed-dividing line in Patagonia's highest peaks pushed diplomats to form an International Boundary Committee in 1896. British colonel Thomas Holdich officiated as the mediator, boasting an impressive career as a boundary broker in the rugged terrain of Central Asia. He brought to the table nearly two decades of surveying experience in Baluchistan (between present-day Iran and Pakistan) and in the Pamir Mountains (which led to the Afghan strip between the then Russian Empire and British India). Perhaps as important, Holdich shared with his Latin American colleagues a belief that "geographical ignorance" lay at the heart of "every boundary dispute."[28] Hence, geographical knowledge could lead to peaceful boundary demarcation. Barros Arana and Moreno served as experts. Each expert coordinated the work of several subcommissions, who surveyed the most litigious areas in the Andes, most of them located in Patagonia. These were places where experts did not agree on the location of the borderline based on their geographical findings. Barros Arana and Moreno compiled this data and prepared their cases to present before an arbitral tribunal in London, which had deputized Holdich. The tribunal's decision would be based on "justice and science" and, therefore, would leave room for no appeals or further discussions.[29] For these scientists, nature defined their nations. However, explorers made observations that not only disarticulated previous conceptions of the Andes but also challenged one another's conclusions. How could diplomats agree on a borderline along the Andes if their experts reported conflicting corpora of knowledge?

LEARNED SOCIETIES IN CHILE

Scientist-explorers who supported the military operations and border negotiations in Chile and Argentina did not work alone. Students, naturalists, geographers, geologists, and engineers found in learned societies and some state institutions spaces of intellectual exchange in service of the state and the advancement of the nation. These institutions contributed to the consolidation of the national territory by surveying the terrain, collecting data, and organizing knowledge. Scholarship about places like Patagonia did not amount to simple encyclopedic annotations on its topographical or ecological characteristics. Reports, maps, and papers outlined the economic potential of natural resources, a crucial aspect of nation-making that scientific societies in Latin America imitated from their European counterparts. Geography, especially, gained the reputation of a stand-alone but quite comprehensive discipline, bringing together fields from geology to anthropology. It provided governments a scientific framework to tell a national history (which often included a natural history) based on archaeological evidence. It also synthesized data to project a development plan based on natural resource extraction. In other words, geography offered a way to construct a national past and create a national future.[30]

Chilean scientific institutions predated military operations in the South against the Mapuche heartland (1861–81) and in the North against Bolivia and Peru (1879–84). Early learned societies supported the formation of Chilean scholars who later participated in the border negotiations while providing a model for collaboration between local and foreign intellectuals. Indeed, Santiago hosted many European and South American scholars in the 1830s and 1840s, from a young English traveler by the name of Charles Darwin to Argentine expatriates and future presidents Domingo Sarmiento and Bartolomé Mitre. Some of these academics were hired by the government to carry out research for the state and teach at the University of Chile and the Instituto Nacional. Among these men was, for instance, French botanist Claude Gay, whom the Chilean government tasked in 1830 to write a natural history of the country. Over the next decade, he embarked on explorations of different parts of Chile, bringing back observations, measurements, and samples. While in the field, Gay published his recurrent accounts in the state's biweekly newspaper, *El Araucano*, where his "personal curiosity" became "a national project."[31] This was probably one of the earliest attempts to sew together a "loosely united group of localities" into a national space through a shared natural history.[32] By the end of the decade, Gay had amassed a collection of plant and animal samples that formed the beginnings of Chile's National Museum of Natural History (Museo Nacional de Historia Natural). Upon Gay's return to France in 1842, most of the collection was ruined, forgotten in a "cramped room," until another foreigner arrived to rebuild the museum in 1853.³³ German naturalist Rodulfo Philippi commissioned explorations, replaced deteriorated objects, expanded existing collections, and started new ones, all while teaching.³⁴

Other institutions centered positivist science in the making of Chile during the following decades. In 1843, the government founded the University of Chile from the embers of a colonial institution, the Royal University of San Felipe (which had replaced a Dominican college in 1747). Its first president was Venezuelan Andrés Bello, tutor to Simón Bolívar and editor of the official bulletin where Gay published his reports. Bello fervently defended the central role of humanist education for the civilizing mission of new Latin American nations.³⁵ His administration hired foreign scholars to teach and do research that would constitute a national science.³⁶ Among these was French geologist Amadeo Pissis, who over the course of twenty years surveyed different areas in Chile. The government especially instructed him to focus on the cordillera "to pinpoint with precision the ridge or culminating line that separates the slopes that go to the provinces of Argentina from those that go to the Chilean territory."³⁷ His crowning achievement was a compendium of Chilean geography and an 1873 national map, which Barros Arana considered a reliable precedent for the border negotiations.³⁸ University presidents who succeeded Bello included Polish geologist Ignacy Domeyko, the "father" of mining engineering in Chile, and José Joaquín Aguirre Campos, physician, governor, and representative in Congress.³⁹ The appointment of Diego Barros Arana as president of the university in 1893 reinforced the alliance between the state and the scientific community, as he was also the chief expert in the border negotiations with Argentina. The university's annual publication, *Anales de la Universidad de Chile*, animated conversations around science, geography, and politics. Many explorers of southern Chile published here their findings for a wider audience, often cross-pollinating knowledge production with policy making.⁴⁰

But perhaps the Chilean agency that best epitomized the alliance between science and state was the navy's Office of Hydrography. Naval officer Francisco Vidal Gormaz negotiated the opening of this division in 1874 with the prime objective of surveying the 8,000 kilometers (4,971 miles) of Chile's continental shoreline plus the many rivers that flowed into the Pacific Ocean.⁴¹ The research center produced maps and charts, statistics on maritime hazards, oceanic currents tables, and weather reports. The study of the coastal terrain enabled technicians to suggest locations for lighthouses, and the general research on the Chilean shores resulted in catalogs of flora and fauna. Thus, though the Office of Hydrography began as a technical support unit for the navy, it soon grew into a research facility focusing on the geography, meteorology, biology, and hydrology of the Chilean coastline. Like its peer institutions, the Office of Hydrography circulated a yearly report on recent explorations. Beginning in 1888, this report included a section describing historical expeditions to Llanquihue, Chiloé, and western Patagonia. With an introduction by Diego Barros Arana, this section threaded a genealogy of explorations back to the eighteenth century, affirming a Chilean presence in the southern territories.[42]

In the second half of the nineteenth century, Chile had established a scholarly tradition in generating national science. Foreign actors played a crucial role in transferring skills, especially German and, to a lesser extent, French academics. Chilean men trained in the halls of the Instituto Nacional and the University of Chile applied this knowledge to advance the interests of the state in multiple fields, from agriculture to medicine.[43] By the early 1890s, when Moreno and Barros Arana began border negotiations, Chile had a robust cohort of locally trained technicians, plus foreign scientists who were still arriving to its shores. These scholars found in the many learned societies places for intellectual exchange and in their journals a platform for disseminating Chilean scholarship.

LEARNED SOCIETIES IN ARGENTINA

In Argentina, similar institutions provided spaces for generating knowledge about the national territory. In 1879, the Argentine Army created the Office of Topography to accompany the military campaign against the Indigenous people of northern Patagonia. This division grouped together engineers and surveyors, who produced cartographic materials based on geodesic measurements. The first maps illustrated Argentina's internal frontier with Indigenous people's land and the military advancement on their territory, such as Manuel Olascoaga's *Plano del territorio de La Pampa y Río Negro*, published around 1880. Olascoaga was the Office of Topography's first director until 1884, when he was appointed first governor of the territory of Neuquén. He joined the military campaign and published an account of operations, which he called *La conquista del desierto* (the Conquest of the Desert).[44] His work influenced geographers, politicians, and historians for more than a century. Indeed, until relatively recently, Argentines accepted the military genocide of Indigenous groups in 1879–84 as the last stage of the conquest initiated by the Spaniards and a foundational moment in the history of Argentina. In schoolbooks, this is described as the point when Patagonia as a whole entered the historical stage to participate in the nation. Therefore, for the authorities, the military raid of 1879–84 indicated a clear entry point for the state's presence in Patagonia.[45]

Like other Patagonian maps published in the next two decades, the *Plano del territorio de La Pampa y Río Negro* located the presence of Indigenous groups, signaling the reach of the state and the incompleteness of the nation. It marked previous explorations, like those of George Musters, Francisco Moreno, or Luis Fontana, and anticipated the "official" advance of the state in the form of military columns. In doing so, it situated cartography as historical narrative, a widespread practice that subsequent maps quickly picked up. Drawing the official contours of Argentina was intrinsically linked to the violent expansion of the state. The *Plano del territorio* showed two lines of towns and forts that acted as buffers against Indigenous attacks. It also included the multiple army columns that had marched into La Pampa and Río Negro, leading up to the Negro River, clearly marked as the new military frontier.[46] Subsequent military maps like the *Plano del territorio* were not only a history of territorial exploration; they were also a look into the future of Patagonia. Conducting fieldwork and gathering data from other maps and reports allowed cartographers to propose locations for towns, railroads, and telegraph lines.[47]

Outside military ranks, the Argentine Scientific Society, the Argentine Geographic Institute, and the La Plata Museum brought together scholars from multiple disciplines to serve national interests. Two students founded the Argentine Scientific Society (Sociedad Científica Argentina; SCA) in 1872 as a research center stemming from the new Department of Natural Sciences of the University of Buenos Aires.[48] In its meetings, students, scholars, and industrialists exchanged scientific knowledge about Argentina and its "applicability to the arts, industry, and the needs of social life."[49] Members visited factories, reported on geological findings, made readings of water levels, and put together exhibits of new technologies.[50] At its heart, the SCA would "serve the Argentine Republic" by bringing "honor and glory to country" through the expansion of science.[51] In this vein, the SCA partly funded twenty-three-year-old Francisco Moreno's first survey of Lake Nahuel Huapi in 1875 and the failed 1877 survey of southern Santa Cruz by explorer Ramón Lista.[52] Through this network, Moreno met other naturalists, geographers, and scientists who would buttress his vision of Argentine expansion in Patagonia and the role that science played in it. However, the SCA's interests in geography and the territorial constitution of the nation waned soon.

The Argentine Geographic Institute (AGI), founded in 1879, instituted a space for generating and disseminating knowledge about the country's space and its people. In the nineteenth century, fifty geographic societies emerged around the world, more than half of them between 1875 and 1880, and most of them in Europe and the United States, including the Royal Geographical Society (United Kingdom, 1830) and the National Geographic Society (United States, 1888).[53] They produced and disseminated knowledge that legitimized territorial expansion. In Argentina, AGI supported expeditions, organized exhibitions, compiled data, and produced "national" geographical knowledge. These surveys carried out a colonizing mission over the perceived Argentine territory, where scientists encountered nature, observed it, measured it, and collected it, but where Indigenous people were hardly seen beyond the artifacts they left behind.[54] Many of the men that participated in its monthly meetings disseminated AGI's scholarship in other state institutions. In part, this cross-pollination was possible because science and politics were a "class-bound activity."[55] More important, national authorities understood the strategic interest in developing a national geography, especially but not exclusively because of the role expeditions played in generating knowledge for the border negotiations of the 1890s. Along these lines, Estanislao Zeballos, AGI's founder and first president, fervently recommended the military occupation of Patagonia "to eliminate the desert and to annihilate barbarism."[56] In the 1890s, he led the Ministry of Foreign Affairs on three separate occasions. Francisco Moreno also participated in AGI's activities, as did other scientists and curious minds. Participating in AGI's activities instilled in these men an expertise that prepared them to construct a national geography and cartography.[57]

AGI's monthly bulletin published expedition reports and research on Indigenous peoples, often conflating anthropology, archaeology, and geography. Scholarship about Patagonia and even the Chaco region, a frontier space in northern Argentina, amplified portrayals of peripheries as Deserts and insisted on the civilizing potential of the Andean valleys.[58] The bulletin's pages were filled with the latest findings from around the world and different corners of Argentina. Beyond new data, articles also commented on the benefits of colonization, nodding to certain European practices that could be imitated here.[59] In 1882, AGI set out to "draw up a map of the Argentine Republic" that would consolidate geographical knowledge and political updates in one multipage chart.[60] The project received enthusiastic endorsements from provincial authorities, mostly in the form of old maps and some money. With this support, AGI authorities compiled the cartographic information they had in an Argentine atlas, "as do all advanced nations, and in South America, Chile, Brazil, Peru, Colombia and Venezuela."[61] Chileans applauded AGI's efforts to compile existing data into one document. However, they complained about the inaccuracies, which they suspected the Argentines had pushed. For instance, the atlas situated Lake San Martín/Chacabuco within the Santa Cruz River basin that drains into the Atlantic Ocean, as Moreno had theorized during his third expedition to Patagonia. A Chilean explorer concluded not only that there was no indication that "lakes San Martín and Chacabuco have any communication with the Santa Cruz [River]" but also that they probably drained into the Pacific Ocean.[62] Barros Arana, the Chilean expert on the International Boundary Committee, resented that the Patagonian boundary was labeled "to be settled" in the Argentine atlas, "as if after the Treaty of 1881 there could be the slightest doubt about the points along which the dividing line should run."[63] Both the publication of the atlas and the protestations about it elucidate the power of maps and the scientific knowledge behind them to draw international boundaries and consolidate territorial claims. Yet they were not the only tools to cement the nationalization of space.

At a time when Indigenous people controlled at least half of what today is Argentina, scholars and authorities vigorously constructed what Jens Andermann calls "an Argentine eternity."[64] This imaginary timeline portrayed a continuous history of the nation from prehistoric times to the present, materialized in museum exhibits. Governing elites shared scholars' urgency to establish museums to educate people in the "ancient and modern history of the country."[65] Particularly, Francisco Moreno envisioned a museum that would centralize a way of seeing the national space and narrating its history. A national museum would be a "monument that commemorates the history of the Argentine people through the centuries."[66] The Museum of Natural History in La Plata, Argentina, embodied the alliance between science and nation-making. The core collections came from a museum that Moreno had founded in Buenos Aires in 1877. When the city of Buenos Aires was upgraded to national capital in 1880, the museum did not enjoy the same fate. Still under the purview of the provincial government, it was transferred to the town of La Plata, under construction just south of Buenos Aires. It was not the "Desert" per se, but constructing an entire city in a couple of years certainly felt like "filling" an empty space on the map with the bursting sounds of civilization.[67] The museum filled the flatness of the pampa with a monumental 170,000-square-foot building. Its elliptical design took visitors through an evolutionary sequence, "from the simplest and most primitive organism to the book that describes it."[68] The "book that describes it" represented the scholarly publications pouring out from the museum's subterranean workshops, labs, and offices, literally supporting the history above them.[69] The main rotunda showcased frescoes that contextualized artifacts from the exhibits, showing them in use in Indigenous everyday life.[70]

The exhibits in the La Plata museum boxed the natural history of Argentina in a primordial time, before the nation-state and before Spanish colonization. Curators also placed Indigenous peoples in this atemporal zone, excluding them from a place in Argentine history. By archaeologizing Indigenous cultures, the curators reinforced the idea of national homogeneity based on Argentina's founding principle: it was a country with no Indigenous people. But in fact, the government had deported them from their lands, separating parents from children. The Mapuche, Tehuelche, and Puelche were distributed among different parts of Argentina following the military campaign of 1879–84. Yet Moreno brought three *lonkos* (chiefs) to the museum—Inacayal, Foyel, and Cañuel—and their families. They lived and worked in the building: the men posed for paintings, and the women made crafts.[71] They were observed, studied, but also neglected. Many, including Inacayal, passed away only a few years after their arrival and within a couple of years of one another. When they passed away, technicians promptly readied their remains for exhibition, a practice denounced by a local newspaper. The paper protested the management of human remains by museum staff, a task typically assigned to the police or the Church. This article attacked methodologies of knowledge that Moreno felt compelled to defend. For him, the study of Indigenous people in life and in death constituted a vital and legitimate step in developing knowledge about the Argentine nation.[72] In the words of Pamela Newkirk, "At the presumed summits of civilization, cruelty was cloaked in civility and a brooding darkness was hailed as light."[73] Other governments, such as Mexico's, were also constructing a historical timeline that included Indigenous peoples only because their absence symbolized a positivist actualization of the nation.[74] In the late nineteenth century, exhibiting remains of Indigenous people proved to scientists and museumgoers that the Indigenous Other was extinct, symbolizing the ultimate triumph of settler civilization over the Desert.[75]

THE DUELING SCIENTISTS

By the early 1890s, the governments of Chile and Argentina had appointed their lead experts (*peritos*) for the border negotiations, Diego Barros Arana and Francisco Moreno, respectively. Both men were embedded in scholarly networks at home, one leading a university, the other directing a museum. But Barros Arana, a prolific historian in his sixties, did not carry out explorations in the same way that Moreno did, who was twenty years younger. While Moreno had in Barros Arana a diplomatic colleague, his scientific arguments found in Hans Steffen, a German-born geographer working for the Chilean commission, an intellectual peer. Moreno and Steffen surveyed the Patagonian Andes, presented at scholarly meetings, and published their findings in national and international journals, crafting intellectual networks that illustrated the multilevel and multipoint creation of knowledge.[76] Steffen belonged to a generation of foreign professors hired by the Chilean government to teach at new institutions of higher education. Under the sponsorship of the state, he led several surveys of the Patagonian Andes in the 1890s. Moreno was more of a self-made man, who aptly took advantage of his acquaintances to gain access to scholarly circles and funding for his expeditions.

Steffen and Moreno, like most technicians working for the International Boundary Committee, believed that nature would reveal a sense of national self. Yet their scientific commitment to finding an objective truth collided with the nationalist interests of their missions to draw the boundary. As Steffen admitted, "From the beginning, these investigations acquired a different character due to the different points of view of the experts of both countries."[77] Each side gave preference to one of the two geographical features that appeared in the Treaty of 1881. Chileans favored the watershed divide, "whether that coincided or not with the crest, or was situated outside, and at a distance from, the Cordillera."[78] This could give Chileans access to fertile eastern valleys, which Moreno vehemently combatted.[79] In contrast, Argentines relied on the main chain of the Andes to draw the boundary, which could give them access to some coastline on the Pacific Ocean. By leaning toward basins or mountains, Chilean and Argentine scientist-explorers provided the other side an argument to access the other side of the Andes, something they agreed would never happen. The only way to resolve this tension was for Steffen and Moreno to provide evidence that they were each correct and the other one was wrong.

Moreno and Steffen engaged in a scholarly debate about their findings, focusing on two fundamental aspects of scientific exploration of Patagonia: toponymy and topography. They joined a cohort of scholars who published locally and internationally based on data collected on expeditions to the Andes. Quite often, these scholars accused each other of bending the truth about the geography of the cordillera to fit the border arguments. Publishing in the journals of learned societies enabled Moreno, Steffen, and their colleagues to earn recognition as experts among their peers in Chile, Argentina, and beyond.[80] Steffen and many of his fellow collaborators in Chile published extensively in German journals, such as the *Geographische Zeitschrift*. It was common for journals to translate German and English to communicate the latest updates in world geography to their readers. Particularly appealing was the journal of the Royal Geographical Society in London, as two members of the arbitral tribunal that would hear arguments for the borderline also participated in the governance of this learned society: Thomas Holdich was vice president of the society and Maj. Gen. John Ardagh was a member of the organization's council.[81] As a result, when Moreno and Steffen traveled to London to present their countries' arguments in 1900–1901, numerous networking activities took place in the context of the Royal Geographical Society.[82]

DEBATES ON TOPOGRAPHY

The basins of the Palena and Futaleufú Rivers (map 1.1) condensed many of the debates around the relevance of rivers or mountains for border demarcation. The Palena River drains from Lake Palena/Vintter toward the east, but it then sharply bends northwest, carving its way across the Andes into the Pacific Ocean.[83] There, locals called it Carrenleufu. Several rivers had different names on one or the other side of the Andes, leading to frequent confusion among scientists. The Futaleufú River originally flowed from a series of four lakes, like beads on a necklace, and it also bends westward to end, via the Yelcho River, in the Pacific Ocean.[84] The topography of these basins presented a dual problem for explorers. If they were to favor the watershed, the entirety of the draining area could belong to Chile, even if this included valleys to the east of the main chain of the Andes. However, if preference was given to the orography (the formation of mountains), then Argentina could claim sovereignty over the western valleys.

If geography was not enough, two other factors complicated the issue. First, Welsh migrants had settled in some of the eastern valleys near the upper Palena River and the Futaleufú basin, both areas claimed by Argentina. Fleeing poverty and oppression, Welsh immigrants began arriving in Argentina in 1865.[85] After an expedition led by Luis Jorge Fontana, first governor of the National Territory of Chubut, the Welsh founded Colony 16 de Octubre at the feet of the Andes. The fertile soil granted the colony prosperity, which in turn "gave way to various settlements in the neighboring valleys."[86] For Steffen, Welsh settlements represented Argentine encroachment into the trans-Andean valleys west of Colony 16 de Octubre along the Futaleufú and Palena Rivers. And it was even more frustrating that Fontana, among other

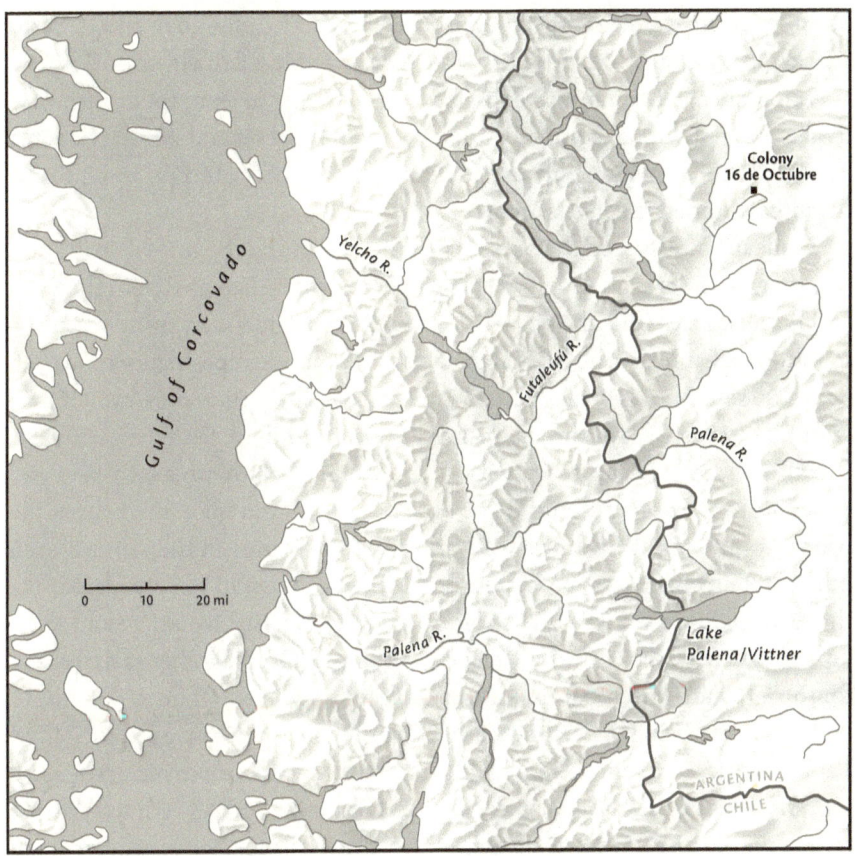

Map 1.1. The Palena and Futaleufú River valleys. Map by Erin Greb.

explorers of the area, did not recognize the Chilean conclusions that the two main rivers that entered the Pacific at that latitude corresponded to the two rivers that bracketed Colony 16 de Octubre.

In addition to the presence of Welsh farmers in what he considered Chilean valleys, a second problem fast-tracked Steffen's expedition to the Palena River valley. In 1892, some farmers from Colony 16 de Octubre found auriferous deposits near the Corinto River, a tributary of the Futaleufú River, and in the upper Palena River valley. These findings joined two other discoveries of gold, one in northern Neuquén and the other in Tierra del Fuego.[87] Argentine law recognized the right to exploit natural resources for those who had discovered them, so the government hurriedly granted land concessions in the Palena-Futaleufú region to promising investors. The excitement was short-lived. The cost of extracting and washing the gold pushed miners to abandon the quest "as soon as they reported their discoveries."[88] Individuals like Alberto Wecker in Chile and Paul Ahehelm in Argentina entered the historical record as recipients of land grants to exploit the mines, only to leave them without a trace.[89] An ephemeral company pooled capital from several Welsh residents in Chubut and in Wales, but lack of understanding of the terrain and internal disputes doomed the investment.[90]

Chileans interpreted the presence of the Welsh and the brief distribution of mining permits as an Argentine push to colonize the Palena River valley. For Steffen, it was clear that the power of rumor had sent intruders from Argentina to settle the Chilean valleys.[91] He suspected Argentine diplomats would then use these settlements to claim further territories down the valley, even if they might be to the west of the main chain of the cordillera. Steffen set forth to show the trans-Andean continuity of the Palena River with a two-pronged expedition. One crew would depart from the east and hike downriver; the other would depart from the west and walk upriver until they found each other. Steffen hoped to demonstrate that "these lands could be considered, like the 16 de Octubre Valley with its colony, as territories that would be under the jurisdiction of Chile."[92] Carlos Reiche (a botanist), Oscar Fischer (a draftsman for the Chilean border commission), and Steffen started off from the mouth of the Palena River in December 1893. The other crew, led by Dr. Pablo Stange, Pablo Krüger, and Pablo Kramer, "all German teachers employed at the local schools," began their journey from 16 de Octubre, the Welsh colony at the foot of the Andes.[93] On February 6, 1894, Fischer, approaching from the west, met Stange, who was coming from the east. Later that day, on the other side of the river, Steffen met Krüger and Kramer, therefore confirming that the Carrenleufu River and the Palena River were one and the same, originating in the east and flowing into the Pacific Ocean.[94] Shortly after, Argentine law enforcement captured "a part of the expedition" together with its transport animals and took them to Fort Junin de los Andes, 250 miles north. Local patrols accused the group of espionage, even though their passports had been visaed at Colony 16 de Octubre. Steffen denounced this incident to Diego Barros Arana and to the international scholarly community. The incident with Argentine local police reveals the spatial ambiguity in which explorers and law enforcement moved. For Steffen, it was a mistake that the upper Palena was "in the possession of Argentina." In fact, he believed the valley lay "to the west of the ranges of the cordillera that form the watershed" and should therefore belong to Chile.[95]

Moreno dismissed Steffen's arguments by supporting his conclusions with evidence from mountain ranges (cordilleras), crests (*serranías*), and hills (*montes*). For Steffen, it was evident that the watershed was to the east of the river, because that is why the Palena River flowed west. Moreno acknowledged this but argued that the crests were "rolling hills and ravines."[96] He concluded, "There is nothing in that region that can be considered a ridge to the east of the river, and what Messrs. Serrano and Steffen have taken as such . . . is only the foothills of the Patagonian plateau."[97] Moreno's focus on orography went beyond the Palena River valley debate. In 1896, the Argentine government authorized the navigation of Lake Lacar and its westward-draining river. It also founded an agricultural colony, San Martín de los Andes, on the eastern edge of the lake the following year. Barros Arana protested this settlement, arguing that the lands to the west of the continental divide belonged to Chile. Moreno objected to this claim, citing Barros Arana's own description of the Chilean territory in his multivolume *Historia jeneral de Chile*: "Two mountain chains that run parallel from north to south, constitute the basis of its [Chile's] orography. One of them . . . is the great and thick mountain range of the Andes, that rises to the east."[98] If the Andes were Chile's eastern border, then Barros Arana could not claim the eastern valleys. Farther south, Moreno also stressed the hierarchy of the main chain of the Andes for boundary demarcation. In his own survey of southern Patagonia, Moreno undermined claims that Mount Palique had a geological correlation with the Andes, which would have made it part of the cordillera. Mount Palique stands alone in a vast low terrain. A river carves its way along the southern slope, flowing east–west, merging with other streams and lakes to eventually arrive at the Pacific Ocean. According to Moreno, these valleys, which "promise easy prosperity to settlers," were located to the east of the Andes, overlooked by "all snow-capped mountains." Isolated peaks like Mount Palique, he contended, "do not belong by any means to the Cordillera."[99]

Chilean explorers favored hydrology and the continental divide to draw the international boundary, while Argentine diplomacy highlighted the authority of the highest peaks at the center of the range.[100] Moreno agreed that the Treaty of 1881 had been worded without much knowledge about the Patagonian Andes, namely that the watershed divide was located to the east of the main chain of the Andes. However, he protested Steffen's claims that the boundary should follow the watershed divide. This would have given Chile access to the fertile valleys on the eastern slopes of the Andes, like the 16 de Octubre valley.[101] The production of knowledge about the topography of the Patagonian Andes was subject to border negotiations, based on fieldwork by local surveyors, and circulated widely to advance the national interests of both countries.

DEBATES ON TOPONYMY

Moreno and Steffen also engaged in a toponymy debate. Both explorers gave name to sites in their travels as a means to affirm their country's right to that place. Toponyms, Raymond Craib reminds us, "matter not solely for the imprint they leave and the impression they make but also for the knowledges they hold, the identities they evoke, the history they convey."[102] Steffen recognized the historical materiality of place-names when he observed that most of the Pacific coast of Patagonia kept Indigenous toponyms. For him and fellow explorers, Native names echoed through time as a subtle hum reminding history of their absence. Colonization, in the form of European and Chilean expansion, erased the original toponymy from spatial references like maps and accounts. Spaniards imposed Christian names "taken mostly from the saints corresponding to the day of their discovery."[103] In the early nineteenth century, English nomenclature along the Patagonian coast evidenced the journeys of the British ships *Adventure*, *Alert*, and *Beagle*, the latter imprinting its name on the southernmost channel of the American continent. Beyond these early explorations of Patagonia, it was the nationalizing force of the Chilean Navy and the boundary subcommissions that displaced old names and established new ones in an attempt to convey a common history. By imposing names on sites and insisting on those names in cartographic documents, explorers "attempted to cleanse imagined territories from inconvenient pasts" to serve the nationalizing efforts of border-making.[104]

One of the most contentious debates between Steffen and Moreno involved the place-names of sites they claimed to discover. In December 1897, after his exploration of the Patagonian plateau and his conclusion that Mount Palique was not a mountain, Moreno sailed north across the Chilean fjords. He arrived at a long sound where a large river—he named it Las Heras—mouthed. He acknowledged that the channel had already been named Calén by a Spanish Jesuit in the eighteenth century.[105] More recently, Chilean sailor Ramón Serrano Montaner had included "Estero Calén" (Calén Sound) in his course annotations of the Chilean fjords (map 1.2).[106] Most explorers, including Moreno and Steffen, recognized Serrano's work and built on his expeditions. Moreno had referenced this toponymy in a presentation to the Royal Geographical Society, which included "photographic slides of that inlet ... and [a] small sketch-map ... [where] these names may be seen."[107] Almost a year later, in September 1898, Steffen led a crew to survey the environs of Calén Sound, only he called it "Baker." As they charted past what is today known as Steffen Sound, the German explorer spotted "a waterway

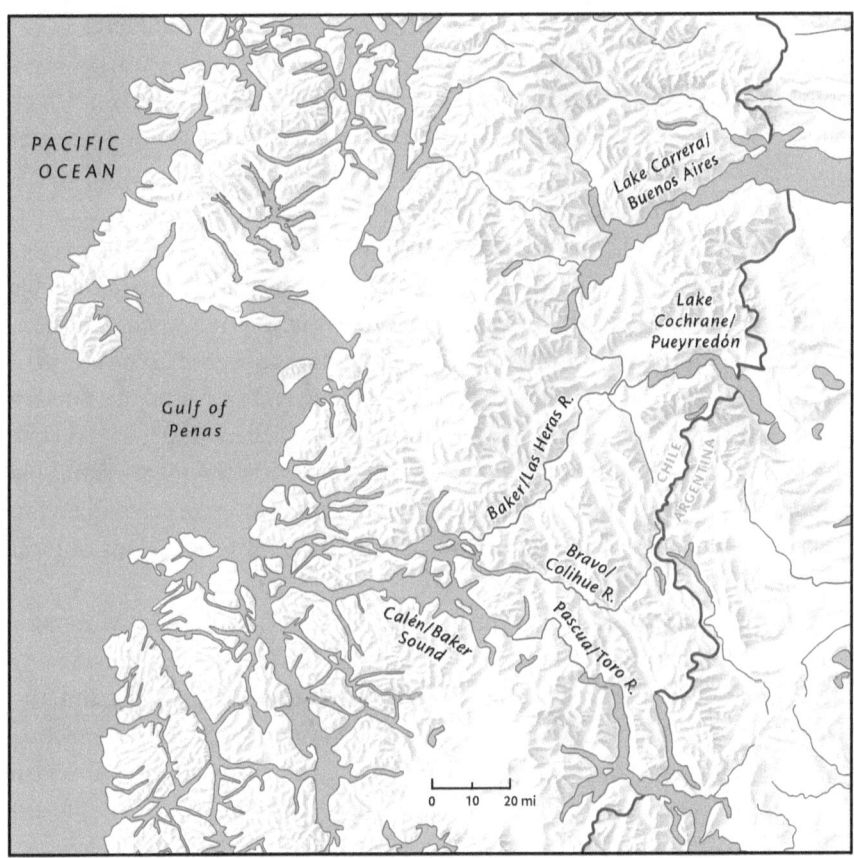

Map 1.2. The Calén/Baker Sound. Map by Erin Greb.

that surely came from the deep interior of the mountains." They did not see any evidence of earlier "prominent" explorations. As a result, Steffen claimed rights of discovery and named this waterway Baker River, because it flowed into the homonymous channel.[108] In doing so, Steffen's crew undermined Moreno's authority. In addition to rejecting Moreno's name for the sound, Calén, Steffen named not one but three rivers flowing into the Baker Channel: Baker, Bravo, and Pascua.[109] In March 1899, one of Steffen's field reports made it to the pages of the *Petermann's Mitteilungen*, a German geographical journal. The notes framing the report praised Steffen for "filling in [an] empty space left on the maps," acknowledging Steffen's work in shrinking the emptiness of Patagonia and expanding the horizon of knowledge.[110]

Moreno protested this attack to his scientific authority in the August 1899 issue of the London *Geographical Journal*. "Under my direction," he wrote,

Science for the Nation   39

"we discovered a large river, to which I gave the name of Las Heras (river Baker of Dr. Steffen); explored in part, river Colihue (that is to say, river Bravo of Dr. Steffen), and the river Toro (river de la Pascua of Dr. Steffen) named by me after the Chilian ship that preceded us." The Argentine scientist also accused Steffen of misconstruing the truth about what he saw: "Surely Mr. Steffen saw our encampments in the [Calén] inlet, in the rivers, and in Lake Pueyrredón."[111] Moreno might have been first to give names to sites scarcely known to Western geographers in 1897, but he was slow to publish his results in the circles that mattered to the experts in the border commission. In May 1899, only two months after Steffen's report appeared in the German journal, the Argentine explorer gave a talk at the Royal Geographical Society and introduced his own toponymy for the channel (Calén) and the river (Las Heras).[112] In a subsequent article he displayed further the place-names his explorations had given to lakes, rivers, and mountains. In an accompanying map, he silenced any other toponymic claims, from either Chileans or local inhabitants.[113]

Chilean and Argentine crews of explorers "almost simultaneously" found a lake in the summer of 1897–98 where the Baker/Las Heras River presumably originated.[114] Similar to what happened with the Baker/Calén Sound, the two sides gave different names to the lake. However, rather than resurrecting previous toponyms, the Chilean and Argentine names imposed on this lake honored national heroes: Chileans called it Cochrane and Argentines Pueyrredón. Thus, unlike in the Baker/Calén case, the Cochrane/Pueyrredón names evidenced how the nationalizing mission of the explorations bled into the scientific border negotiations. Moreno contested the *Geographical Journal*'s use of Chilean toponymy: "The name of Lake Cochrane was given to Lake Pueyrredón by Chilean explorers posteriorly to its discovery and survey by my assistants."[115]

Needless to say, in his 550-page report to the British arbitral tribunal, Moreno did not recognize any of Steffen's toponyms. In addition, he claimed Argentine right of discovery to at least seven lakes, including the contested Lake Cochrane.[116] Unsurprisingly, Diego Barros Arana, the Chilean expert before the tribunal, did not use Argentine names. Today the Chilean-Argentine boundary cuts through five lakes, each of which have different names on each side. However, the dual names do not represent remnants of the toponymic debate between Steffen and Moreno. By the time of the tribunal decision, explorers and authorities agreed on the names of known lakes in their cartographies. The exception remained Lake Cochrane/Pueyrredón, the only of the five binational lakes encountered by Chilean and Argentine explorers almost simultaneously. In the case of Lakes Buenos Aires and San Martín, these were names given by Argentine expeditioners Carlos Moyano (to the former) and Francisco Moreno (to the latter) in the late 1870s. Those who followed in their footsteps into the Andes, like Steffen, recognized their work by incorporating their toponyms in their maps. Eventually, Chileans changed the names of Lakes Buenos Aires and San Martín to honor national figures, becoming Lake General Carrera and O'Higgins Lake, respectively.[117] This change honored two heroes of Chilean independence, José Miguel Carrera and Bernardo O'Higgins, both of whom sought Argentine assistance in an hour of need. Thus, while the lakes' toponyms might have pushed against the spatial traces of Argentineness, they might also evoke their common history across the Andes. Different toponymy for the same bodies of water in the Andes persists to this day.

CONCLUSION

The valley-centered approach to border-making appeared in Barros Arana's objections to Argentine arguments presented to the arbitral tribunal. The treaties of 1881 and 1893 stressed the orographic and hydrographic nature of the border, but Argentines only partially accepted this. The Chilean expert contended the boundary should run through the highest peaks while also dividing the watershed. Confronted with the geography of the Patagonian Andes, Barros Arana protested that Argentines went looking for "absolute maximum altitudes," disregarding the watershed principle.[118] However, the Chilean objections were mild at best. Preoccupied with diplomatic treaty interpretations, the authorities overlooked the social dynamics in the Palena and Futaleufú River valleys. Their shortsightedness "allowed the spread of Argentine colonization to the litigious valleys." By the time Barros Arana submitted a boundary proposal to the arbitral tribunal, the Chilean borderline "would have meant an unfortunate intervention not only in new settlers' peaceful existence as farmers and ranchers but also in the relations that had been created with the eastern coast."[119]

Moreno, on the other hand, exploited opportunities to gain territory for Argentina. Ahead of the International Boundary Committee's inspection of the area in 1902, he planned to show off Argentina's "positive and efficient colonization policy."[120] A little admiringly and a little indignantly, Steffen observed that Moreno instructed his assistants to check in with settlers of the Palena and Futaleufú valleys. In the event of questions "from the delegates of the arbitration award, they would declare themselves to be loyal Argentine citizens."[121] Furthermore, he arranged for settlers to express their desire to belong to Argentina when he visited the Welsh Colony 16 de Octubre. On April 30, 1902, "many settlers" in the colony allegedly signed a petition asking Holdich to resolve the border dispute promptly "so that they could focus without worry [on] the cultivation of their fields [and] the construction of [their] houses."[122] With this petition, the agricultural Colony of 16 de Octubre created a foundational myth about Argentine presence in the Andean valleys. To those present in Colony 16 de Octubre, the proclamation was less a desire to recognize Argentine authority and more a concern over their land titles issued by that government.[123] Holdich favored the Argentine claims. While he did not appreciate imperial subjects fleeing the inconvenience of hard work for the unruly Patagonian pampas, he admitted that the Argentine government was well equipped to "reconcile them permanently."[124] The ruling demonstrated how historical contingency circumscribed scientific negotiations. Chilean complaisance when Argentines allowed the colonization of the litigious valleys (even though it was forbidden by the treaties) and indifference to scientific reports resulted in their loss of fertile farmlands in the Palena and Futaleufú River valleys.

The debate about the Patagonian mountains elucidates the prevailing ideas about geography, science, and authority in the late nineteenth century. Scientists and authorities envisioned Patagonia as a non-place, outside of national histories. Drenched in imperial views of the world but animated by local interests, they justified the deployment of violent nationalizing policies, from military raids to border demarcation. The universal ambition of positivist science flooded the specificity of Patagonia. To some extent, this explains how effortlessly the Argentine government appointed an explorer of the Chaco, Luis Jorge Fontana, governor of Chubut and how the British government chose Thomas Holdich, who had participated in the Russo-Afghan and the Pamir Border Commissions, to mediate between Chile and Argentina.[125] Holdich assumed that "geographical ignorance" was the root cause of every dispute; thus scientific knowledge allowed "civilized communities" to agree on a boundary.[126] Border negotiations were a scientific matter.

The topography of the Patagonian Andes, however, limited scientists' universal ambition. What worked in other mountainous borders did not work here; what worked in other sections of the Andes could not be replicated south of the thirty-ninth parallel. Explorers filled their notebooks with scribbles, measurements, and sketches of the Patagonian landscape. It was in the publication and circulation of these reports that their findings had an impact. Letters to authorities, academic articles, and cartographic materials incorporated new data and disseminated new conclusions. The debates between Steffen and Moreno illustrate how scientists integrated new knowledge into their work but also how they were strategic about what information they accepted. Hence, both in circulating their arguments and in responding to others' conclusions did Steffen and Moreno contribute to a transnational conversation about geography, nation, and borders in service to their nations. In cooperation with colleagues in their countries and abroad, Steffen's and Moreno's examples highlight the multiple sites of production and reproduction of knowledge. The Andean valleys they so frequently considered empty and peripheral proved central to the making of the nation.[127]

Chapter Two

SETTLING PATAGONIA

1890–1915

This chapter is about settlement. Authorities in Chile and Argentina, like others across the Americas, used state-sponsored (or, at least, state-encouraged) settlement as a means to incorporate distant territories into the orbit of new nation-states. Based on the knowledge generated by explorers and surveyors, authorities on both sides of the Andes tried to frame the colonization of the South according to the way they imagined the nation. They could be understood as men that did not fully comprehend this space or the people in it, and their ideals were very far from reality, resulting in conflict. However, here I propose a different approach, away from focusing on the distance between the law and what was happening on the ground. I argue that even though these efforts sought to eliminate the Desert, that emptiness of history and identity, they created one, or at least they redefined it. Legislation distributed land in southern Chile and Argentina among nonlocals, especially among foreigners. Understandably, newcomers arrived with their own ideas of space, of community, and of nation. The attempts to eliminate the Desert ended up perpetuating the sense of emptiness of history and detachment from the nation that authorities obsessed over, which resulted in conflict among neighbors, landholding companies, and authorities.

Attempts to colonize Patagonia certainly rhymed with settler colonial practices elsewhere. Chilean and Argentine governments sought to incorporate Patagonia into the nation through the dispossession of Native inhabitants from their lands, the extermination of their culture, and their replacement with an immigrant population. Hence, settler colonialism structured a postcolonial vision of spatial organization that rested on the creation of private property, a trend common across the Americas. Indeed, ruling elites of the second half of the nineteenth century resorted to myriad policies to consolidate the territorial expansion of the state and the capitalist exploitation of resources, at the expense of Native inhabitants. These strategies included legislation on landed property, military campaigns, and colonization at the hands of individuals or companies. For instance, the Brazilian Land Law of 1850 "forbade the acquisition of public land through any means but purchase," while the Lerdo Law of 1856 in Mexico sought to privatize communal lands and "mobilize them in a commercial economy."[1] Farther south, Argentine authorities accompanied expropriation laws with a military campaign against nomadic and semi-nomadic communities in Patagonia and Chaco.[2] From the coffee plantations in western Guatemala through the ranching estates in southern Brazil and the sugar mills in northern Argentina, the steady privatization of lands, especially at the hands of few companies, cornered peasants into selling their land and joining the workforce.[3] In essence, argue Juan Castro and Manuela Lavinas Picq, this method of state formation rested on a long history of land grabs from Indigenous communities, which, ultimately, structured the modern state.[4] Land dispossession and redistribution, then, lay at the center of policies of "incorporation" of different territories into the nation. Settler colonialism provided a configuration for the nation, but it did not yield results.[5]

I begin this chapter by examining the legal framework that legitimized the removal of Mapuche communities from their ancestral lands in Araucanía, Valdivia, and Llanquihue in Chile. In the mid-nineteenth century, the Chilean government created three new provinces south of the Bío-Bío River that would receive legislation on landed property: Arauco (1852), Llanquihue, and Magallanes, the latter two both created as colonization territories and then upgraded to provinces in 1861 and 1929, respectively. Legislation defined Arauco as "Indian territories south of the Bío-Bío River and north of Valdivia

Province."[6] Llanquihue was carved out from its northern neighbor, Valdivia, and the continental area of Chiloé Province in the South. These new districts implied that Indigenous communities lived in specific places—that is, in the now province of Arauco—with serious consequences for those living farther south.[7] Over the course of the second half of the nineteenth century, Arauco suffered multiple changes to its administrative boundaries, all resulting from policies that sought to incorporate this Mapuche heartland into the Chilean nation.[8] For this reason, I follow other scholars in referring to the historic region between the Bío-Bío and Toltén Rivers as Araucanía. In the 1860s, scholars estimated far more than 100,000 Indigenous people living in Araucanía. This number decreased in the following decades as a result of military advances and displacement by settlers, to about 90,000 Mapuche in the early twentieth century. In the provinces of Valdivia and Llanquihue, scholars estimated about 30,000 inhabitants around 1845, where the majority was probably of Indigenous background.[9] Because of this demographic composition, it was urgent for Chilean authorities to target landed property legislation for Araucanía and parts of Valdivia and then adopt (and adapt) it to the provinces farther south.[10] With this, state bureaucracy forced Indigenous populations under the jurisdiction of governors, inspectors, and judges, effectively bringing them under national sovereignty and suppressing their autonomy.[11]

Land conflict in southern Chile and Argentine northern Patagonia originated in the multifaceted violence against the Mapuche.[12] Beginning in the 1850s, demand for foodstuffs around the globe pushed farmers to cross into the fertile lands south of the Bío-Bío. Individuals bought more and more land, slowly and surely threatening Mapuche livelihoods in Araucanía. Despite the fraudulent nature of many of these purchases, new settlers would claim ownership later on.[13] Since the 1860s and through the early 1880s, Chilean military forces plundered, intimidated, and advanced on Araucanía, while Congress legislated on landed property to legitimize the state's distribution of lots. Following a Mapuche coordinated uprising in 1881, the Chilean army launched a series of counterattacks, collectively known as the War of Occupation (1881–83), forcing Indigenous leaders to surrender. In early 1883, the government refounded the colonial city of Villarrica, which Spaniards had been forced to abandon in 1603, symbolizing the effective occupation of Araucanía.[14] In Valdivia and Llanquihue, the situation was very different. Immigrants arrived in the midcentury, settling in the old cities of Valdivia and Osorno and the new towns of Puerto Montt and Puerto Varas. Newcomers established businesses and farms, displacing the Mapuche-Huilliche (Mapuche of the South) from the best land and into the rugged terrain of the Pacific coast or the foot of the cordillera.[15] Overall, this strategy was rooted in what Jorge Pinto identifies as an ideology of occupation that portrayed Indigenous peoples as incorrigible enemies of the nation and, therefore, deserving of military action.[16] Creating public lands in the South allowed the national government to "set a precedent to grant itself legal and military sovereignty south of the Bío-Bío River," especially after the War of Occupation in Araucanía. Additionally, new legislation provided the state with a means to disarticulate persistent colonial-era hierarchies in the countryside.[17]

In Argentina, minister of war Julio A. Roca launched a military campaign to occupy northern Patagonia in 1879. Some documents, especially those referring to military awards, called this operation the "Expedition to the Negro River," a term I have mostly seen in documents about land distribution among veterans of the campaign.[18] But the designation that took hold was "Conquest of the Desert," mostly intellectualized by military officers and politicians. It condensed a foundational narrative for the national state anchored in dispossessing Indigenous lands and making them available for non-Indigenous settlement.[19] Beginning in 1879, several military columns advanced from different forts toward the Negro River and the cordillera. They arrived at Lake Nahuel Huapi in 1881, where Argentine forces hoisted a flag to symbolically secure sovereignty from that point to Patagonia. Through 1885, the Argentine army cornered different Indigenous communities against the cordillera or in the South, away from their ancestral lands. The army flocked them in gathering points and redistributed them in multiple directions with the express purpose of disarticulating Indigenous organization and reeducating them to become part of Argentine society.[20] But redistribution of Indigenous people was slow, and soon these meeting places turned into effective concentration camps. The state chose able-bodied people from among those who were relocated and sent them to labor in other parts of the country in sugar fields, households, or even the Museum of Natural History in La Plata, per Francisco Moreno's request.[21] The government also set up a disciplinary camp in Martín García Island, in the River Plate, to educate, Christianize, and inoculate prisoners, including school-aged children. Finally, orphans were put up for adoption and widowed mothers were requested for domestic service in urban households.[22]

New legislation created avenues for land tenure for people of different national origins, favoring European immigrants more than any other group and vilifying Indigenous presence. I use reports from Chilean land inspectors to illustrate how authorities reconciled their expectations of settlers with the challenges newcomers faced when moving to a new place. These difficulties combined with a heightened urgency to bring settlers into the context of the

Chilean-Argentine border negotiations of the 1890s. As a result, the Chilean government transferred the colonizing efforts to private capitalists, distributing large estates in spaces they deemed empty, reconstructing the ways applicants positioned themselves as makers of the nation in southern Chile. But settlement did not occur in a vacuum. Indigenous and mestizo Chileans found themselves tormented by new land grants to holding companies. Hence, I examine how *agricultores* (title-less farmers) resisted the violence of removals often by framing their agrarian practices as means to build the Chilean nation as well, which garnered sympathy from local elites.

I finish the chapter with an overview of land distribution in Argentine northern Patagonia to pinpoint similar genocidal policies against Indigenous peoples. The legislation that distributed lands in the aftermath of the military campaign failed to populate the South. Against the expectations of the law, land surveyors and provincial authorities noted the slow progress of settlement. With it, they inadvertently prolonged the Desert they were trying to eliminate. As a result, the incorporation of Patagonia to the nation remained incomplete.

THE LAW THAT FRAMED THE LAND

A person's access to landed property in southern Chile depended on their ethnicity and the location of the land they applied for. The Colonization Law of 1845 allowed foreigners to obtain land in Llanquihue and Valdivia, while in Araucanía and Chiloé, foreigners had exclusive access to state-managed agricultural colonies.[23] In areas without colonies, such as Llanquihue and Valdivia, foreigners could lease available lots. The Chilean government, displaying a concerted effort to incentivize immigration, subsidized travel for overseas comers, tools, and seeds for new settlers and did not collect taxes for the first twenty years. Foreign settlers received a plot of land on the condition they lived there with their families and worked the land. To encourage settlement, the government earmarked fertile expanses of land for newcomers, making them available at a nominal cost, which often displaced Mapuche communities to smaller lots or reservations typically far from where they had lived for generations. Additionally, maritime vessels ferrying immigrants to Puerto Montt did not pay anchor and tonnage fees and were permitted to load produce and timber for export without incurring additional charges.[24] *Colonos*, or settlers, who proved residency by building a home and working the land within the first three years could be granted a property title.[25] The government harbored a preference for immigrants who exhibited several desirable attributes, including being male, preferably of European or American origin, and possessing rural skills.[26] On occasion, the government did permit the settlement of single or widowed men, often integrating them into labor infrastructure initiatives in Llanquihue before offering them a lot. This policy met with limited success, as most immigrants remained in Puerto Montt or left for the mining north.[27] Finally, immigration status did not preclude newly arrived settlers from taking up residency in unoccupied lots, which precipitated conflicts, both among settlers themselves and between settlers and the state. These tensions persisted throughout the turn of the century, shaping the dynamic landscape of southern Chile's land distribution and settlement.[28]

Mestizo Chileans, which the documents usually referred to as "national" settlers (*colonos nacionales*) sparsely populated Valdivia and Llanquihue since before the arrival of German-speaking immigrants in the mid-nineteenth century. Mestizo Chileans in Araucanía, Valdivia, and Llanquihue existed at the margins of non-Indigenous and Mapuche societies, unable to claim a place in either one. Some emigrated to Argentina, pulled by the Avellaneda Law of 1876. This legislation underpinned immigration policy for the following decades, assigning to newcomers the responsibility to colonize the Argentine countryside.[29] In both countries, the mestizo population had little access to land leases that could result in permanent titles. In the eyes of the state, they settled illegally, rented from others, and worked in local estates, sometimes moving seasonally between countries. They were not at the center of policy making in Santiago, but land tenure policy did impact their livelihoods. Since many resided in Valdivia and Llanquihue before the government granted land leases, their presence triggered legal conflicts for newcomers who encountered people living in their assigned lots.[30]

Indigenous groups had uneven access to land through new legislation. The Mapuche in Arauco and Cautín could only apply for land titles through the Comisión Radicadora de Indígenas (Indigenous Settlement Commission). This commission was established as part of the Law of 1866, which regulated the foundation of towns, auctions of public land, and transactions between Indigenous and non-Indigenous individuals in "Indigenous territory." In the context of the Law of 1866, it was safe to assume that "Indigenous territory" typically referred only to Araucanía.[31] The Law of 1866 stipulated that Indigenous peoples could only get land titles (*títulos de merced*) through the Comisión Radicadora, which only began its work in 1883, after the military campaign in Araucanía. Claimants needed to show evidence of their Native status and multiyear residency in the lot they applied for, forcing Mapuche *lofs*, or extended family groups, to settle and register their property. To support its work, the Comisión Radicadora often rallied Capuchin and Jesuit missionaries for help approaching Indigenous communities about settlement.[32] After witness testimony of an applicant's Indigenous status and long-standing residence, a surveyor would grant the title in the presence of neighbors and a protector of Indians, a government-appointed officer in charge of advising Indigenous communities on legal matters.[33] While Indigenous land access through the Comisión Radicadora segregated Mapuche property from other settlers', it also addressed the rampant corruption among local authorities that forged sales or forced people to sell their lands. The Comisión Radicadora first focused its work solely on Araucanía, and only in 1907 did it begin its work in Valdivia and Llanquihue.

The Law of 1866 reflected a statist approach to land leasing, reducing Mapuche communities to smaller territories and forcing them to register their lands.[34] The law reinforced the role of state agents in certifying sales or leases to halt illegal occupation. Failing to prove ownership enabled the state to create the notion of vacant land as public property, which the government could then sell or lease.[35] The creation of public property through the delineation of private lots contrasted with the colonizing strategy in the United States, Canada, and Argentina, where immigrants could buy land directly from Indigenous people. Additionally, Chilean authorities coupled legislation and military occupation to exclude Indigenous peoples from the republican project.[36] Furthering this ideology of occupation, in 1883 the government forbade transactions between Indigenous people and other settlers in Arauco, a ban that was extended to Valdivia, Llanquihue, and Magallanes ten years later and through 1903.[37] Anchored on a discourse that vilified Indigenous peoples, the occupation of Mapuche territory would unify the Chilean geographical space while incorporating fertile lands into the capitalist economy.[38] By creating a legislation that regulated land tenure and taxation, the Chilean government forced the Mapuche to apply for titles to their own lands.[39]

Legislation on land tenure initially seemed straightforward, but it ultimately resulted in chaos. Relatively low land prices in Valdivia and Llanquihue in the mid-nineteenth century encouraged speculators to accumulate and sell land.[40] However, as land prices surged in the 1870s and 1880s, this prompted various Chileans from different backgrounds to settle without obtaining leases, exaggerate their land boundaries in an existing lease (if they held one), or deceive their Indigenous neighbors into selling their land.[41] The state bureaucracy was soon distracted from its colonizing mission by complaints, petitions, and trials between settlers with leases and settlers without them but who claimed landownership by working the land. In 1883, a law tried to patch the problem by prohibiting the Mapuche from selling or renting their land for ten years; the law was then extended in 1893. At heart, this law forced Indigenous communities into small lots and agriculture, a strong contrast from their cattle-herding, foraging, and fishing traditions in large spaces.[42] Additionally, the state tried to fight illegal occupation or purchases of lands by foreigners and mestizo Chileans alike, revealing the weakness of rural titles.[43] State officials resented that the government was missing out on an opportunity to collect taxes as people settled on empty lots or as they bought land from the Mapuche.[44] Beginning in the 1890s, the Chilean government granted land to holding companies with the responsibility of bringing immigrants to settle. With this, authorities addressed the slow colonization of the south of Chile, especially at a time when border negotiations with Argentina were at their most strained.[45]

AGAINST NATIVE CULTURE

The urgent drive to accelerate the colonization of Llanquihue is better understood against the backdrop of authorities' anxieties in Araucanía, often associated with Indigenous culture. Distributing land to foreigners carried the implicit goal of transferring knowledge and morals to the Chilean population, including Indigenous people. In the context of the colonization of southern Chile, morality evoked Western values of social, economic, and political life. While these were not outlined in land contracts, we find them in yearly reports of land inspectors and in contrast to Indigenous traditions. Polygamy exemplifies a tradition that Chilean authorities viewed as immoral and, therefore, undermining the nation they were trying to build.

In the Mapuche world view, having multiple wives signified social and economic status and also served to weave alliances.[46] Incorporating women from other families could extend a *lonko*'s (chief of family) network. Additionally, multiple wives signified wealth and status among peers. In the Mapuche world view, polygamy was not only accepted but desired. Nineteenth-century Chilean writers, however, criticized polygamy as an attack on civil society and Christianity and as the root of moral poverty among Indigenous groups. Authors like Benjamín Vicuña Mackenna protested that polygamy enslaved women to sow the fields and saddle the horses of idle spouses.[47] By portraying Mapuche women as victims of male practices, governing elites used the trope of Mapuche polygamy to delegitimize the authority of Indigenous men as heads of families.[48] A friar worried in 1896 that "as a consequence of the infamous and repugnant vice of polygamy, in which [Indigenous people] have lived and live at present, parents do not always know their children, nor do the latter recognize them as such, nor do they respect or obey them."[49] The reports that arrived at the Ministry of Foreign Affairs in 1899 complained that polygamy was "the greatest obstacle that now stands in the way of the conversion and civilization of the [Mapuche] Indians."[50] Accounts made sure to illustrate that smaller, non-Mapuche Indigenous groups of western Patagonia, now either disappeared or forced to accept Chilean laws, were not polygamous at the time Spaniards encountered them, appealing to a sense of (past) good savage. A navy captain described groups on Cailín Island, southeast of Chiloé: "They did not know drunkenness or polygamy, so deeply rooted in other reductions."[51] But the anxieties about Mapuche traditions served as a mirror for authorities to see their work. As happened all around the world where Western powers colonized territories, high-ranking officials saw in the presence of Indigenous peoples the failure of their work. "The Indian," wrote the head of the Office of Lands and Colonies in 1899, "is the screen that serves to mock the work of the [national] State in its own lands."[52] Indirectly, these inferences attempted to portray Mapuche *men* as morally corrupt.[53]

Gendered attacks on Mapuche political organization around a polygamous *lonko* undermined their communal power. Disarticulating such loyalties would free up space for two other patriarchal forces: the state and the Catholic and Protestant Churches.[54] For the Comisión Radicadora de Indígenas working in Araucanía to settle Mapuche families, the law that initially allowed traditions such as polygamy to persist needed amendment. "The Indigenous person does not respect our laws," argued a proposal, to which the commission submitted "matching their civil status to that of the rest of the population."[55] In layman's terms, this meant forcing Mapuche people to accept Western understandings of family unity and marriage. Specifically, disbanding polygamy had very concrete uses for land distribution, as Chilean law recognized only one line of succession for heirs. Private property, as a result, crippled Mapuche civilization because it destroyed the multispousal family unit and its landed property. Under the guise of moral superiority, the Office of Lands and Colonies believed that "settling Indigenous people" and "subject[ing] them to our social community" would disappear Indigenous culture. Brazenly, Agustín Baeza Espiñeira predicted that "the stronger race absorbs the weaker race."[56] The Mapuche confronted the national state with its own skeletal power in Araucanía. In turn, this further motivated authorities to erase Indigenous heritage.

INSPECTING THE NATIONALIZING EFFORTS

Two state agencies worked exclusively south of the Bío-Bío River to accelerate the settlement of "desirable" populations. The Office of Lands and Colonies operated under the Ministry of Foreign Affairs and at times bore the term "colonization" in the title, reflecting its original mandate to coordinate the immigration of foreign settlers and assist consular representatives abroad. Other responsibilities included surveying lots, compiling documentation, granting titles, and frequently representing the interests of the state in courts. The office functioned based on perceptions of southern Chile as an internal frontier that would be eliminated through land distribution and the settlement of non-Indigenous, preferably foreign farmers. In addition to the Office of Lands and Colonies, the Chilean government also created the Office of Land Surveys in 1907. Its immediate goal consisted of setting landmarks on the boundary between Chile and its neighbors. It also carried out geodesic measurements to create maps, draw administrative boundaries, and address land disputes that arrived at its desk.[57]

Land engineers and land inspectors spearheaded the colonizing mission south of the Bío-Bío River. Engineers measured lots and distributed leases or titles, thereby creating a body of spatial knowledge that solidified a vision of rural Chile centered on landownership, as discussed by Amie Campos.[58] Land inspectors oversaw settlers in state-sponsored colonies, providing insight into how the expectations of European colonization and environmental transformation manifested in southern Chile. Their assessments of state colonies in Llanquihue and Chiloé serve as a window into how authorities distinguished between successful and failing settlers. Sub-inspectors of lands and colonization stationed in three jurisdictions (Temuco, Valdivia, and Llanquihue and Chiloé) enhanced their narrative reports to the inspector general with detailed tables of every family leasing a lot in a state colony. These records included the name of the household's head, typically a man, along with a breakdown of the number of male and female family members and their ages. For instance, a 1902 report featured Juan Jönsson's family, consisting of five male members, between seven and fifty years old, and four female members, the youngest being an infant. With the exception of one Chilean member, the entire family hailed from Sweden and had arrived at the Chacao colony on Chiloé Island on September 22, 1895. The inspector assessed the property, as migrants were required to meet specific thresholds to demonstrate successful settlement. Even though the Jönssons had cleared less than 10 percent of the land for agriculture, the inspector deemed their efforts acceptable, given that they cultivated wheat, potatoes, and oats and possessed two oxen, one cow, two calves, three horses, one pig, thirty sheep, and poultry.[59]

Observations occasionally transcended mere data recording. Amid the linguistic diversity in state colonies, one sub-inspector marveled at immigrants living "in peace and good harmony," idealizing the commitment to hard work and land among settlers.[60] Authorities openly favored German and British settlers, praising their contributions to the prosperity of the region.[61] Their productivity and good behavior made them "good" settlers.[62] But some authorities also believed immigrant tensions disrupted the idyllic narrative. Immigration recruiter Nicolás Vega argued for Chilean authorities to mediate disputes, portraying immigrants as both part of the "civilized" community and slightly superior.[63]

Legislation often failed to anticipate rural challenges, hampering colonization. Some newcomers struggled in their first year despite government support. *Colonos* had to clear land for crops and often lacked proper roads to markets. Weather, particularly in winter, hindered travel. Language and cultural differences posed obstacles. Authorities aimed to establish concentrated communities for "religious services, instruction, [and] small businesses essential to farmers."[64] However, efforts to integrate foreign settlers into the national fabric, like by providing Spanish instruction to "English, French, German, and Dutch settlers" in places such as Huillinco, proved challenging in practice.[65]

Many immigrants did not passively accept land. Some settlers opted to relocate, either due to changing preferences or unmet expectations in southern Chile. They filed their grievances with the diplomatic offices, citing issues about the quality of the lot, unfavorable weather conditions, or the spending power of the Chilean peso, which often irritated the immigration agents in Europe.[66] For instance, Norwegian Cristian Pedersen refused his lot until more fishing families arrived. Otto Rehren, the sub-inspector of land in Llanquihue and Chiloé, reported that out of the 336 foreign-born families (comprising 1,801 individuals) who arrived to settle in state colonies between 1895 and 1901, about 70 percent eventually left for urban centers, mining regions in the North, or rural areas with better infrastructure. This trend was consistent across nationalities, with all Russian and Australian immigrants in Chiloé abandoning their lots, as did all but one Belgian family.[67] Arturo Whiteside Toro, a naval officer who accompanied another land inspector, concluded that "either the colono is useful for work, but useless for the one who requires this colonization; or else the colono is useless for all work and only apt to live off others, he is a parasite."[68] Rehren echoed Whiteside Toro's assessment, suggesting that certain foreign *colonos* were "completely incapable of resisting the sacrifices imposed on them by settler life."[69] Both officials believed that not all ethnicities were prone to succeed, perpetuating xenophobic notions of difference.[70]

Immigrants also faced health challenges that hindered agricultural production in southern Chile. Conditions such as diabetes, heart disease, sores in the extremities, and rheumatism prevented newcomers from working the fields. For instance, Cornelio Branje, a fifty-year-old Dutchman who relied on crutches, believed he had contracted an ailment (we do not know which) during his voyage to Chiloé. His daughter was paralytic "and need[ed] to be carried," and his grown son "cannot work because he is not strong enough."[71] Another family, the Rosses, settled in Ancud in the mid-1890s. Shortly after their arrival, the mother succumbed to cancer, and the father, plagued by rheumatism, struggled to tend his land. Tragedy struck when one of his eight children accidentally ignited their wooden home. With the father bedridden, the children remained inside, unaware of the danger. Fortunately, a passerby rescued them, but they lost their possessions. Finally, the father's failing health put him in the hospital, leaving all his children up for adoption.[72] The incident raised the question of how ailing immigrants could contribute to the nation.[73]

LARGE ESTATES FOR THE NATION

At the end of the nineteenth century, the Chilean government transferred to private capitalists the expense of bringing foreign families and building infrastructure. Border negotiations with Argentina raised concerns among Chilean authorities about the scanty populations in Llanquihue and Valdivia. Despite these worries, the 78,315 inhabitants in Llanquihue Province (excluding Chiloé) in 1895 surpassed all the population of Argentine Patagonia (54,955 people), who lived mostly on the Atlantic coast.[74] Concerns had to do more with the type of population than the numbers, which is why inspectors assessed colonies not only in terms of amount of people or produce but also in terms of their attitude toward work.[75] Additionally, food security also underpinned the creation of private property in Llanquihue, as the government sought food independence from Argentine imports. In the late nineteenth century, local markets in southern Chile depended on Argentine cattle and grains, but industrialist organizations in Chile, like the Sociedad de Fomento Fabril (Society for Industrial Development) and Sociedad Nacional de Agricultura (National Agricultural Society), lobbied to change this and give land access to productive enterprises. As a result, in 1893 the government extended to Valdivia and Llanquihue an Araucanía law that allowed the expropriation of properties with "dubious" origins, where people (Native, Chilean, and foreign) had been working the land before the state could grant titles. In essence, argues Jorge Muñoz Sougarret, the new legislation reversed land from private to public with the almost exclusive purpose of redistributing it among private businesses that could make it profitable. In the eyes of the governing and economic elites in Santiago, neither Indigenous inhabitants nor settlers working a small farm were apt for the job.[76]

Under the auspices of the Office of Lands and Colonies, several individuals received large land concessions with a requirement to bring European immigrant laborers. This practice facilitated the concentration of land in the hands of individuals and, subsequently, holding companies, as grantees often sold these leases at profitable prices. For instance, Juan Contardi, former secretary of Magallanes Province, received in 1903 a lease for "the environs of Baker, Chacabuco, [and] Salto Rivers, and Lake Cochrane" in the present-day Aysén Region, on the border with Argentina. His contract mandated settling forty foreign families of Saxon origin and farming profession and providing them with necessary resources.[77] Pablo Hoffmann, in another case, received a grant farther north near the mouth of the Bodudahue River, opposite Chiloé Island. He was tasked not only with bringing in "fifty Saxon settler families" but also establishing a monthly shipping service between Bodudahue and Puerto Montt or Valdivia.[78] These individual land grants amounted to an impressive 4.7 million hectares in southern Chile, including Magallanes.[79] Many of these large land concessions eventually evolved into holding companies that attracted investments from Santiago, Germany, and Great Britain.[80] For example, Juan Contardi sold his lease to Juan Tornero, who established the Compañía Esplotadora del Baker (Baker River Company).[81] Teodoro Freudenburg's lease in the Bravo River valley was transferred to the Sociedad Esplotadora de Río Bravo (Bravo River Company).[82] Part of Tornero's Llanquihue land became the Compañía Comercial y Ganadera Chile-Argentina (Chile-Argentina Trading and Cattle-Breeding Company).[83] Notably, Amadeo Heirenmans received one of the largest concessions, totaling 40,000 hectares, especially significant in the context of Llanquihue Province, where available state land was limited. This lease encompassed an area nearly equivalent to all the Mapuche-Huilliche lands along the Pacific coastline of Llanquihue and exceeded all prior concessions to foreigners.[84] These vast land grants extended into the cordillera and included the Chilean side of a trans-Andean pass. At a time when high-level border negotiations had

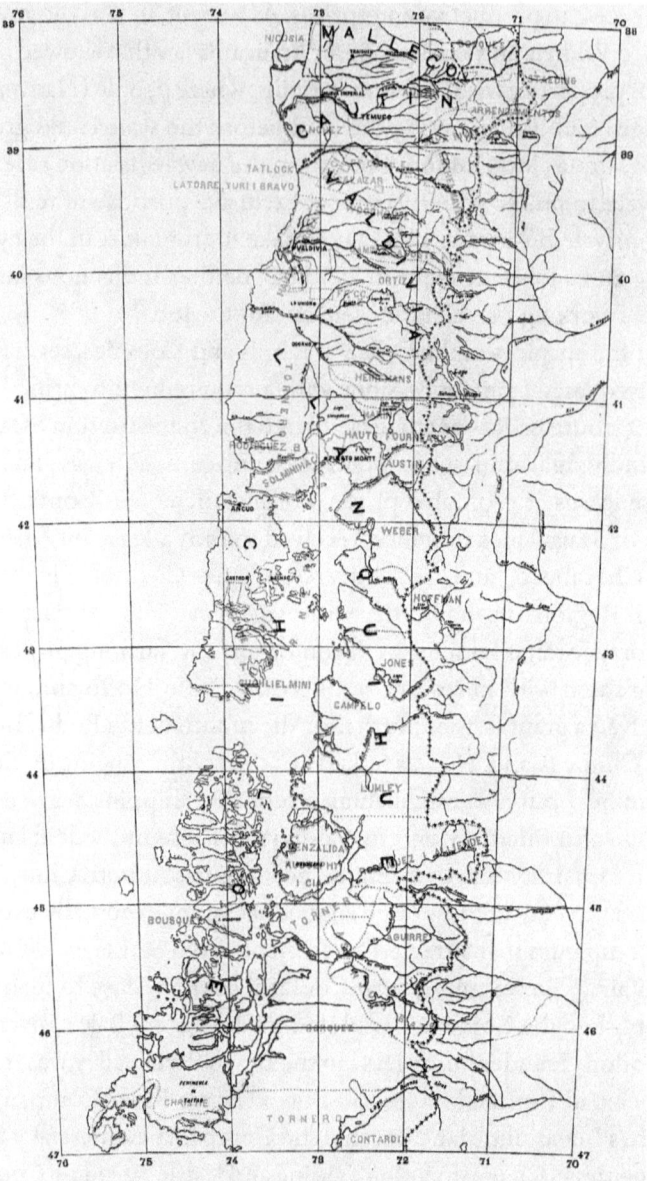

Figure 2.1. Map of land concessions in southern Chile, 1908. Sixty-four maps relating to the boundary dispute between Argentina and Chile. Portfolio 1 contains sheets numbered A1–A27; portfolio 2 contains sheets numbered A28–A64. Each portfolio also includes a glossary and a descriptive list of the maps within it. Originally annexed to the Argentinian Memorial to the Argentinian-Chilean Boundary Commission, 1966. Map portfolio, Selected Maps Representing the Long 19th Century, 1966 1893, MPK 1/406: South America, Nineteenth Century Collections Online, The National Archives, Richmond, United Kingdom.

intensified, Chilean authorities sought to establish a history of property and distribute these lands to what they deemed desirable inhabitants, essentially European immigrants.

The Office of Land Surveys produced the "Map of [Land] Concessions" in 1908 (fig. 2.1). It aggregated all the concessions signed over to holding companies since the late nineteenth century, even though many had changed hands since then. Mapped together, they show the aspiration of a populated Chilean territory. Between the Toltén River and Ancud, the boundaries of each estate are relatively precise, often skirting long-established urban centers like Valdivia or Puerto Montt. The decrees granting these lands mirrored the accuracy on the map. For instance, Jorje Woodhouse received 80,000 hectares (309 square miles) where the northern boundary amounted to "the line that starts at kilometer 28 of the railway line to the east, until crossing the line that goes from north to south at kilometer 24, continuing four kilometers north of this line until crossing the line that goes from east to west at kilometer 24 and its continuation to Lake Villarrica."[85] Such precision contrasted with the loose and certainly more sizable land grants farther south. For example, Tornero's name appeared in several sections from southern Chiloé Island, other islands in the archipelago, and across the Chonos Archipelago (except for Melchor Island, clearly marked for a person named Bousquet). The land concession gave Tornero and his associates dominion over the empty lots on Chiloé Island and beyond, bounded by "north, Parallel 42; east, gulfs and channels; south, Taitao Peninsula; west, Pacific Ocean."[86] The map combined the reality of land grants (at least on paper) with the hopes of populating the South with productive enterprises.

EXPECTATIONS AND CHALLENGES OF LAND CONCESSIONS

In the vast land grants to individuals such as Tornero, Contardi, or Heirenmans, settlers needed to comply with "age, nationality, and morality requirements," meaning they had to be young enough to have a family, old enough to work in the fields, and European.[87] For instance, in 1894, Chile agreed to bring in Norwegian families. In particular, conversations revolved around subsidies for newcomers so as to better compete with the United States.[88] Perhaps ironically, this arrangement was brokered between two non-Chileans acting in the name of Chile: Frenchman Carlos Colson, representing the Chilean Immigration Agency in Europe, and Justus Hansen, the Chilean consul in Norway. But then again, Chileans (much like their other hemispheric neighbors, including the United States) chose foreign settlers to nationalize their territory. A general consensus on the use of bringing foreign settlers to Llanquihue and Valdivia did appease some objections. Indeed, Francisco Fonck, a German physician who had settled in southern Chile, recommended to the minister of foreign affairs that "German settlers [were] the best and perhaps the only ones who would be willing to populate the regions disputed by the Argentines [in the South]." In his view, people from Basque Country, France, England, or Scandinavia were "less suited [for settlement]" because they didn't "have the cosmopolitan and accommodating genius of the Germans." Fonck had surveyed extensive parts of southern Chile since the mid-nineteenth century, witnessing its transformation at the hands of mostly German migrants. He viewed their emerging towns as "beacons of morality" in a barren land, perpetuating tropes that Indigenous life was backwards and uncivilized.[89] Fonck's arguments certainly aligned with end-of-the-century views about southern Chile. Still, the Chilean government sought to attract immigrants from all over Europe.

It seemed likely that applications to bring settlers from Europeans who had lived in Chile for a number of years would receive energetic approval from state agencies. For instance, in 1902, the Chilean Immigration Agency supported an application by a Danish trading house to bring families. The trading house offered to cover the settlers' travel expenses, which would save the Chilean government a good amount of money. Additionally, the agency hired Danish lieutenant colonel Víctor Lindholm, often described as *asimilado* because he joined the Chilean army, to recruit fishermen from his home country.[90] A report contested that Lindholm single-handedly changed the allegedly negative press about Chile in northern Europe and secured an unknown number of migrants bound for Chilean ports.[91] Despite these glowing evaluations, most of these enterprises did not come to fruition. "In reality," wrote the land inspector general, "no individual enterprise has yielded results; all of them have failed to this day."[92]

Despite their capital advantage, not all grantees had the funds to pay for immigrants' travel, tools, and initial support while they settled. Consequently, some sought subsidies. The case of Hungarian Pablo Tuza illustrates the budget needed to make colonization a reality. While Tuza's store in Arica had a value of no more than fifty Chilean pesos, his plan to bring 100 Austrian families to Castro (Chiloé Island) necessitated a significant budget.[93] This was not a land concession, as he would not be establishing and administering an agricultural colony. Tuza would coordinate the recruitment and transportation of European families to their destination, and he would provide land, seeds, and tools to his *colonos*. He requested about CLP$917.50 per family through the Office of Lands and Colonies.

This was indeed an extraordinary amount for the 100 families he planned to bring. His proposed budget allotted about CLP$55 (6 percent of the total) for transportation costs from Valparaíso to Castro, including temporary lodging while traveling, meals, and all the shuttling to and from ports and to the immigrants' new lots. Another CLP$255 (27.8 percent) was earmarked for settlement expenses, including new lots with clear farming areas, demarcation of lots with tracks spanning two and a half miles, and a small rectangular house of about nine by four and a half *varas* (about 304.6 square feet). The remaining CLP$607.50 would cover seeds, animals (two cows or oxen and one horse), a monthly allowance of CLP$30 for one year, and tools, including saws, chains, yokes, and plows.[94] It seems the application for funds was approved, as the inspector of land and colonies, Ricardo del Río Pinochet, reported in 1902 that seventy-two families recruited by Tuza would arrive any day and anticipated that "double or even quadruple of that amount would come [the following] spring."[95]

ENVIRONMENTAL KNOWLEDGE FOR PRODUCTIVE LANDSCAPES

The commitment to bring immigrants in the name of the nation did not suffice for receiving land concessions. In addition to land distribution and immigration, knowledge played a substantial role in securing a grant. Knowledge, in this case, had less to do with scientific descriptions of the geography and more with the capitalist potential of a territory. In 1902, Ciriaco Álvarez and Mauricio Deffarges applied together for a land concession in the estuary of the Palena River. Álvarez and Deffarges formed an odd partnership, the former a Chilotan (who later established a successful lumber company) and the latter a French explorer who discovered a depository of calcium phosphate in the environs of Lake Huillinco in Chiloé. Furthermore, dexterous in "using a compass, dividing lots, and drawing plans," Deffarges worked as an aide for the land inspector of Llanquihue and Chiloé, Jerardo (or Gerardo) Lindh, and eventually earned a promotion to land engineer.[96] Therefore, Álvarez and Deffarges honed their combined experience to show that they were well prepared to develop a colony at the mouth of the Palena River. In essence, their project epitomized a recurring pattern of filling the Desert, where authorities viewed unutilized but potentially productive territories as detached from a sense of shared mission to bring progress.

Settling Patagonia

In the eyes of nineteenth-century governments, progress looked very much like felling trees to grow crops, fencing lots to visualize private property, and building ports, roads, and railroads to connect sites of production with sites of consumption. In order to undertake such enterprises, it was crucial to understand the environment, or so argued the Spaniard Agustín García in 1900. García applied for a land grant to "hire in Galicia, Spain, fifty families of fisher people and farmers" and to settle them in El Volcán, a site located north of the Aysén River estuary.[97] A producer of canned goods, García inspected several possible sites to establish a seafood-canning factory and decided to set up shop in El Volcán, which offered a good balance between land apt for modest agriculture and access to plentiful waters.[98] For García, the math was simple: for a colony to be prosperous, the recruiter of immigrants should be knowledgeable of the geography of the site, like he was, so that settlers would adapt seamlessly to the environment. In his view, only people who knew what they were doing could spark prompt economic activities. A Spanish businessman in Chile who had collected data on possible locations for a canning factory could obviously mobilize Galician families more effectively than anyone else, or so he claimed. He described Galician people as skillful at land and sea and "of good and long-suffering character, honorable, and work- and family-oriented."[99] The argument, weak in the eyes of authorities, maintained that the failure of other colonies, especially state-run colonies in Chiloé, had been due to the lack of knowledge of recruiters in Europe about the territories they assigned to immigrants.

Environmental knowledge and the skills to use it did not suffice to push the colonizing mission over the finish line. For García, the successful colonization of southern Chile depended on the state's commitment to it: "These [conditions] should favor the settler in the number of hectares given to him, in the seeds, tools, machines and farming equipment that are provided, in the animals that are granted to him, in the means of communication that have been or may be established, in the moderate freight prices and in the salary that he should receive monthly for two years to attend to his urgent needs."[100] A canny businessman (no pun intended), García garnered support from the minister of foreign affairs, Rafael Errázuriz Urmeneta, the Chilean Immigration Agency in Paris, and the Sociedad de Fomento Fabril, an organization that grouped business interests. The Chilean Immigration Agency unequivocally supported García's proposal to bring Galician settlers to southern Chile, mirroring the common misconception that it would be as simple as uprooting a plant. To the agency, the explicit lack of European immigration "has plunged [Chile] into the saddest industrial backwardness and in an endemic afflictive economic situation." Like García, the agency thought that the best solution to "cure these evils" was to support colonizing projects that brought European immigration. Benjamín Dávila Larrain, president of the board of the Sociedad de Fomento Fabril and former immigration agent in Europe, knew the value of García's project. His letter of support echoed the belief in the simplicity of resettling "carefully selected" people to establish an industrial colony in southern Chile. The success of such an endeavor would "attract others" in a "continuous flow of Spanish and Chilean [settlers]." García also had the enthusiastic support of Minister Errázuriz Urmeneta and Agustín Baeza Espiñeira, the inspector general of lands and colonies, after a trip they made to the proposed site, El Volcán, invited by the Spaniard. This in-person experience translated into unwavering approval for García's proposal and a positive evaluation three years later, in 1902. In fact, Baeza Espiñeira praised García's efforts to bring families to Chile while chastising the poor performance of Chile's European Immigration Agency. Per the report, García had only recently come back from Europe "with the first 36 families" and fifty more ready to embark.[101]

The internal communications between the inspector general and the minister, however, dim García's commendable efforts. In April 1902, twenty-six Galician families were stood up in Ancud (Chiloé), awaiting transport to El Volcán. No lodging facilities were available on-site, so the newcomers were on hold until the government built a temporary warehouse. Additionally, the local inspector, Otto Rehren, requested a vessel to transport the migrants with the animals, timber boards, and tools provided to them. A few months later, in July, the settlers were heading back north and refused to return any of the equipment they had received. A fiscal agent (*promotor fiscal*) prosecuted them, voiding their leases and demanding they return their tools. We know from other documents that tools, seeds, and animals were among the most valuable items newcomers received, not only because they could resell them (or reuse them) but also because they represented at least 50 percent of the budget allotted to each family. Just a year later, in 1903, Alfredo Weber, the first land inspector for Chiloé, reported that the colony of El Volcán "has been a failure, we do not know why... and the settlers have withdrawn from there."[102] The faith placed in individuals like Agustín García to narrow the gaps between distant territories succumbed before the human challenges posed by the reality of distance, weather, and above all, lack of profit.

Despite García's ultimate failure, he did raise a crucial point about the value of environmental knowledge. Such an argument pushed back against the all-too-common imperial ideas of space that transplanted assumptions from one place to the next. Rejections to land grant applications evidence a hint of resistance toward European businesspeople who went about South America with the certainty that, as a Chilean official put it, "their knowledge is universal and always superior to any specialty in those backwards regions."[103] Recently arrived Swede John Öhlander embodied this sense of entitlement. In 1899, he applied for a permit to bring Swedish immigrants. Like other applicants, he expected the government to provide marketing materials and compensation for his work. Unsurprisingly, he sang the same tune that portrayed southern Chile as an empty space in need of labor, stating, "I think Chile should favor [Swedish] immigration for populating its vast and uninhabited southern territories and exploit their wealth."[104] In the same stroke, he asked for government support, especially in terms of funding. Although his petition resembled Tuza's, he provided no budgetary details. Officially, the Ministry of Foreign Affairs rejected Öhlander's request out of a zealous sense of jurisdictional authority of the Chilean Immigration Agency in Paris against him and "many other foreigners from European nations who think that in countries like ours, they are capable of anything."[105] Certainly, his unapologetic ignorance of the location, soil condition, and size of the terrains he would promote in Europe did not help. Neither did the fact that he had never been to southern Chile and offered to go there "to personally examine the country" only on the government's dime.[106] The agency evaluated Öhlander's application and concluded that it was out of place for a European man to request from the Chilean government what he would not from his own state.

PROTESTING REMOVALS

Land legislation served the mission of civilizing southern Chile by removing Indigenous people and non-Indigenous *agricultores* from fertile farmland. The 1896 land grant to Carlos Colson on Coihueco Island best exemplified this goal. The resistance it spurred over the years illustrates how different communities used legislation, protests, meetings with authorities, and outright physical violence to challenge the validity of new grants in inhabited territories. Colson, a French businessman, received this grant with the mission of bringing 5,000 families over the course of eight years. President Jorge Montt personally supported this contract because Colson enjoyed connections "with multiple colonization and transportation companies" that could help sow "reliable factors of progress, such as a gradual immigration of well-chosen people based on their nationalities, their habits of life and their work skills."[107] The president's words illustrated a shared view among Chilean political elites in territories inhabited by Indigenous people or non-Indigenous peasants as inherently backward. In 1904, Colson transferred his grant on Coihueco Island to Amadeo Heirenmans, who then passed it on to the newly constituted Rupanco Colonizing, Farming, and Cattle-Breeding Company (Sociedad Colonizadora Agrícola y Ganadera Rupanco), or Rupanco Company, of which he was a shareholder. Heirenmans's land grant for Coihueco Island obliged him "to respect Indigenous settlements that might come in his land lease and the rights of settled residents."[108] The law also required current residents to register their land so that the grant holder could differentiate between legal settlers and squatters. However, this would require a resident to be aware of the law, travel to Osorno (and forgo a day of work), pay the registration fee, and risk the possibility of being challenged on the legality of his property. This requirement might have had a deadline, making the burden of proof a little heavier for families with limited resources. Such legislation provided fertile ground for landholding companies like Rupanco to forcibly remove farmers.

Indigenous and mestizo *agricultores* in Coihueco Island protested the encroachment onto their land in courts, to authorities, and through demonstrations. In part, resistance to removal responded to an immediate need to protect one's property. Additionally, generational experiences at the receiving end of violent state policy also informed individual and community resistance. Such was the case of the Currieco family. The Curriecos lived on Coihueco Island for at least three generations. As a young boy, Juan Currieco served Vicente Pérez Rosales and Francisco Fonck in their explorations of the area surrounding Lake Llanquihue in the mid-nineteenth century.[109] As a reward, Currieco received land on Coihueco Island, a marshy area bounded by the homonymous river and the Rahue River in southern Chile.[110] Despite Fonck's letters to the national government in support of Currieco, his land title never came.[111] The Curriecos built a home, bred animals, and grew crops on their land on Coihueco Island, all symbols of effective residence.

In 1895, more than 200 Indigenous and non-Indigenous farmers objected to the auctioning of lots in northern Llanquihue Province, including on Coihueco Island. José Domingo Currieco, Juan's grandson, joined others in petitioning the minister of foreign affairs to either recognize their property or compensate them for their land if it was sold.[112] Such protests prompted the governor's secretary Rafael B. Pizarro to call for a town hall meeting in Cancura, a township on the western edge of Coihueco Island and in relative proximity to Osorno.[113] Pizarro knew people might not produce land titles, but he was ready to recognize their residence status if they could prove they had lived on their lots for at least ten years. Only nine signatories showed up to the town hall, all of whom claimed that Vicente Pérez Rosales had promised land to their families decades earlier, much like in the case of the Curriecos. No one could provide "written evidence" of such a promise. Despite this, Pizarro acknowledged the improvements they had made on their land as evidence of residency: "They have cleared a part of [their lots,] and some of them had plantations."[114] But all other signatories, argued Pizarro, received multiple orders to either leave or appear before a judge, and only those with verifiable land titles did. The protests made it to the desk of then minister of foreign affairs, Luis Barros Borgoño, under whose purview fell the Office of Colonization that distributed land in southern Chile. Barros Borgoño, a future shareholder of the Rupanco Company, resolved to displace farmers and compensate those who were "truly affected" with some land near the town of Cancura.[115] At heart, the protests of land concessions showed not only that people lived on that land but also that they had the legal recourse to fight back.

The pressure to find arable land increasingly put farmers without land titles at risk. The Currieco estate survived the auctions of 1895, but in late 1907, the Ñuble-Rupanco manager Francisco Hechenleitner arrived on the Curriecos' land to remove the whole family. He was accompanied by nine other men, armed with revolvers, shotguns, axes, and whips to make sure the Curriecos complied. Unintimidated, Ceferino Currieco, José Domingo's father, resisted the attack and was fatally shot.[116]

Local law enforcement often endorsed violent removals. Hechenleitner's brother Federico used his acquaintances in the police force to remove Francisco Huaiquipan, Manuel Díaz, and José Antonio Cumacure from their land near Osorno. This was a site where Federico Hechenleitner had acquired an estate that overlapped with some auctioned lots.[117] The farmers had maintained rights of possession by upkeeping "buildings, tenants, crops, animal husbandry, and other works." Despite this proof of more than twenty years of residence, law enforcement allowed Hechenleitner to remove people and animals; he also "reduced their homes to rubble and burned them to the ground."[118] The resolution is unclear, but it does not seem that the farmers who had lived there for decades got to keep their land.[119] Similarly, in 1911 a judge brought a crew of known perpetrators of other attacks to the house of Candelaria and Genaro Saldivia in the region of Coihueco Island to remove them from their lot. They swung their axes against the wooden posts that supported the kitchen, causing the roof to collapse under its own weight and shatter everything beneath it. They rammed the house with a beam tied to the back of their saddles, smashing a wall. The couple barely made it out of their home before it collapsed. With the judge as their witness, the attackers gathered whatever could be salvaged into carts and dropped off the elderly couple at the edge of Coihueco Island, where their son-in-law found them sick the next morning.[120] Perhaps one of the most vicious attacks on farmers was the repression of a protest in Forrahue, a Mapuche-Huilliche hamlet northwest of Osorno, in 1912. As police officers approached, locals armed themselves with shotguns, a rifle, knives, axes, machetes, and sticks. Children wielded clubs; women had basins of boiling water ready. Police officers brandished Mauser rifles, a symbol of the power of modernization that contrasted with the improvised weapons of the villagers. The clash between people and police resulted in thirteen Mapuche deaths, including an eleven-year-old boy, all of whom officers buried in a mass grave in the local cemetery.[121] Violent removals of marginalized communities, most of Indigenous descent, often occurred with the visible support of local authorities.

Some Indigenous communities presented themselves as Indigenous Chileans (*indios chilenos*) to resist encroachment by landholding companies. In doing so, they evoked their historical right to the land and their right as citizens of Chile. In 1908, cacique José Esteban Canuipan filed a petition for land his father had received fifty-three years earlier. The estate was located on the 1855 version of the border with Argentina: "the mountain range where the streams part eastwards and westwards, that is, the *divortium aquarum* [dividing watershed]."[122] Canuipan argued that those lands belonged to his community "by nature" (*por la naturaleza misma*) and it was up to the government to give them what was theirs. By identifying themselves as "the sons of this beloved but wretched Chile," they simultaneously asserted their identity as people of the land (Mapuche) and acknowledged the important role the state played in verifying that identity.[123] Canuipan wondered, "Are not we Chileans and, as such, [entitled to] shelter under the Patriotic Laws dictated by the wise for halting any threats from the ambitious [men] and the citizens?"[124] In other words: the land gave Canuipan and his community their identity as Mapuche(-Huilliche), and it was both challenged by and reinforced by state laws. It was because they were Indians *and* Chileans that they could continue to claim those lands on the border.[125]

Even in the face of violence, Indigenous people used their ethnicity to claim right of possession. In 1912 and in light of the recent massacre at Forrahue, Ceferino Catrilef pushed the state government to protect him against the Ñuble-Rupanco Company. He heard that the company, "exercising its overpowering action," attempted to "remove [him] from his land in Coihueco Island and take it."[126] Catrilef argued that the Ñuble-Rupanco Company had him confused with Sixto Catrilef. Hence, Ceferino desperately petitioned the intendente of Llanquihue not to "facilitate or give orders to the police force to dispossess him from his property." This would have been a serious mistake "given my quality of Indigenous person," aggravated by "the recent disgraceful events in Forrahue."[127] The following year, a whole community filed a complaint against "the people occupying the lands in Rupanco and Coihueco," because they belonged to them, "the natives that come to claim our blood loss in the battlefield of our fatherland."[128] Recalling the many times Indigenous people aided Chilean forces against the last Spaniard strongholds in the South, the petition drew similarities between colonialist forces that had displaced their Indigenous ancestors and the corruption in the highest ranks of government. Under the purview of President Germán Riesco (1901–6), argued the document, authorities had granted a questionable lease to the Ñuble-Rupanco Company, while others had jumped at the occasion to make some profit: "Everyone in the House [of Representatives in Congress] owns this company and even President Ramón [Barros] Luco [(1910–15)] ... out of fear became a shareholder in this company."[129] Indeed, shareholders included future senator Luis Barros Borgoño, political rising star and future minister of finance Javier Ángel Figueroa, former senator Carlos Walker Martínez, and representative Francisco Rivas Vicuña.[130] Sadly, the document has no signatures and no acknowledgment that the petition was sent up the ranks of government to be further investigated.

OPPOSITION TO REMOVALS

Violence against Indigenous and mestizo residents of Coihueco Island did not go unnoticed. In 1902, Alberto Stegmaier, a first-generation Chilean and future mayor of Valdivia, reported to the land inspector that government officials had tried to remove Indigenous people from their lands. His concern stemmed from a paternalistic view of Natives that typified them as "credulous and timid," noting that "many people in the frontier want to take advantage of these weaknesses."[131] As early as March 1905, only two months after the creation of the Ñuble-Rupanco Company, the newspaper *El Llanquihue*, published in Puerto Montt and distributed in the entire region, denounced the extensive acreage granted by the government. "It is truly incredible," its article read, "that the government has ignored the rights of the many inhabitants who live there and has ceded it to a private individual."[132] A German newspaper in Valdivia reported in September 1905 that some residents of Coihueco Island had formed an "Anti-Rupanco Society." Journalist Fritz Gädicke argued that some "shyster lawyers" organized the group with the express purpose of collecting dues from locals, holding dramatic meetings, and publishing hysterical pasquils in Puerto Montt. Even though the anti-Rupanco group did not achieve anything and does not seem to have lasted long, it evidences a shared sentiment against the encroachment of the Ñuble-Rupanco Company.[133] Another newspaper, *La Prensa*, from farther north, stated that "Coihueco Island is not a state-owned property that can be disposed of."[134] A magazine from Santiago similarly drew attention to Ceferino Currieco's murder on Coihueco Island: "It seems the [Ñuble-]Rupanco Company considered itself entitled to the properties of the Indians" and "sent armed people from its service" to destroy the house Currieco was building.[135]

Incidents with the Rupanco Company made their way to the Chilean Congress. Alarmed by the volume and severity of denunciations, the lower chamber put together a special commission to survey southern Chile to investigate "grievances against residents and national settlers in some colonization regions."[136] At its heart, it sought to reconcile the current legislation with the reality on the ground and propose solutions where there was dissonance. To that end, the parliamentary commission held numerous town halls across the southern provinces. During the month of February 1911, they convened at La Unión, Osorno, Puerto Montt, Puerto Varas, and Puerto Octay, near Coihueco Island.[137] The meeting at Puerto Octay received the largest number of applications to address land disputes (138), where more than half (seventy-five), including José Domingo Currieco's, requested protections (*amparos*) against the Ñuble-Rupanco Company.[138] The commission advanced two proposals in an attempt to remove squatters, provide land titles to long-term farmers, and protect the interests of the state. None of them moved forward.[139]

A land lease to Amadeo Heirenmans and, a few months later, to the Ñuble-Rupanco Company in 1905 created legal paths for the violent removal of Indigenous communities and non-Indigenous settlers in Coihueco Island. This lease enjoyed the support of a legal system created to displace Indigenous people from the Chilean national space and populate that space with non-Indigenous settlers. The conflict of Coihueco Island at the beginning of the twentieth century illustrates a long history of violence against a geographical area construed as vacant by legislation, military action, and occupation. Indigenous and non-Indigenous *agricultores* resisted violent acts of removal mostly by appealing to the courts, participating in local town halls, and, in a handful of instances, standing their ground armed with what was at hand. The numerous attempts to remove Indigenous and non-Indigenous *agricultores* from Coihueco Island illustrate the state's anxieties around land possession. Authorities favored European and US American settlers who would import their farming techniques and work ethos. Little did they know that the rising number of landowners and their enclosures would spark conflicts among residents who use roads to move from farm to market.

LAND DISTRIBUTION IN SOUTHERN ARGENTINA

Imagining Patagonia as empty—even though it was not—allowed the Argentine Congress to organize space and distribute land through legislation. For the Argentine government, settlement needed to follow the military campaign of 1879–84 to effectively transform the Desert into a productive region within the national economy. Land distribution constituted a central aspect of the incorporation of Patagonia into the nation. Yet its legal framework did not foresee possible obstacles, such as people who moved without permission and people who did not move, because of lack of infrastructure. It did prove successful in removing Indigenous survivors from their lands and settling them in agricultural colonies. Yet at the end of the day, the legislation pertaining to landed property in Argentina did not eliminate (and perhaps augmented) the Desert.

The Argentine government partially funded the military campaign of 1879 by selling land certificates to the territories still under Indigenous control. For those who applied for titles in northern Patagonia, such a feature shaped their own perception of their role in nation-making in what was deemed a frontier space and their relationship with the national government. To attract settlers, Congress issued the Homestead Act in 1884, which sought to create rural colonies in Patagonia. In some cases, new colonies like Catriel and Valcheta (in the territory of Río Negro) and Cushamen and San Martín (in the territory of Chubut) prioritized granting land to Indigenous survivors of the military campaign, provided they met the requirements for anyone receiving land: age, settlement, and a commitment to working the land.[140] For instance, Valentín Sayhueque and his people went to San Martín, and the Nahuelquir group settled in Cushamen.[141] Walter Delrio argues that creating these agricultural colonies effectively forced Indigenous communities to identify as Argentines in order to qualify for land titles.[142] In 1885, Congress passed the Military Prizes Act, rewarding veterans of the campaign of 1879–84 with land titles.[143] Enlisted or discharged soldiers received land, which varied in size depending on rank. The family of Adolfo Alsina (who had initiated the defense tactics against the Indigenous communities in the 1870s), for instance, received 15,000 hectares, while regular troops received about 100 hectares. Additionally, higher-ranking officers could choose where they wanted to settle, while the general troops had to accept a location on either bank of the Colorado or Negro Rivers or on "other sites suitable for the for the purpose for which they are intended." As an incentive, applicants received ten animals, farming implements, and "a bushel of wheat and a bushel of corn," but they were not allowed to leave for at least a year, ideally leading to long-term settlement.[144] The military campaign had created the imagined Desert, and subsequent legislation sought to fill it.

National authorities expected beneficiaries of the Military Prizes Act to move to their awarded land promptly. However, lack of infrastructure and relatively little support discouraged most recipients. As a result, many awardees sold or transferred their prizes to land speculators, who had no intention of moving there.[145] Noticing that many certificate holders had not participated in the military campaign, the minister of the interior, who coordinated the surveying of lots for awardees, appointed in 1891 a commission to purge those who had not served from the list of applicants. He also repeatedly moved the deadline for people to claim their prizes, with the hope that people would. In addition to transferring their prize certificates to land speculators, many awardees also transferred them to family members in their wills before they could apply for land.[146] Authorities tried to hurry veterans to claim their prizes by giving awardees lands in La Pampa, Río Negro, and Neuquén instead of faraway western Chubut, where next of kin could receive land.[147] Additionally, national authorities claimed that most of this area had been surveyed. The national government offered several deadlines by which to apply for their prizes. On more than one occasion, a decree stipulated that the awards had all been claimed, only to provide another date later to allow more people to request their titles and move to Patagonia.[148]

Like in Chile, land distribution in Argentine northern Patagonia relied on the work of surveyors. Grantees could not receive land if surveyors did not measure it first. Measuring lots constituted a symbolic way of exercising control over the vastness of frontier spaces in Patagonia and the Argentine Northeast. Surveys collected data about what land remained in the hands of the state and started the clock on settlers to transform barren land (*terrenos baldíos*) into productive farms. The work of surveyors should be understood within the tradition of data collection in northern Patagonia that flourished in the context of the border negotiations with Chile. Similar anxieties about decoding the chaotic nature of the Patagonian Andes fueled the rush to produce geodesic knowledge of the "new" territories. In theory, grantees needed to employ their own surveyors, which delayed settlement even more. Frequently, the national government slowly hired land engineers to plot several lots in one go.[149] Yet surveyors were slow in carrying out their work, partly because of the difficulty accessing the assigned lots and partly because of the size of the land to cover.[150] In more practical terms, surveyors eyeballed initial measurements, compromising future appraisals when the Argentine government opened up land in Patagonia for sale to the public.[151] In 1888, surveyor Telémaco González, working on the south bank of the Limay River toward Lake Nahuel Huapi, requested permission to extend his appointment and finish his work. Delaying the measurement of lots, after all, delayed settlement.[152]

In the eyes of the national government, surveyors would demonstrate that land perfectly fit for agrarian production was empty, symbolically bringing the work of the military and the law full circle. But this approach backfired. Surveyors and land inspectors reported that people had settled in Patagonia without land titles. That was the case of Ildefonso Linares, Juan María Saavedra, and Guillermo Iribarne, who in different situations showed they had been raising cattle since at least 1880.[153] Incomplete or imprecise measurements meant that leaseholders did not know the exact dimensions of their lots, resulting in disputes with people living there or other grantees.[154] Such was the case of Col. Lino de Roa, who applied for unsurveyed lots in Río Negro only to find out that one of them had been sold to Ildefonso Linares just months earlier.[155] Additionally, impromptu settlers, including Indigenous people who were forced to relocate, preceded any type of law enforcement, so there was no real way to deport them. For example, José María Paichil (or Paisil) from Salinas Grandes (La Pampa) settled on the northern shore of Lake Nahuel Huapi in the late 1880s. In 1902, when the Argentine government created the colony of Nahuel Huapi and the town of Bariloche, it legalized Paichil's occupation of lot 9. Historical documents acknowledged that Paichil had bred cattle and grown crops since at least 1895, which would have met the requirements to right of land through labor. However, this did not prevent the Argentine government from displacing his descendants and building the tourist town of Villa La Angostura in the 1930s. Subsequently, a historic Mapuche cemetery was relocated, making way for the construction of Plaza San Martín. This square now serves as the town's central hub, though its development has regrettably obscured the rich Indigenous heritage that once thrived in this very place.[156]

THE GRID ON THE MAP

Geodesy measures and maps the earth's surface, heavily relying on multiple instruments and meticulous calculations. Trained in these techniques, Argentine surveyors geometrically divided some of the national territories into sections, fractions, and lots to imprint a rational grid on an irregular surface. In 1896, a decree from the Argentine Ministry of the Interior founded the colony Coronel Barcalá. The formulaic decree declared that lots 2–9 and 12–19 in section 33, fraction B, of Neuquén Territory would be reserved for a future agricultural settlement. Fifty thousand square leagues (roughly 150,000 square miles) would be divided into lots of 625 hectares for new settlers. Without a map that clarified sections, fractions, and lots in each national territory, no one would have been able to locate the colony, and these maps seldom accompanied the transcribed decrees that survived in the printed volumes of the Ministry of the Interior's legislation. One of the best-conserved documents is a map from 1900 published by the Geodesy Division of the Office of Colonization, *Plano demostrativo del estado de la tierra pública en los territorios nacionales del Sud* (fig. 2.2). At the time, this office had transitioned from within the Ministry of the Interior to the new Ministry of Agriculture, where accredited surveyors and support technicians measured lots and granted land titles.[157] Although decrees in published reports did not include maps, the frequency of these formulas suggests that the involved offices did have access to cartographic references. This may have particularly been the case in the Office of Lands, Colonies, and Agriculture (later Lands and Colonies). The *Plano demostrativo* superimposed a mathematical matrix (a western form of epistemology, of making sense of the world, of ordering it) onto a mostly unknown, beyond-control space.

The *Plano demostrativo* declared the authority of a national state over a territory it did not effectively control. The straight lines and right angles concealed the specificity of the Patagonian terrain. Let us go back to section 33 of the National Territory of Neuquén, where the Ministry of the Interior founded the Coronel Barcalá colony. The *Plano demostrativo* flattened the rugged topography with color-coded grids for future land grants, agricultural colonies, forestry exploitation sites, and mining zones. In general, the grid followed longitudinal and latitudinal axes, such as in the Coronel Barcalá colony and the area between Chos Malal and Pichachen. An exception to this checkerboard was the Neuquén Territory south of the homonymous river, where lots were aligned diagonally, facing the Limay River. Noticeably, the sections in this part of the territory were smaller, as were the lots in each of them, signaling, at the very least, authorities' hope that many newcomers would settle in what was once the Mapuche heartland.

Other maps from the turn of the twentieth century also show more aspiration than reality. A map from the Post and Telegraph Office (1904), *Carta de las comunicaciones postales y telegráficas*, illustrates what Carla Lois calls a

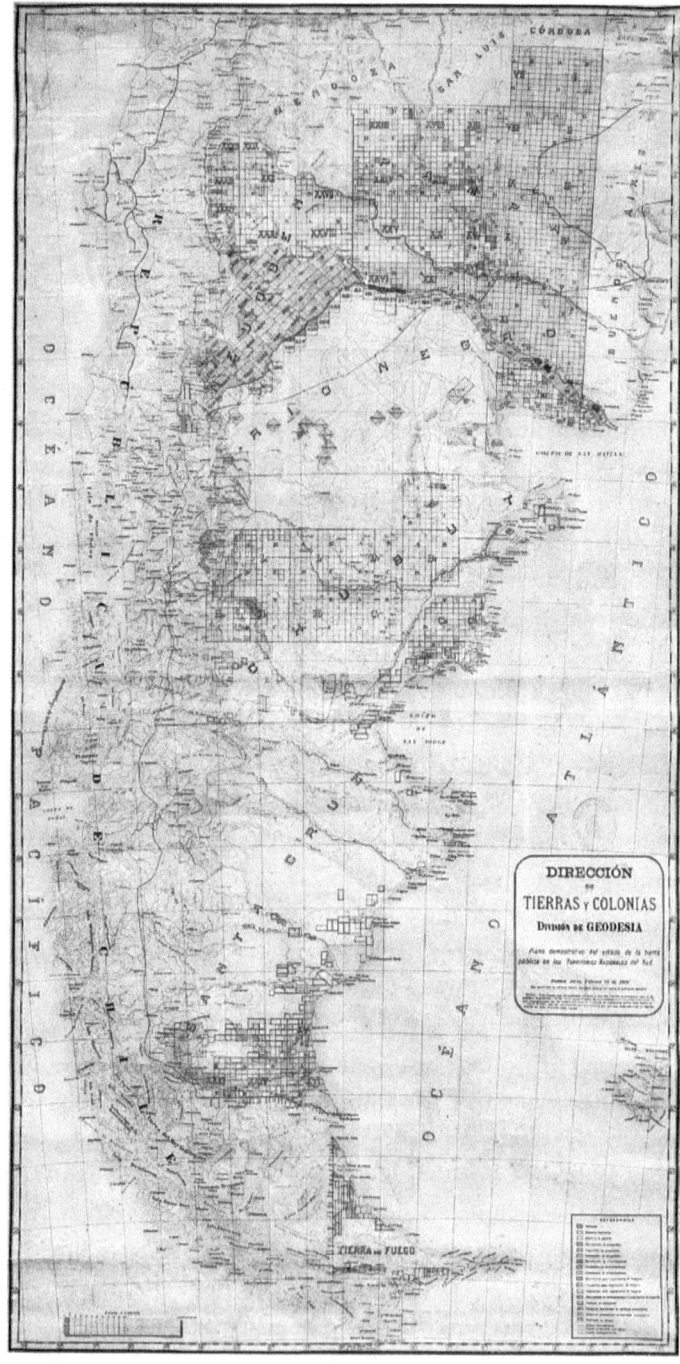

Figure 2.2. *Plano demostrativo del estado de la tierra pública en los Territorios Nacionales del Sud*, 1900. Courtesy of Biblioteca Nacional, Argentina.

"show for Argentine modernity."[158] In it, the department poured onto the map not only operational telegraph, postal, and rail lines but also those that were "projected" or "under construction."[159] In addition to Argentina's network of communications, cartographers included postal shipping lines with their destinations and frequency. The British sailed to Southampton twice a month, to Liverpool four times a month ("twice to the Pacific Ocean and twice to Europe," from Montevideo), and to Walvish Bay (in present-day Namibia), Natal, and Cape Town monthly. Two German vessels departed to Bremen and Hamburg three times a month total, as did ships to Genoa, and five monthly departures were bound for Bordeaux and Marseille. Once a month, ships left for Rio de Janeiro and New York and less frequently to Desterro (present-day Florianópolis), Rio Grande do Sul, and Paranaguá, all in southern Brazil. Finally, twice a month a vessel sailed to the Patagonian ports, anywhere between Bahía Blanca and Punta Arenas, Chile. Overall, the result is an image all too familiar for anyone who has seen cartographic visualizations of early twentieth-century Argentine communications. The first impression is the imbalance of lines across the country, with provinces like Buenos Aires, Santa Fe, and Córdoba crossed by countless tracks and other places like Chaco, Misiones, and Chubut with bare surfaces. Additionally, routes are generally oriented toward the port of Buenos Aires, the central node of the network. Despite these recognizable traits, the *Carta de las comunicaciones* overlapped Argentina's communication network with at least two major topographic markers. In the north of the country, low trees symbolizing the bush fill in the emptiness of the territories of Chaco and Formosa on the map, as well as the northern and eastern parts of the provinces of Santiago del Estero and Salta, respectively. More clearly, mountain ranges dramatically appear in the west of the country and in some areas in the center (Córdoba) and in parts of Patagonia. In the central and northern Andes, unfortunately, the efforts to illustrate altitude backfired and made it impossible to read the names of places, as critiqued by the *Geographical Journal*, the paramount publication on the subject at the time.[160] In most of Patagonia, geographical features, including *mesetas* (plateaus), not only filled in the vastness of the Patagonian steppe but also distracted readers from projected telegraph and rail lines. Thus, cartographers used topography to counteract the stark imbalance between means of communication in the national territories and in the rest of the country. The promise of incorporating Patagonia into national routes represented a quality proper of modern nations at the turn of the century.

Despite the authoritative divisions and the parceling on the *Plano demostrativo*, the color-coded grid blotted it with ambiguity. In the mid-1890s, authorities revealed the consequences of improper land measurements taken a decade earlier. As one high-ranking officer put it, the spirit of land-granting legislation following the military campaign of 1879–84 might have been well intentioned, but it fell short in its execution. It was unclear whether land surveys for dividing lots had been done "in accordance with the regulations" or whether surveyors "had worked only on paper . . . and mapped, sometimes, streams and rivers."[161] Incompetent measurements of lots resulted in poor markings of lot boundaries, which delayed the transfer of land titles. Additionally, surveyors' reports sometimes exaggerated (or simply reimagined) soil conditions for agricultural development, which resulted in overpriced sales of unproductive land.[162] In 1895, the governor of Río Negro, Liborio Bernal, reported that the locations and sizes of land granted as prizes to the veterans of the military campaign of 1879–84 "will bring incessant litigations" because the measurements were done without the proper tools (*a ojo*).[163] At its heart, the clumsy work of surveyors delayed colonization efforts in northern Patagonia.

The tidiness of the grid in the *Plano demostrativo* clashed with the uneven topography that could affect land value and productivity. As Benjamín Zorrilla, the minister of the interior, contended in May 1894, "There are vast surfaces of public land which population and productivity need to be the subject of thorough studies because they are a key element for the nation's prosperity and growth."[164] It is not clear whether authorities objected to this poor surveying on ethical grounds, but they certainly did on financial ones. Like their counterparts in Chile, Argentine officials cringed in response to anything that would delay the settlement of newcomers, mainly because it would incur a steep financial cost to the state. Indeed, recipients of infertile land in Argentina refused to pay further fees and "requested a change of location to places where it would be possible for them to practice the industry they intended to engage in."[165] While the Office of Lands and Colonies was forced to hear these petitions, it also resisted assigning more valuable land. At the heart of this tension between the expectations and the reality of the terrain lay obsolete legislation. In Zorrilla's view, laws passed in the context of the military campaign against the Indigenous communities of northern Patagonia (or, as the minister camouflaged it, for "the incorporation of immense areas occupied by the Indians into the nation's domain") had achieved their goals and were now overdue. While the "[legislation did] not obey a homogenous plan," the *Plano demostrativo* depicted a national, coherent policy toward land distribution in Patagonia.[166] While the map might have concealed the contradictions from a still-young state apparatus, it simultaneously illustrated the authorities' aspirations, at least at a national level, of how they hoped legislation would interact with each other.

CONCLUSION

The regulation of land access in Chile and Argentina contributed to a larger plan to legalize inclusion in and exclusion from the nation. For governing elites, legislation provided a roadmap for those whom they saw as being "outside" the nation to become a part of it, especially for Indigenous people to transition from the category of "savage-Indian" to one of "national-Indian." Military raids, legislation, surveys, and ultimately settlement displaced surviving Indigenous communities. To paraphrase Alberto Harambour, turning Indigenous territories into public land meant accepting the imagined emptiness as a legal reality, an optimal condition for the advancement of capitalism.[167] The spatial discourse about landed property constituted a central condition of nation-making in southern Chile and Argentina.

This chapter has examined how national governments attempted to incorporate Patagonia into an imagined national territory. Land-granting legislation organized the Patagonian space. National authorities in both countries created jurisdictions, sanctioned laws, and founded agricultural colonies with the hope of seamlessly incorporating the territories in the South. The national governments of Chile and Argentina granted land titles and resource-extraction permits, opened roads, erected telegraph lines, and founded colonies. They drew all this on maps, literally filling in the "empty" Patagonian space. However, local authorities struggled to reconcile the law, a symbol of a modern state, with the reality on the ground. Both environment and people challenged the imagined nationalization of Patagonia that lawmakers in capital cities envisioned. The clarity of the law that national authorities sought to impose in Patagonia backfired into a chaotic plethora of ambiguous legislation resulting from minimal state presence, little knowledge of the physical terrain, and disinterest in people's local needs. Incorporation, then, was incomplete at best.

Chapter Three

THE MATERIALITY OF SPACE

1895–1925

In 1905, rancher Augusto Minte filed a complaint to the district authorities of Llanquihue Province against a new company, Empresa Cochamó. Every year, Minte herded his cattle from Ensenada, where he lived, to his brother's ranch in El Bolsón across the Manso Pass.[1] However, Empresa Cochamó "prevented him from taking his cattle along the public road that leads to the Argentine Republic."[2] Roads materialized one way in which day-to-day travelers experienced geographical space. What had been a common route for herding cattle across the Andes was now a private road that Empresa Cochamó would either charge a fee for or close. This forced cattle herders to cross the cordillera along dangerous routes, seriously harming their business and putting their lives at unnecessary risk.³ The dispute between the Empresa Cochamó and Augusto Minte offers a window into three key aspects of settlement in the northern Patagonian Andes that I examine in this chapter: the use of roads and railroads to organize space, the imagining of space as organized either because of or in spite of roads and railroads, and the impacts that aspiration and practice had at the local level.

Consolidating the frontiers of Patagonia into a nation relied on people's ability to move. Movement signified a crucial aspect of settlement. Landowners needed to herd their cattle from ranches, like Pilcaniyeu, to trading hubs, like Bariloche. Authorities governed through correspondence, both telegraph and postal service. Stores brought merchandise like tobacco, garments, or canned food. But also movement constituted a crucial aspect of modern nations. Until the mid-nineteenth century, the Mapuche sustained *rastrilladas*, or trails they carved out with heavy usage.⁴ Indigenous traffic, which often crossed the Andes, decreased significantly after the military campaigns against them and the subsequent border negotiations, and it was slowly replaced by the comings and goings of new settlers. Roads and railroads could make settlement quicker, more efficient, and more permanent.

This chapter argues that movement emerged as a linchpin for consolidating nations in the Patagonian Andes. Yet authorities, ranchers, and farmers interpreted this in different ways. For those designing policies, moving people and goods along planned routes served as a tool for organizing space and cementing progress. For those using the trails and railroads, movement represented their individual aspirations as citizens, enabling production and trade. The opening of roads in southern Chile reflected both planned state interventions and informal trails shaped by local convenience, illustrating the dual nature of transportation networks. Opening roads usually meant clearing vegetation to create paths between places. Additionally, some roads were *caminos de conveniencia* (desire paths) people used for their everyday business. Hence, the term "roads" engulfed state-built tracks connecting cities, towns, and ports as well as trails *agricultores* used to get to rivers, forests, and public paths. This chapter showcases the impact of state planning on everyday life, the role of midrank authorities in executing these plans, and the responses from residents of the northern Patagonian Andes to such projects. Ultimately, it highlights the intricate relationship between movement and the making of Chile and Argentina south of the thirty-ninth parallel.

On both sides of the Andes authorities attempted to use roads and railroads to organize space and make it legible for them.⁵ The first three sections delve into how the opening of roads in southern Chile accompanied settlement. Because roads carried so much meaning for authorities (that of order) and for travelers (that of movement), it is not surprising that they also represented a tool to exercise control. The first section shows how state-built roads symbolized the advancement of the nation-state, infused, as James Snead puts it, with "the intent of authority" rather than the need of those who used them. The second section traces how informal country roads, like the one Minte took to herd his cattle, evolved more out of convenience than planning. For locals, such trails provided a chance to consolidate wealth and torment (and maybe displace) neighbors. The planning and use of roads often resulted in conflict. The third section shows how conflicts exemplified tensions between the legal framework that supported landed property and long-standing road use.

At the turn of the twentieth century, the railroad arrived in southern Chile and Argentine northern Patagonia. With it came the hopes of national authorities to expand the reach of global capitalism to the corners of the continent and to materialize the promise of progress in wagons filled with products heading to the ports. Additionally, the railroad and the space it occupied also symbolized the power of the national government in the south of both countries. It shortened the distances between political centers and the South while bounding these once "remote" regions to a national territory.[6] In the fourth section, I discuss how national authorities in Chile and Argentina imagined the railroad would reorganize their territories south of the thirty-ninth parallel. Chilean governing elites worried about territorial fragmentation and thus insisted on threading urban centers along the country's north–south axis. In Argentina, concerns swirled around connecting the productive Patagonian valleys with Atlantic ports. To that end, construction of railroads followed an east–west orientation.

Despite attempts to pull away from the Andes, market pressure from Puerto Montt continued to attract trans-Andean trade. Land distribution in southern Chile pushed cattle herders to increasingly take their animals to graze in the Argentine valleys. Demand from Chilean tanneries, as well as meat-salting, candle, and soap factories, prompted Chilean capital to expand eastward, exerting pressure to "put Argentine lands into production."[7] These demands put Argentine authorities in a delicate situation. They bought into the idea that the growth of the Andean valleys depended on the ability to take their products to ports on the Atlantic Ocean. Yet governors also acknowledged the attraction of Chilean markets both in terms of demand and in terms of accessibility. In the fifth section, I provide some examples into how governors navigated this tension, illustrating that the idea of an Atlantic-oriented Patagonia was a bit easier said than done.

The pull of trans-Andean markets represented a unique opportunity for Chilean capital. From Neuquén to Santa Cruz, landholding companies purchased ranches in Argentina to breed cattle or sheep and export them west of the Andes.[8] In the sixth section I analyze the role of one of these companies, the Chile-Argentina Trading and Cattle-Breeding Company, the epitome of trans-Andean trade. Originating from Chilean investments, the Chile-Argentina Company monopolized the Pérez Rosales Pass between Lakes Llanquihue and Nahuel Huapi, which allowed it to grow animals in the environs of Lake Nahuel Huapi and take them on the hoof to Chile. More to the point, the opening of a store on the shores of Lake Nahuel Huapi attracted settlement, and soon the town of Bariloche was born. Although the trading company experienced a relatively short life-span (its activities waned in the early 1910s), it exemplified the trans-Andean pull of Chilean ports for agricultural production from Argentina. Even after its demise, as we shall see in chapter 5, the influence of the Chile-Argentina Company reverberated through the environs of Lake Nahuel Huapi decades later.

The last section of this chapter breaks down Argentine attempts to make northern Patagonia legible, that is, to organize it in such way that authorities in Buenos Aires could make sense of it. A new generation of governing elites, influenced by the experts of two decades earlier, understood the territories of Río Negro and Neuquén in terms of their available resources. Led by minister of public works Ezequiel Ramos Mexía, the director of national territories and the governors attempted to build a network of roads and railroads that would shift the orientation of the Andean valleys from the Pacific to the Atlantic Ocean. Indeed, for these authorities trails connecting "isolated villages" helped surmount "impenetrable hills and seasonal floods."[9] In 1908, Congress passed Law 5559, which spatially reorganized the national territories around projected railroads connecting productive valleys with Atlantic ports. Beginning in 1914, surveying crews worked across Río Negro and Neuquén, mapping the terrain to build tracks and roads. Yet, the apparent orderliness of state-planning collided with what Michael Bess identifies as "disorder in everyday construction efforts."[10] Scarce funding and terrain difficulties dampened the initial momentum of the 1910s, which completely halted by the end of the 1920s. As it unfolds, this chapter taps into how authorities at different levels envisioned movement as it pertained to the making of Chile and Argentina south of the thirty-ninth parallel.

Authorities in Santiago viewed state-built roads (*caminos públicos*) as the backbone of a unified territory. State roads usually followed older tracks that had been used by Indigenous peoples or that had been created during the late colonial period. In Chile, the first legislation on road-building passed in 1842, and for the next sixty years Congress added laws on funding, protocols, materials, and maintenance, which amounted in their view to a guarantee of rapid prosperity.[11] "Without good stable roads," stated a Chilean official, "no colonization is possible.... Bad roads can isolate the settler and kill the colonizing [efforts]."[12] Additionally, the expansion of a transportation network, in the authorities' view, evidenced the unwavering presence of the state.[13] Consequently, roads (both old and new), the telegraph, the postal service, and, in the 1910s, the railroad realized the aspirations of a unified territory arranged along the north–south axis. With this, Andrés Núñez argues, authorities pushed back against the remnants of the colonial east–west configuration that followed river valleys and sought to realign what they saw as peripheral regions to the centers of political and economic power, Santiago and Valparaíso.[14]

In southern Chile, modernist aspirations of orderliness through road-building often succumbed to the messiness of everyday construction and maintenance of roads.[15] Legislation stipulated that roads needed to be between sixteen *varas* wide (forty-four feet) for hilly areas and twenty-six *varas* wide (seventy-three feet) for flat terrain. Roads in plain regions would have ditches to either side. Authorities anticipated using the dirt excavated from these to improve the road itself. The rainwater collected in this side drains could flow into rivers or irrigation channels.[16] In southern Chile, public roads did comply with width requirements. Yet, costly and infrequent maintenance handicapped public road service. Roads "demand[ed] the constant attention of the Supreme Government inasmuch as the progress and development of the colonies depend on them."[17] Local law enforcement and judges found it hard to "fulfill their duties," even during the dry months of summer, because they had to cover "great distances" often on "difficult roads."[18] The abysmal conditions of public roads also obstructed the work of the postal service. During "the prolonged winter of these regions... the incessant and heavy rain turns [roads] into real marshes and swamps."[19] The Ministry of Public Works delegated to municipalities the repairs of most roads, but it provided financial support to pay for them. It seems monies were thin and slow to come by, but local governments were quick to request more help. In turn-of-the-century

Puerto Octay, for instance, heavy rains "completely interrupted traffic [to Cancura,] and the postal service [workers] to Frutillar [made] their travels risking their lives."[20] Years later, the provincial engineer reported that the road between Puerto Montt and Puerto Varas was "in such [poor] conditions that if it remains this way until the start of next winter," the cost to repair it would have been much greater.[21] Such views, no matter how hyperbolic they might seem at times, captured a shared concern among local elites that roads should be maintained and that the national government should shoulder those expenses. Further, histrionic portrayals of environmental obstacles to the advancement of modernity led to even more theatrical depictions of the obstacles that state authorities and foreign settlers had to overcome in the name of Chile. As one administrator put it: "Only the unbreakable will and tenacious perseverance of the European [immigrant] can triumph over this difficulty, capable by itself of making the most enterprising nationals faint."[22] Roads were a good idea to connect the scattered populations and make space legible for authorities, but they required constant upkeep.

Authorities, thus, looked for land grant applications that offered the capability of opening and maintaining roads. For example, in 1900 investor Roberto Christie requested a lease in the Tres Montes Peninsula. The Tres Montes Peninsula branched off the larger Taitao Peninsula. Tres Montes is on the northern edge of Gulf of Penas, a point where vessels exited the channels westward to circumnavigate the Taitao Peninsula and the Chonos Archipelago. Christie proposed making Tres Montes not only a productive estate but also a refuge. He pledged to bring immigrants "acquainted with sea life," provide the state with cheap timber, be prepared to aid shipwrecks, and maintain harbors on the inner bays so ships could avoid sailing into the Pacific Ocean to reach Chiloé, which was deemed dangerous.[23] He planned to build a road to a projected lighthouse on Cape Tres Montes, connecting the Puelma and Newman Fjords across the rugged terrain of the Taitao Peninsula. In Christie's view, "this road would be of great utility for the entire area of the Gulf of Penas and the places further south" because travelers and merchants would avoid circumnavigating the Taitao Peninsula and could simply cut through it along his envisioned path.[24]

ROADS OF THE COMMUNITY

Private or communal roads (*caminos vecinales*) also carved through the Chilean landscape. Country trails emerged from constant use as farmers needed to access rivers, forests, grasslands, or public roads to town. While public funds maintained public roads, private trails depended on the people who used them. Traffic—sometimes constant and sometimes infrequent—coupled with heavy rains resulted in poor road conditions. Engineers measured the importance of a specific private road for a given community based on its *servidumbre*, a term that encapsulated both who used it and how frequently. As more people received land leases in Llanquihue Province, they gated rutted tracks cutting through their lots "so that passers-by and animals would not destroy [their] crops."[25] Hence, those who had been using country roads for years might have found themselves suddenly unable to do so. The Chilean government attempted to disentangle, at least partially, the conundrum of informal trails by making it the responsibility of settlers to "open and keep clean the public road that passes through their land for a width of 20 meters [66 feet] and for the length of [the lot]." If a public road did not exist, a surveyor could draw a new one. If it was unnecessary, settlers still agreed to open and maintain a path at the edge of their lots.[26] This legislation embedded road work into the responsibilities of lease recipients, together with building a house, working the land, and enclosing the lot. In doing so, the government reinforced the idea that progress looked like agricultural productivity in southern Chile.

The materiality of public and private roads, however, challenged the legal framework that supported it. Because communal roads crossed lands that the state later leased to others, road access and use grew into a bargaining tool among rural settlers. In March 1912, Juan Antonio Cárdenas and Manuel Oyarzún complained that Juan Linde closed a road opened twenty years earlier, disrupting their nearly decade-long use of this path. They used the private trail to access the public road to Las Quemas, the nearest town. What made things worse was that Linde allowed another neighbor and his employees to use this road. For Cárdenas and Oyarzún, "Linde's clear purpose [was] to bore [them] so that [they would] be forced to sell [their] farms for half the price."[27] Linde refused to open the road. If they wanted access to the public road through his lands, they should buy those lands from him. The feud landed on the desk of land inspector Francisco Steeger, who concluded that Linde did not have a reason to block the private road but neither did Cárdenas and Oyarzún to use it. The root of the conflict was that all these lots were subdivisions from two larger land concessions passed on to Linde, Cárdenas, and Oyarzún through inheritance. Gaspar Münstermann had owned an estate and leased some lots in it to several farmers, one of whom was Ramón Tellez. Tellez's lot was located "in the back of Münstermann's estate," so Ana Menge, Münstermann's widow, had given Tellez permission to open a trail from his land to the public road.[28] Sixteen years later, the land engineer concluded that Menge's permission granted to Tellez did not compel her second husband, Juan Linde, to allow passage through his land. Cárdenas's and Oyarzún's lots were located on a neighboring estate and they used Tellez's road. The land inspector concluded that the authorization given to Tellez was shaky enough, but it certainly did not imply that two farmers leasing on another estate could automatically use a road that cuts through an entirely different property. Instead, they should either use an older, longer, but open track across other neighboring lands or open a new trail to the public road, at their own expense, on the edge between both estates.[29]

This explosion of names illustrates the collision between the legal trajectory of landownership and the practical, improvised, but still effective solutions to everyday needs. On the one hand, the land engineer saw the clear materialization of land legislation. Two men received large estates, and they both rented pieces of them to others. Because these leased lots were inside their estates, they could be expected to allow people to open trails across them to a public road. In Steeger's view, Cárdenas and Oyarzún had no right to use a trail across Linde's estate because their lots were not located there. On the other hand, the reality of everyday mobility drove Cárdenas and Oyarzún to use a trail that was easier for them to access and had a shorter distance to the public road. To them, the nesting of their lots in one estate or another did not (or should not) affect how they moved in and out of it. Simply put, this case illustrates a conflict between the state's idea of space and farmers' everyday life. Francisco Steeger rendered space as a blank canvas that authorities could divide and subdivide as they stamped and signed land titles, leases, and transactions. Cárdenas's and Oyarzún's experiences proposed a different way of ordering space, where the location of their lots, the topography of the trail they used, and its distance from a public artery directly informed how they used the space and how they prioritized accessibility. Furthermore, while legislation sought to provide an atemporal framework for regulating land tenure and use, residents' views of the landscape rested on a locally situated and historically contingent experience of geographical space.

LOCAL CONFLICTS ON MUDDY ROADS

The formality of landed property was often at odds with the informality of private roads. Inevitably, this led to conflicts of all sorts. Some people did not go to the extent of blocking public roads but did bring their wire fences very close to them, often "obstructing traffic flow."[30] On rare occasions, farmers

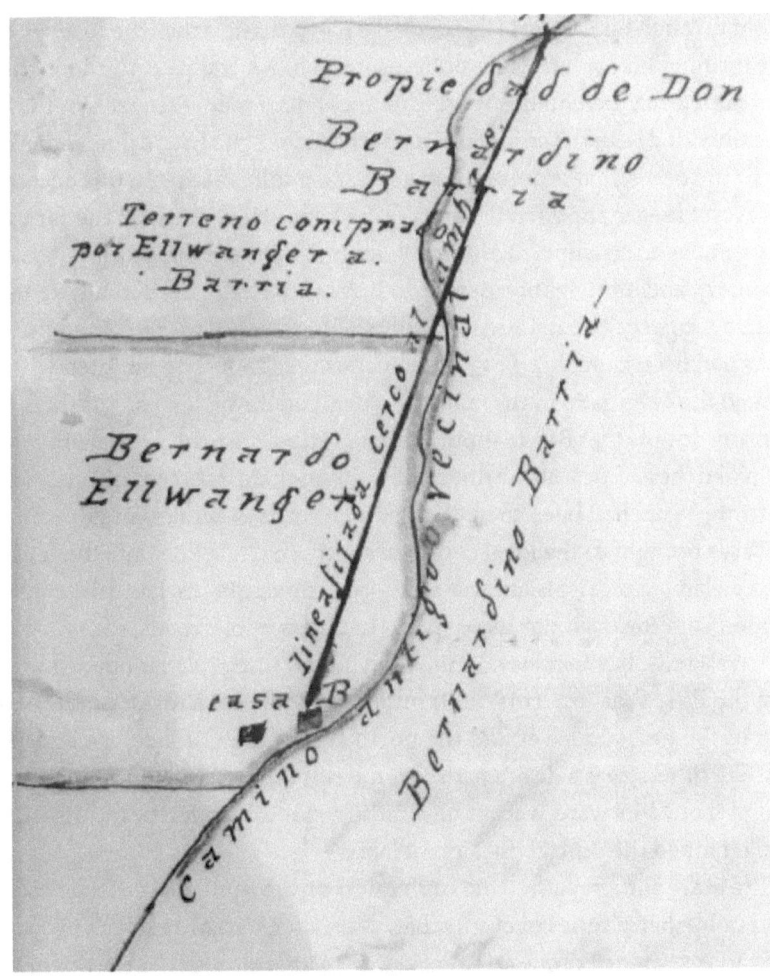

Figure 3.1. Old and new roads on Bernardino Barría's property. Bernardino Barría, "Bernardino Barría solicita el ingeniero de provincia se informe de lo que espone," October 1912, Fondo Intendencia de Llanquihue, Solicitudes varias 1912–13, vol. 227, Archivo Histórico Nacional, Santiago.

got ahead of the conflict and offered to open an alternative track outside of their lots in exchange for closing paths that cut through the middle of them.[31] But most conflicts arose when people blocked one another's way along these private roads. For instance, three farmers from Pelluco complained in 1912 that their neighbor Carlos Larrain closed the road that gave them access to Puerto Montt. Land engineer Francisco Steeger examined the case and concluded they had another road that connected them to town. Another neighbor, Bernardino Barría, objected to the farmers' use of this second road because

The Materiality of Space   87 it went through his own lot (fig. 3.1). This was partially true (the road wound a bit through his lot); thus he built a gate to block traffic. A third neighbor complained that he could not use this secondary road. Steeger would have typically sided with the complainant, because the state had to guarantee, one way or the other, that everyone had access to public roads. On this occasion, however, Steeger supported Barría's blockade. He dismissed the farmers' complaint as an attempt "to annoy the Intendencia, the Office [of Lands and Colonies], and his neighbor Bernardo Barría." Barría's gates actually protected state property. Some insulators on telegraph lines that ran along the rail tracks had been vandalized by "mischievous neighborhood children who go through the open gates to the railroad to play and throw stones."[32] If there had been any doubts that Barría should be allowed to close the path, such attacks dissipated them. Local authorities, then, found themselves confronted with a question: What had been there first, private roads or public lands?

Cases brought to the local courts of southern Chile illustrate the tension between land tenure legislation and the use of private roads. Land distribution collided with the everyday use of paths to access public roads, rivers, forests, and grasslands. For instance, in the early 1880s, three *colonos* opened a path from the Ensenada–Puerto Montt public road to a larch forest on the south bank of the Pescado River and the north side of the Calbuco volcano. This trail, like others, was a desire path that did not observe lot boundaries (even though many lots were "vacant," according to local authorities). Immediate need trumped the official division of land.

The innocent dirt path to the larch forest on the south bank of the Pescado River unleashed a series of events that, years later, became a severe headache for land inspector Francisco Steeger. Since at least 1892, but probably for longer, Indigenous farmer Francisco Huenchuman had leased a lot on the north side of the Pescado River. He built a house, grew crops, sowed grains, and bred cattle, all signs of permanent dwelling that documents refer to as "improvements" (*mejoras*).[33] In other words, he complied with the state requirements for land tenure that legislators had introduced to force Indigenous people to abandon nomadic practices, a widespread strategy among colonizing states. He also "built roads and trails," which he "maintain[ed] in good condition with [his] own work and resources, without the help of the Treasury or the help of [his] neighbors."[34] He later bought a lot from Pedro Gallardo on the north side of the river, but it seems Gallardo had never owned the land and thus had no right to sell it.[35] For the next fifteen years, more people settled around him and more travelers used the road that went through his land connecting the larch forest to the public road.

A dispute broke out in 1907 between Huenchuman and some of his neighbors about the use of the rutted road to the larch forest. In August, Guillermo and Lorenzo Mödinger led a small crew to log a wooded area on the north side of Huenchuman's land, "which [he] had reserved for his oxen to spend the winter."[36] The Mödinger brothers argued in court that their neighbor had closed the private road they used, so they had to cut through his land. They instigated others to join them, including Augusto Minte, who used the private road to access his own land. In what the sources described as an outburst, Huenchuman blocked the communal path and prevented his neighbors from using it. But what had preeminence, the road or the property? Land inspector Francisco Steeger, whose meticulous research populates the archives, followed the legality of landownership vis-à-vis the historic use of the road. He discovered that Gallardo's sale had been illegal because he had simply "[taken] possession [of the lot] on his own, without authorization, or legal rights."[37] While the inspector based his conclusion on the genealogy of land tenure, Huenchuman claimed that the very improvements he introduced ("*derechos de cultivo y mejoras*") granted him real landownership. By transforming a vacant space into a productive farm, Huenchuman implied that his rights of possession were undisputed and, indeed, acknowledged by the state.[38]

A local judge agreed with Huenchuman and qualified his "positive" transformations of the space as effective acts of "peaceful possession" of the land for more than two decades.[39] Minte and other settlers agreed to move the communal road to the edge of Huenchuman's land and pay for the labor. That way, they would not cut through his property. Despite the decision in Huenchuman's favor to move the road to the edge of his lot, Huenchuman's life did not get easier. People still used the old path, and construction of the new road brought oxen-pulled carts to his lot along with men of "questionable backgrounds" who came and went "as they please[d]." Workers "armed with machetes and revolvers" taunted the landowner to call the police on them. Huenchuman saw this as a blatant trespassing, so he gated the path and requested the immediate intervention of provincial authorities. With the same pen that he called for state officials to intervene, he also rejected the police chief that had ordered him to allow people through until the new road was ready. Steeger seconded this order, arguing that the old path had seen "uninterrupted service for thirty years."[40] Steeger applied the use of the road as evidence that Huenchuman should let people come and go. The land inspector did not accept Huenchuman's argument of right of possession through long-standing occupation and improvements. For Steeger, this also meant that Gallardo did not have the right to sell the lot to his neighbor;

Huenchuman's mercurial protests had no standing because he had no property rights to the territories he claimed and because it prevented the continued use of space by others. The land inspector's opinion, hence, rejected Huenchuman's arguments for land tenure—improvements in the name of agricultural productivity and settlement—but used them to support the use of the private road by others.

Roads epitomized the production of space sometimes as a shared and sometimes as an individual experience.[41] Hence, roads, trails, and paths reckoned with the particularities of southern Chile. They also evoked a more universal concern among landowners and *agricultores* with the promise of access to roads against the uncertainty of exclusion.[42] Farmers' interests collided in the use and maintenance of private roads. Often, land inspectors needed to mediate.

RAILROAD PLANS AWAY FROM AND ACROSS THE ANDES

As in other parts of the world, the extension of railroad lines in southern Chile and Argentina reconfigured geographical space. National authorities across the Western Hemisphere viewed railroads as symbols of progress and modernity. They connected sites of production with trading hubs and the world market at great speed, allowing for economic expansion across the South American continent. The railroad simultaneously shrank and expanded space, allowing products to reach ports from farther away and enabling people to travel to corners of their country they had never visited before. It cemented a technology-centered modernity that celebrated the efficiency of movement of people, capital, and goods.[43] Undoubtedly, governing elites in Santiago and Buenos Aires, like their neighbors, subjected the development of the railroad to economic goals of resource extraction.[44] However, examining the development of the railroad in southern Chile and Argentina through the lens of movement—both how people moved and how they hoped to move—illuminates the contested understanding of geographical space, one geared toward the centers of political power and port cities and one grounded in trans-Andean trading networks.[45]

In Chile, a main track sought to connect the country along its north–south axis, situating Santiago and Valparaíso at its center. The first railroad (1851) linked the silver mines of Copiapó with the still-rudimentary Caldera Cove on the Pacific Ocean.[46] From then on, private capital expanded the rail network in the North, akin to the export-oriented model that typified Latin American democracies of the late nineteenth century. It was primarily foreign companies that constructed tracks in mining districts, while the Chilean government either maintained ownership stakes or eventually purchased them.[47] Between 1880 and 1920, the Chilean network quadrupled from 1,777 kilometers (1,104 miles) to 8,210 kilometers (5,101 miles). The majority of this expansion occurred north of the Bío-Bío River. By the mid-1910s, state-owned railroads in Chile amounted to more than half of the network, and a decade later they surpassed that. They also accounted for more than half of the public debt.[48] In 1884, the Chilean government created its state railway company, Empresa de Ferrocarriles del Estado, which operated mostly in the center and south of the country. In Argentina, railroads developed radially from the port of Buenos Aires and, subsequently, from the other secondary ports into the fertile plains of the North, West, and South.[49] Foreign investments (mostly, but not all, British) had propelled Argentine railroad growth since the first train journey in 1857. By 1890, the network boasted 9,254 kilometers of track (5,750 miles), and in less than thirty years it had more than tripled to 34,534 kilometers (21,458 miles) in 1916.[50]

In Chile and Argentina, the railroad arrived relatively late south of the thirty-ninth parallel. In both cases, investments mostly came from state agencies, with the exception of Magallanes in Chile and the Bahía Blanca–Neuquén line in Argentina.[51] Private firms did not see impressive profit projections for lines in Patagonia. Perhaps conveniently, some authorities believed that the civilizing mission of railroads should remain at the hands of state companies, especially in a region like Patagonia that had triggered so many debates about the border. In Chile, the train crossed the Bío-Bío River in 1876, when it arrived in Angol. President José Manuel Balmaceda (1886–91) personally pushed a north–south main railway with secondary routes branching eastward and westward. Amid border negotiations with Argentina, while still in the aftermath of both the War of the Pacific and the occupation of Araucanía, the president saw the railroad as a powerful tool to cement the unification of the Chilean territory, with Empresa de Ferrocarriles del Estado as the main state actor. Such a latitudinal axis undermined preexisting east–west understandings of space that for centuries had organized how the Mapuche moved along riverways and Andean valleys.[52] But for Chilean authorities, the railroad brought people and capital, and it vanquished desolation and fear.[53] Steadily, construction advanced southward and the main branch arrived in Valdivia before the end of century (1895), in Osorno a few years later (1906), and in

Puerto Montt, the last station before the ocean, in 1913. Rail transportation reoriented trade away from the port, anticipating the effects of the opening of the Panama Canal in a port that relied on transoceanic trade. Import fees collected in Puerto Montt decreased more than in any other port in Chile, by 78 percent in the span of two years (1913–15).[54]

Some farmers in southern Chile perceived the construction of the railroad as an annoyance. Such was the case of Carlos Wiederhold, founder of a trans-Andean trading company with land in Chile and Argentina. Though in the late 1910s he sold his assets, he did not retire from doing business in the environs of Lake Llanquihue. He founded a linen company that quickly grew and needed a warehouse in a central location. To that end, he applied for a land title in El Desagüe, a hamlet at the point where Lake Llanquihue drains into the Maullín River. Wiederhold applied for a lease of a lot on the north side of the public harbor, which would have connected his business with other ports around the lake. Additionally, it was strategically located across from the new train station. He offered to pay the same annual fee as the business on the south side of the harbor. Land inspector Francisco Steeger applauded the initiative, a clear example of the industrious activities that would bring "life, progress, and future" to Llanquihue. But a railroad technician objected to the application. He cited studies that projected an increase in freight and travel across El Desagüe, anticipating an expansion of the train station possibly into the lot that Wiederhold requested. Fortunately for Wiederhold, this opinion did not sway the governor, and his application was approved.[55] For other people, the railroad came to be more than a nuisance. In 1913, nineteen rural residents from Pelluco, east of Puerto Montt, complained that the construction of the Osorno–Puerto Montt railroad branch had left them with no road. They had accepted such dispossession "only temporarily" and used a neighbor's lot to access the public road. The "high public interest" of the extension of the railway to Puerto Montt was clear to them, but they feared the railway company would not reinstate the "long-existing" road that gave them access to their properties.[56] Other people perceived the railroad extension workers to be dangerous. People who traveled near the railroad construction sites, such as Daniel Maldonado of Puerto Montt, requested permits to carry weapons for "travels on roads that are nowadays dangerous due to the construction of the railroad."[57] Indeed, farmers reported "disorderly conduct and theft" near construction sites, portraying strangers as dangerous and thus implying that the railroad undermined their safety.[58] The office of the intendente (provincial governor) received all permit applications but the police ultimately approved them. Police officers considered applicants' concerns about railroad construction sites as sufficient for warranting permits to carry weapons. They also did not comment on reasons altogether but rather commented on an applicant's character.

In Argentine northern Patagonia, two main lines branched from the Atlantic coast to the foot of the cordillera, the Southern Railway Company (Ferrocarril del Sud), funded by British capital, and the State Railway (Ferrocarril del Estado). Connecting farmers and ranchers with ports on the Atlantic, the Southern Railway Company arrived in Bahía Blanca in 1884 and the following year inaugurated its own dock in the port, symbolizing the financial opportunity the pampas represented for British investors.[59] From Bahía Blanca—a hub between the western plains, the Negro River valley, and Buenos Aires—the Southern Railway track extended westward to Río Colorado (1897) and Choele-Choel (1898), both in the territory of Río Negro, and then to Confluencia (1899), in Neuquén. Confluencia was a hamlet at the confluence of the Limay and Neuquén Rivers, which form the Negro River. Partly because of railroad access, in 1904 the capital of the Neuquén Territory moved from Chos Malal to Confluencia and was renamed after the railhead, Neuquén. Until the late 1910s, when the State Railway advanced across the Río Negro plateau, people and correspondence would travel along the Neuquén route from Viedma or Buenos Aires to Bariloche. In Chubut, a railroad line was inaugurated relatively early, in 1889, connecting the inland city of Trelew with the port of Puerto Madryn, seventy kilometers away. After the state nationalized this line in the early 1920s, it sought to reach west toward Esquel and build a junction to Jacobacci in Río Negro. However, this ambitious plan of an interconnected western Patagonia track could not overcome the slopes of the plateau a third of the way through Chubut.[60] None of the projected rail ends in the Andes made it beyond the planning stage, except Bariloche, concluded in 1934.[61]

Amid the border negotiations of the 1890s, building a railroad to the cordillera represented a strategic interest for Argentine authorities. Diplomatic negotiations, as we have seen, did not prevent a naval race between the two countries, suggesting that an armed conflict was imminent. If an armed conflict with Chile loomed, the Argentine military believed it urgent to construct a railroad to the Andes. A rapid extension to the foot of the cordillera was a military priority for national authorities, including Francisco Moreno, anxious about a possible Chilean invasion in northern Patagonia.[62] Indeed, Chilean military exercises a bit farther north showed clockwork coordination between rail services and military operations.[63] The issue of imminent danger was not hard to sell to Congress. For representatives, a railroad to Neuquén would "complete [Argentine] sovereignty in that vast region, enforcing its definitive possession." Replicating tropes of emptiness and fear, senators agreed that railroad tracks, a symbol of the nation-state, would "guarantee complete security to the frightened spirit of the scarce and fearful settlers."[64] For military and civilians alike, immigrants (preferably European) would lead the settlement of the transnational Andean valleys. Manuel Olascoaga, a veteran of the raid in northern Patagonia and former governor of Neuquén, voiced a slightly different project, which, though it did not prosper, added texture to a seemingly homogenous view of the Andes. In 1901, he proposed a railroad that, mirroring Chile's model, would run north–south. He echoed military concerns about a possible war with Chile and shared the military's view that a railroad parallel to the Andes from Mendoza to Chos Malal (Neuquén) would bring about settlements. Yet for Olascoaga, such a string of villages played a more vital, long-term goal in peace-building in the Andes. These future towns, he argued, would represent the first line of defense because they would "turn this long zone of danger into the best and most durable bond of international fraternity." For him, people in the borderlands living in close proximity to one another would be moved to "discover and know how to take advantage of the mutual respect that conciliates well-being."[65] Perceived danger could be deflected with real connections.[66]

Even when the border negotiations concluded and the fear of war passed, hopes for a trans-Andean railroad did not wane in Patagonia. In 1908, the Ministry of Public Works approved the extension of the railroad from Neuquén City to Chile.[67] That same year, Francisco Moreno, now a respected figure among the governing elites, proposed a railroad that would connect the Atlantic ports in Bahía Blanca and San Antonio Oeste, in Argentina, to Concepción, in Chile.[68] At a time when the Panama Canal was still under construction, shortening the journey between the two oceans and overall avoiding the Strait of Magellan was always an appealing argument. Moreno's observations carried a striking similarity to governor of the National Territory of Chubut Franklin Rawson's from a decade earlier, arguing that the best route for the extension of the Southern Railway Company would be west of Chos Malal. Such a railroad would serve densely populated areas in both Neuquén and southern Chile.[69] For this reason, and probably informed by his own experiences in the border negotiations, Moreno warned that in case of armed conflict, a trans-Andean railroad would benefit Chile more than Argentina because it would give Chileans quick access to "rich regions south of the Aluminé [River]." Moreno knew all too well that the origin of the Aluminé River in the homonymous lake was located at the thirty-ninth parallel, where the Andes begin to significantly decrease in altitude and allow anticyclone winds from the South Pacific to irrigate the valleys.[70] In part, the inauguration of the trans-Andean railroad from Los Andes, Chile, to Mendoza, Argentina, in 1910 encapsulated Chilean-Argentine collaboration, echoed by a 1922 commitment to build other cross-border rails in the North (Antofagasta–Salta), active since 1948, and in the South (Curacautín–Zapala), which remains unfinished.[71] All in all, the search for trans-Andean connections certainly overcame the fears of war.

THE PULL OF TRANS-ANDEAN TRADE

At the turn of the twentieth century, trans-Andean markets pulled production from Neuquén, western Río Negro, and western Chubut to Chilean markets more than to Argentine trading circuits.[72] More people lived in the province of Llanquihue in 1895 than in all of Argentine Patagonia in 1914.[73] In addition to a larger population, southern Chile offered a more direct connection to Europe. Thus, the trans-Andean pull to the ports of the Pacific put the governors of the national territories in a delicate position. On the one hand, authorities recognized the importance of trans-Andean trade routes; on the other, they also hoped to reorient the routes toward the Atlantic. For them, such reorientation was the key to transforming the "latent wealth" in the national territories "into products tailored to the needs of domestic and international trade."[74] How did the authorities advance the nationalizing mission in Patagonia?

Franklin Rawson, governor of the National Territory of Neuquén, walked this thin line between the pull of trans-Andean trade and the hopes for Atlantic-facing traffic. Since at least 1893, he had been requesting funds to open roads connecting trans-Andean passes, such as Pichachen, with larger towns in the northern Patagonian plateau.[75] In early 1895, Rawson petitioned the minister of the interior for a new road between the then capital of his jurisdiction, Chos Malal, and Acha, the capital of neighboring La Pampa Territory. Poor communication had been the governor's primary concern since the creation of the national territories in 1884.[76] Yet at the heart of his letter to the minister was another, more pressing petition: Rawson insisted on an extension of the Chos Malal–Acha road to the west up to the "dividing line of the Andes."[77] When he wrote this letter, the official borderline did not exist. Despite this, Rawson shared the official view that the boundary roughly coincided with the main chain of the Andes, making the exact location of the official borderline irrelevant. Far from the Andes being an insurmountable wall, there were at least thirty known trans-Andean passes in Neuquén alone that had been used for centuries; about five of them were accessible all year round, and two of them, including Pichachen, were apt for wheeled vehicles such as carts or *catangas*.[78] An international road would serve at least the 61 percent of the *neuquina* population who identified as Chilean.[79] In Chos Malal alone, then capital of the territory, non-Argentines outnumbered Argentines two to one.[80] Thus, a clear road to Chile represented a popular need.[81] Additionally, following "customary law" (*ley de la costumbre*), a high volume of products circulating in Neuquén were imported from Chile. In other words, Rawson recognized a long-established tradition of trans-Andean circuits that dated to the prenational period.[82]

For Governor Rawson, trans-Andean exchanges were of national interest. A transnational road would advance both commercial and patriotic interests because it would address "the needs . . . of the population . . . to make transportation . . . easy."[83] Probably influenced by his military training, Rawson attached to his request a concrete plan for where the road should run, weighing the possible routes, all sketched on a hand-drawn map. This map, entitled "Chos Malal a Pichachen" (From Chos Malal to Pichachen), casts a web of streams to the east and west of the border (map 3.1). There is no reference to countries or other jurisdictions. The plan accompanying the map reminds the minister of public works to foresee possible rivers rising during the melting season. The meticulous rendering of waterways along the valleys contrasts with the lack of reference points for someone who is not familiar with the area. The map, then, shows a precise place in an imprecise space. State ambivalence toward the border, the geography, and the people combined into an equivocal form of authority. We do not know who put together this report, because some pages are missing and the territorial government did not "have an engineer or an office that could survey maps or undertake technical studies."[84] Despite this, it seems to have come from the governor's office, and we can discern how the author evaluated possible sites for the road.

The proposed routes from then capital Chos Malal to the pass in Pichachen needed to contemplate the topography of Andean valleys and the means of transportation most people had at their disposal. The most common vehicle in this part of the country was the *catanga*, an improvised cart of "small wheels, one wooden board, axis of the same material, and rustic handrail."[85] For each portion of the projected road from Chos Malal (marked on the map as point A) to Pichachen (marked as point D), which would run on an existing trail, the report assessed the cost of clearing the route and maintaining it. From Chos Malal, the mapmaker traced two possible routes to Pichachen, one that went through Malal-Caballo (point B) and one that went through Mount Carcayen (unmarked on the original map but marked in my rendering with a

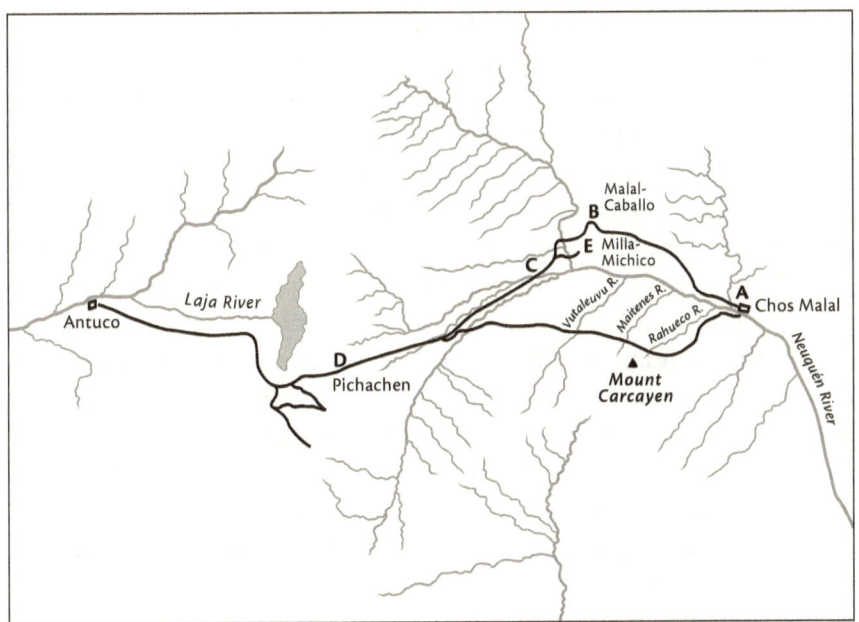

Map 3.1. Simplified sketch of northwestern Neuquén Territory. Capitalized letters on the map mark points along different routes. Map by Erin Greb.

small black triangle). The road from Malal-Caballo (point B) to Milla-Michico (point E) was one of the most traveled in the area, especially with *catangas*. Despite this, the first of these options would have required sixty laborers to clear it and M$N13,500 for maintaining the bridges, without taking into account materials and tools to carry out the job.[86] The second option would require less investment, but it would probably be more dangerous because it cut through uphill terrain. Travelers typically encountered boulders on the improvised trails, which could injure traction animals or damage the wheels of the carts, slowing down their journey. People rarely used this second path because *catangas* could not cut through the narrow gorges of the Vutaleuvu, Maitenes, and Rahueco Rivers. Locals seem to have used more frequently the leg between Milla-Michico (point E) and Pichachen (point D), which goes along a wider valley. Despite this, the report anticipated heavy work needed due to the "geography of the soil."[87] The topography of sharp ravines, especially leading up to the Pichachen valley, could increase the cost of road-building and maintenance because the roads would require numerous drains for the spring months, when melting snows increase the flow of water in otherwise tranquil streams.

The Materiality of Space

Any road from Chos Malal to Pichachen required substantial investment. Governor Rawson requested support from the Ministry of the Interior, which oversaw virtually anything pertaining to the national territories; the Department of Public Works (formerly called the Ministry of Public Works), which assessed infrastructure needs across the country; and the National Treasury, which evaluated the budget for the project. Works like the road to Pichachen appealed to a sentiment of nation-building that moved authorities in wood-paneled offices in Buenos Aires to authorize the works. The military commander at Fort General Roca, 183 miles southeast of Chos Malal, echoed these sentiments about the pass at Pichachen. For him, better international roads "would significantly improve exchanges" (no surprise there) and, equally important, allow for more efficient movement of troops. Good communication routes enabled better surveying of the terrain and policing of people, two cornerstones of land redistribution efforts in northern Patagonia. As Rawson himself put it, "Your Excellency [the minister of the interior], interpreting the yearnings of a population, perhaps the most distant of the republic, that wishes to join the concert of progress whose note predominates in all areas of the country, should seek from the high powers of the Nation the necessary authorization to build this road, useful and necessary."[88] Other departments endorsed the construction of the road, "provided funding was available."[89] Funding was available, but not in the sums the governor was expecting.[90] While Chileans were investing about M$N60,000 on the pass on their side, Neuquén received from the national government the "meager amount of $9,000m/n." With the half-diplomatic, half-complaining tone that was not unusual from governors, Rawson noted that "[the road] will be done in any event and in the best possible way."[91] The governor was partially mistaken: the road would not be done in any event. The contractor faced "snowstorms, rains, and other setbacks" in early 1896 and requested a subsidy of M$N3,000. Maybe he was unprepared, maybe he ran into bad luck, but in any case, his petition was denied. It does not seem that the work was halted for too long, however, because on February 15, 1896, Francisco Moreno, who was visiting Chos Malal, applauded Governor Rawson's efforts in keeping the construction going.[92]

Trans-Andean collaboration acknowledged the persistent trading partnerships that have existed since at least the eighteenth century.[93] Rawson understood the vital role trans-Andean routes played in his territory's economy and social life. In 1895, he not only proposed reliable roads to connect Chile with the Argentine pampas, but he also projected three options for a rail extension from Confluencia to Chile. For each proposed route, Rawson weighed the topography, the available population to provide a workforce, and the productive potential (either mining, logging, or farming). The governor concluded that the best option was the one that went through then capital Chos Malal and from there to the Chilean port of Concepción, which he assessed, perhaps a bit hyperbolically, as being "one of the most important [ports] in South America."[94] Although Rawson's vision did not pan out, it did linger in the minds of other authorities. Immediately after Chilean and Argentine authorities agreed on a borderline in 1901, they geared up to resuscitate a clause in one of their earliest treaties for the creation of a free-trade zone along the Andes, the *cordillera libre*.[95] Its cornerstone was the elimination of tariffs for Argentine cattle imported to Chile and for Chilean wines exported to Argentina. Yet Chilean ranchers and Argentine winegrowers fiercely opposed the treaty. Overall opposition to the trans-Andean railroad and mutual skepticism around "reliable and accurate statistical data" shelved the project.[96]

MAKING A TRADING ROUTE ACROSS THE ANDES

At the dawn of the twentieth century, trans-Andean trade overshadowed the international boundary between Chile and Argentina. Cattle breeders of southern Chile herded animals to the Argentine valleys. The expanding economy of the mining North put pressure on farmlands in the South in the last decades of the nineteenth century. As a result, some Chilean landowners opted to herd their cattle for the summer months in the Andean valleys and even across the cordillera.[97] An 1897 law allowed ranchers to move their cattle to graze in the Argentine valleys and then return them to Chile free of charge. Animals acquired or born in Argentina, however, needed to be declared at a port of entry and were subjected to a tariff.[98] Herders branded their animals and showed their travel documents to account for them. The Chilean customs system revealed serious flaws. Andean ports of entry (*resguardos*) were far from the supervision of customs officers, located in the Pacific ports. In many cases, officials allowed cattle through for a small bribe, significantly less than what ranchers would have had to pay in taxes. The remote location of such outposts coupled with low salaries resulted in staffing issues. In 1902, there were fewer officers than ports of entry along the Andes. The cattle ranchers also did not help. Some declared adult animals as calves, alleging that they had been born in Argentina to Chilean animals and thus did not have to pay tariffs. Authorities suspected that many herders falsified receipts (*tornaguías*) for the cattle they purchased in Argentina, costing the state thousands of pesos.⁹⁹ By some exaggerated calculations, the cattle contraband amounted to three times more than what was declared at the cordillera ports of entry.¹⁰⁰

Ranchers from the plains of Osorno herded cattle into Argentina along the Puyehue Pass. Transit was so frequent that Jorge Hube (or Huber, according to some sources) established a small store and rest station in Rincón, on the northwestern shore of Lake Nahuel Huapi. There, he bred 1,500 heads of cattle and grew wheat, barley, potatoes, onions, and beans.¹⁰¹ Another Chilean rancher who set up a store in the environs of Lake Nahuel Huapi was Carlos Wiederhold. Wiederhold was a first-generation Chilean whose German parents had established a family business in the 1870s. Their soap factory, beef jerky factory, and distillery benefited from the economic bonanza caused by the rising exports of nitrates in the last third of the nineteenth century.¹⁰² The expanding mining industry increased the demand for foodstuffs from the farmlands in the Central Valley and farther south. Between 1877 and 1905, wheat production grew by 350 percent in Valdivia Province and 450 percent in Llanquihue, more than in any other part of Chile, echoing a trend in terms of farmed area. Similarly, the production of barley and potatoes witnessed an upward trend.¹⁰³

Business owners like the Wiederholds prioritized agriculture on their land and thus herded their cattle farther east, including in the Argentine valleys. Carlos Wiederhold brought his cattle frequently across the Puyehue Pass. He befriended some local ranchers, including José Tauschek, a Bohemian immigrant to Chile, and Jarred Jones, from Texas, and soon opened a store and rest station similar to that of Jorge Hube's. On February 2, 1895, he inaugurated La Alemana to serve the community of 200 people in the environs of Lake Nahuel Huapi, which soon became one of the fastest-growing and most transformative businesses in Patagonia. Traffic around La Alemana grew quickly. It attracted other businesses and residents, resulting in a small hamlet. In 1902, Argentine president Julio A. Roca recognized the village of San Carlos and reserved 6,200 hectares (15,320 acres) around it for an agricultural colony.¹⁰⁴ The president might not have known this at the time, but the name "Carlos" honored Carlos Wiederhold as the founder of the small community. Within a couple of years, the town adopted the full name of San Carlos de Bariloche.

Carlos Wiederhold's store, La Alemana, owed its success in part to the inconvenience of the Puyehue Pass. It was hard to move cattle, timber, and grains along the faint trail to Rincón, especially during the winter months. Looking for other possibilities, Wiederhold and his friend Tauschek surveyed the Pérez Rosales Pass, which required multiple lake crossings. This was the same pass that Franciscan priest Francisco Menéndez had used in the late eighteenth century, whose decayed rafts Fernando Hess and Francisco Fonck found in 1856. Nearly forty years after that journey, Wiederhold and Tauschek sailed on a small boat to the end of Blest Branch in Lake Nahuel Huapi. From there, they opened a trail using machetes to cut through the dense vegetation. They crossed the watershed and went downhill, encountering "pouring rain" that forced them to use large leaves of the *pangue*, or *nalca*, plant (*Gunnera tinctoria*) to shelter themselves, "giving name [to the site,] 'Casa Pangue.'" The following day, they "continued along the Peulla [River] valley and we were forced to cross this river more than twenty times until [Lake] Todos Los Santos." The next morning, they made a raft from dry wood and *boqui*, a tree-climbing plant they used to tie the logs together. Probably because of the regular headwind, it took them two days to cross the lake to its draining river, Petrohue.[105] They followed the river valley up to the point where it bends southward. Wiederhold and Tauschek continued moving west to Lake Llanquihue. From the shore of Ensenada, they sailed across the fourth lake to Puerto Varas, finalizing the trip that transformed the pass into a trading route.

La Alemana's business took off spectacularly once it relied on the Pérez Rosales Pass. In 1899, Wiederhold partnered with his son-in-law Federico Hube (who was not related to Jorge Hube from Rincón) and Adolfo Achelis, a German investor settled in Bremen, Germany.[106] Hube rebranded the small business as the Chile-Argentina Trading and Cattle-Breeding Company, evoking its trans-Andean scope.[107] Federico Hube forged the Chile-Argentina Company as a truly trans-Andean business and positioned himself as a desirable partner for the Chilean and Argentine governments. He applied for subsidies to bring timber from Valdivia to build a steamboat for a mail service on Lake Nahuel Huapi, which was nonexistent. A steamboat would rely on its own power and not on the temperamental winds on the lake. This more modern technology would guarantee frequent and timely trips around Nahuel Huapi. Several Argentine state agencies applauded Hube's entrepreneurial initiative. In the context of "constant development and progress in the National Territories, the [national] government should support industrialists" like Hube.[108] The Argentine government showed its support for Hube's endeavors by appointing him consular agent in Puerto Montt.[109] Hube's steamboat application hit a temporary roadblock at the Post and Telegraph Office. Its experts concluded that the project "lacked purpose," as it did not connect Lake Nahuel Huapi to other parts of Argentina.[110] For them, business interests should accompany the development of the state and not vice versa, a clear contradiction of earlier letters of support for Hube's plan. To unblock the application, Hube withdrew the request for funding, which led to the approval of the project in 1901. In general, government officials understood that advancing the interests of "industrialists," especially foreign ones, would buttress the advancement of the state.

Hube's permit applications corresponded with two moments in the Chilean-Argentine border negotiations. The first one, in 1899, coincided with the submission of arguments by Chilean and Argentine experts to the arbitral tribunal in London. The second application, which was approved in 1901, overlapped with the moment when the border tribunal announced its ruling about the international boundary. In these two years, while Hube developed his business, the tense relationship between the two governments evolved into a formidable naval arms race. By the turn of the century, Chile and Argentina ranked first and third, respectively, for armed naval tonnage relative to population.[111] Yet the anxieties about the border that kept diplomats busy did not immediately become apparent in everyday life in the environs of Lakes Llanquihue and Nahuel Huapi. Hube's trans-Andean vision aligned with trade expectations across the cordillera. Until that moment, his company had exported from Bariloche 300 tons of raw wool and imported about ninety-three tons of goods such as furniture, garments, and lamps. In 1900 alone, the dock in Bariloche boarded 500 tons of wool and leather bound for Chile over the course of sixty trips.[112] In 1905, Chile and Argentina lifted customs barriers for all products (except wine) along the Andes, cementing their orientation toward the Pacific in northern Patagonian valleys.[113]

The Chile-Argentina Company began to decline in the early 1910s. The managers made a few unsound decisions. These included purchasing vessels, attempting to build a trans-Andean cable car, buying a costly machine to make wood briquettes that broke down on the first try, and investing in 10,000 sheep that died in the first, unusually snowy winter. Local context certainly did not help. The arrival of the railroad in Puerto Montt in 1912 offered a more effective connection to Santiago, redirecting trade to Valparaíso. Shifts in international trade further crippled the company. Tonnage in Puerto Montt/Calbuco declined after the Panama Canal opened in 1914, from 394,640 tons in 1910 to less than half that in 1916.[114] Finally, the beginning of World War I interrupted trade with German ports, the company's principal destinations, and it slowed down the Argentine export sector.[115] In 1918, the Argentine government increased export taxes to benefit from international wool prices. However, the price of wool plummeted that year, causing many workers in the Patagonian steppe to lose their jobs. Those who kept them saw their salaries fall.[116] With investors fleeing, the Chile-Argentina Company was forced to liquidate its assets. In Chile, Ricardo Roth, a Swiss employee, bought the trans-Andean transportation business in 1914 and transformed it into an early tourist agency.[117] Soon after, Roth partnered with Augusto Minte and purchased the company's lands in Ensenada, Peulla, and Casa Pangue, all strategic points between the Pérez Rosales Pass and the ports on Lake Llanquihue. In Argentina, the company's lands were sold to other holdings with competing interests. In Bariloche, employees acquired different branches of the business, like shoes, soda water, drinks, books, and cigarettes. The more commercial and industrial part of the business, the timber production, and the docks were transferred to Primo Capraro, an Italian resident in Bariloche. In Nahuel Huapi, rising tariffs and risky ideas fatally wounded the Chile-Argentina Company in 1918.[118]

Wiederhold's concern with finding a cost-effective trans-Andean pass and Hube's vision for the company's lake service correspond to what Andrés Núñez identifies as horizontal understandings of geographical space in northern Patagonia.[119] As discussed earlier in this chapter, Chilean national authorities of the late nineteenth century viewed their territory as fragmented and sought to unify it with a north–south integration. This integration materialized in the form of roads, the telegraph, the postal service, and the railroad. Núñez argues that the north–south visualization of space sought to displace earlier spatial renderings that portrayed the Chilean territory in a west–east orientation, following river valleys from their mouths in the Pacific Ocean to the depths of the cordillera. Explorations of the eighteenth and nineteenth century all followed this pattern, from the oceanic coastline to the valleys of the Andes, as did trans-Andean Mapuche movements. Hence, the Chile-Argentina Company's trans-Andean organization of space was not an innovative form of spatial rendering, but it did go against national expectations of how the territory should be organized.

DOUBLING DOWN ON THE ATLANTIC ORIENTATION

Beginning in the late 1900s, when the Chile-Argentina Company operated at its peak, the Argentine government renewed its efforts to reorient the Andean valleys toward the Atlantic Ocean using roads and railroads. Ministers and governors were perhaps emboldened by favorable national and international contexts, which fractured trans-Andean commerce. In Buenos Aires, the conservative governing elite worried about the growing number of immigrants that came to the city and did not leave to populate the interior.

At the height of conservative rule in Argentina, a progressive group within the administration, the Liberal Reformists, defended a structural economic policy in the national territories that would transform them into desirable places for immigrant settlement. Within this caucus was minister of public works Ezequiel Ramos Mexía, who spearheaded legislation in 1908 to overhaul the development of infrastructure in the interior of the country, especially Patagonia. Law 5559 sought to consolidate economic policy in the national territories, delineating an active role of the government in planning and funding works, especially as they pertained to roads and railroads.[120] World War I also disrupted Argentine-Chilean trade across the northern Patagonian Andes. The armed conflict in Europe interrupted transoceanic routes out of Puerto Montt, which decreased demand from the environs of Lake Nahuel Huapi. To increase revenue, the Argentine government imposed new tariffs on Argentine products, deterring even more exports to Chile.[121] National and international contingencies helped authorities reorient Argentine trading routes toward the Atlantic Ocean.

Until the late 1900s, dirt roads cut through Patagonia more out of necessity than careful planning. They did not make legible for authorities the ways in which local residents used space but rather forced authorities to understand them. Governors and other state employees had to adapt their expectations about, for instance, ranchers, small merchants, and laborers opening paths on land and using barges to cross rivers where no bridges were available.[122] Authorities also hired private companies with established trading routes to manage official postal service. Such was the case of the Chile-Argentina Trading and Cattle-Breeding Company on Lake Nahuel Huapi and of the San Martín Cattle-Breeding Company on Lake Lacar.[123] Similarly, the Argentine Southern Land Company, a British landholding corporation, built a telephone line between its estate and Pilcaniyeu, with the understanding that authorities would be allowed to use it (especially in the case of armed conflict).[124] Governors also encountered challenges to the way they were supposed to inspect their territories. The governors of Río Negro usually traveled to towns close to Viedma, like Pringles (present-day Guardia Mitre), Conesa, San Antonio Oeste, Valcheta, and in some cases even Maquinchao.[125] The western third of Río Negro, and even the northwestern area of Chubut, usually fell under the travel networks departing from Neuquén City. Expense reports show frequent official trips from the confluence of the Limay and Neuquén Rivers south to Mencué, Pilcaniyeu, and Bariloche.[126] Topography, weather, and infrastructure dictated how governors experienced their territories on official business.

Map 3.2. Proposed division of the national territories in Patagonia, 1914. Map by Erin Greb.

Perhaps the spatial experience of governors moved the Ministry of the Interior to propose dividing Patagonia into smaller territories. The project, illustrated in map 3.2, clustered together regions where topography and infrastructure supported relatively shared spatial orientations. For instance, Neuquén would absorb the upper Negro River valley and the eastern bank of the lower Limay. With the Southern Railway as the main axis, the capital would sit at the intersection of rivers, roads, and tracks. Río Negro would keep its center-eastern region, engulfing the Negro River in the North and the State Railway up to then-railhead Maquinchao. In southern Neuquén and western Río Negro, a new territory, Los Lagos, would center around the environs of Lake Nahuel Huapi and its sphere of influence, from Lake Tromen (on the thirty-ninth parallel) to Cholila. Chubut would suffer a similar fate. The southern section would be merged with the area north of Santa Cruz to form the new territory of Patagonia, arranged around the ports of Comodoro Rivadavia and Puerto Deseado and their respective railroads that traversed the steppe. The center-eastern part of Chubut and Río Negro would inherit the area surrounding their respective capitals, Rawson and Viedma. Mirroring Los Lagos, a new territory would comprise the Andean valleys, acknowledging spatial practices along the cordillera dating back at least three decades, if not more.

Minister Ramos Mexía understood, as did others in the administration, that railroads comprised the main arteries of long-distance communication. They offered a faster alternative to roads and certainly could overcome the challenges that the environment posed for vehicles. Within the framework of Law 5559, he deployed surveyors to study the terrain with an eye toward constructing railroads that would connect the Patagonian ports in the Atlantic Ocean with the Andean valleys and north–south connections converging in Bariloche, echoing Manuel Olascoaga's vision for the cordillera.[127] Ramos Mexía was particularly committed to building a state railroad that connected Nahuel Huapi with San Antonio Oeste, the Atlantic port, and San Antonio with Viedma, the capital of the territory of Río Negro. If stakeholders of the Southern Railway protested the competition from the state, Francisco Moreno counterargued that northern Patagonia was big enough for more than one company.[128] By the end of 1910, the Ministry of Public Works had surveyed half the distance between San Antonio and Nahuel Huapi, had designated a budget for about half of that, and had built a rail for about half of that budget, inaugurating the first train ride for an eighth of the total distance.[129] Work on a railroad to the west was coupled with work on the port of San Antonio, to "prepare the dock to serve the movement of cargo to be shipped by the state railroad."[130] This pace looked promising. "The locomotive," reported Ramos Mexía, "has already begun its civilizing march, traveling the first leagues in the Patagonian territory, on its way to the fertile Andean colonies, filling those immense regions with the auspicious promise of new life and prosperity."[131] The civilizing mission was on track.

In this context, in 1911 Ramos Mexía appointed the Hydrologic Studies Commission (Comisión de Estudios Hidrológicos, CEH) to survey freshwater sources across the Río Negro plateau through Lake Nahuel Huapi. Mapping rivers and streams in an east–west crusade would allow the national government to better plan the route of the forthcoming railway connecting the port of San Antonio (south of Viedma) on the Atlantic coast with Bariloche. The CEH was presided over by Bailey Willis, a US American geologist who had surveyed the projection of the Northern Pacific Railroad in the United States.[132] Assisting Willis was Argentine Emilio Frey, a topographer with vast experience in the Patagonian Andes from his work for Francisco Moreno's border commission almost two decades earlier. The CEH certainly looked like a newer generation of explorers, with better tools and more data. It relied heavily on foreign expertise, as most of its twelve to fourteen members were US American. The CEH also gathered expertise from different fields, including topography, geography, civil engineering, and geology.

The CEH embraced Ramos Mexía's developmental vision for the valleys of northern Patagonia, from Aluminé in Neuquén to Esquel in Chubut, with Nahuel Huapi as the epicenter. Consequently, the commission proposed constructing an industrial city near the lake at the crossroads of "raw materials, abundant electro-motive power, and the great market of the agricultural provinces of Argentina." Progress could not rely on old strategies of distributing lands, building infrastructure (if at all), and hoping settlers would do the work. Instead, it depended on concrete plans based on "appropriate investigations."[133] To this end, in 1913 Willis identified a site on a plain on the easternmost tip of Lake Nahuel Huapi to build an industrial city, as illustrated in figure 3.2. The location sat at the crossroads of communications, connecting the railroads from Neuquén and San Antonio (neither of which existed at the time), the telegraph between Neuquén and Trevelín (this was not new), and several roads in all directions. Additionally, the city would sit close to a gorge on the Limay River, where Willis projected a small dam "to supply more than a hundred thousand horsepower" to support city life and businesses.[134] Even in the early stages of railroad planning for Nahuel Huapi, which came relatively late compared to the rest of Argentina, the triad of power, productivity, and progress that Dolores Greenberg identified for the Age of Steam now reverberated in the corners of Patagonia decades later.[135] A moraine between the lake and the city would protect dwellers from wind, and the flat topography would make the connection to railroads and urbanization feasible. Bariloche's location exposed residents to leeward winds, but the slopes prevented any realistic plan for a rail station. Echoing Francisco Moreno's vision of a future national park in Nahuel Huapi, Willis also chose a site where "buildings and activities may not intrude upon the observation of tourists or destroy the beauty of the scenery."[136] The linchpin to the industrial settlement was the creation of a national park. In part, it built on Moreno's donation of land on the border with Chile, inscribing the work of the CEH into a larger historical vision for northern Patagonia. More specifically, it sought to protect the Andean forests that regulated water flow into the rivers. Rivers were crucial not only for freshwater access for irrigation and consumption but especially for hydroelectric power. Industry depended on forests. Though never published, Willis's plans for the city shaped the Ministry of the Interior's proposal

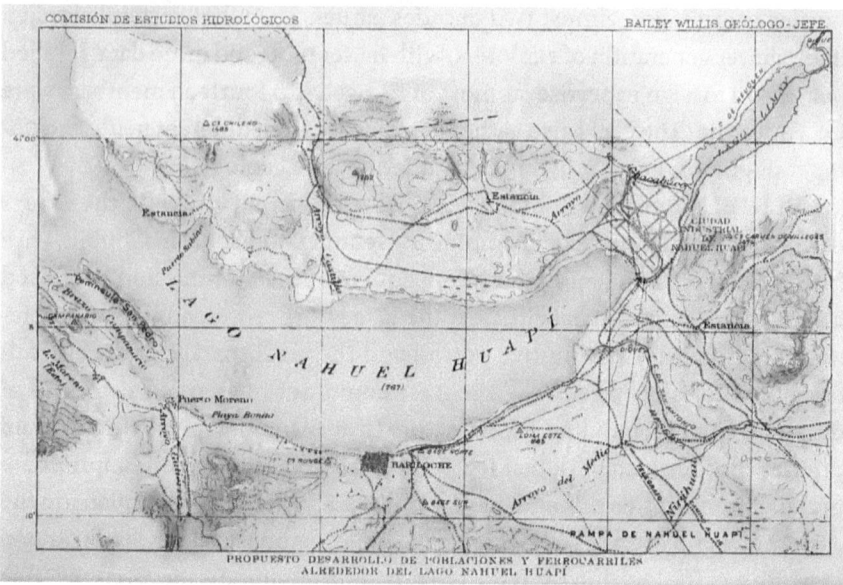

Figure 3.2. Settlements and railroads proposed by the Hydrologic Studies Commission for the environs of Lake Nahuel Huapi. Comisión de Estudios Hidrológicos and Bailey Willis, *El norte de la Patagonia*, vol. 1, *Naturaleza y riquezas*, trans. Pablo Lacalle (New York: Scribner, 1914).

for dividing the national territories. For Willis and his crew, the future of Patagonia lay in an "Andean Province," a "site of an industrious population, in stark contrast to the exclusively agricultural Argentina."[137] The proposal for an industrial city in Nahuel Huapi did not prosper. The CEH was left without funding, and when a new administration was elected in 1916 the commission was completely shelved.

CONCLUSION

The construction of the State Railroad to Nahuel Huapi moved relatively swiftly to Pilcaniyeu. The work of the CEH had yielded topographic data that helped engineers carve through the plateau. While in the rest of the country railroad construction waned, in Río Negro work peaked in the 1920s, serving a growing demand. Each new railhead in the Patagonian steppe turned hamlets into villages. As populations grew, though modestly, residents requested more state services, such as justices of the peace, telegraph stations, and police officers. For example, in 1917 the governor of Río Negro moved the justice of the peace from thirty-five kilometers away to Nahuel Niyeo, a new railhead and the "most important population center in the county."[138] Passenger traffic through San Antonio, which represented about two-fifths of its activity, almost quadrupled between 1922 and 1926, from 4,964 to 17,104 people.[139] For the 1925 fiscal year, it garnered more profits than any other railroad in a national territory, all of which also covered a shorter distance.[140] Yet this relatively heavy traffic, warned then minister of public works Roberto Ortiz, was spread out across the 423 kilometers (263 miles) of track. In other words, the growing use of the railroad across the Río Negro plateau, both in terms of passengers and cargo, was concentrated in the railheads, San Antonio (the port), and Ingeniero Jacobacci, where people connected with car services to Bariloche, 160 kilometers (99 miles) west.[141] Stagnant funding and a rugged terrain halted the construction in Pilcaniyeu, just fifty miles from Bariloche. The construction of railroad lines on both sides of the Andes illustrates how national authorities envisioned the orientation of the national space. In Chile, the government prioritized a north–south connection that disarticulated the east–west riverine understanding of the space. In Argentina, authorities sought to connect the productive Andean valleys with the Atlantic ports (which only occurred in 1934) and they also projected, but never built, a north–south railroad that would move military resources quickly in case of attack.

Movement was central for settlement. Chilean roads helped authorities organize space. They made clear how local farmers of different backgrounds used space and expected others to do so. Conflicts about roads cutting through people's land showed that people were settling but that settlement was messier than anticipated. In some documents, we can almost read land inspector Francisco Steeger's frustration with some feuds. His meticulous research showed that not everyone settled in the same way or at the same time. Some occupied lots with permission, worked the land, lived there for years, but never got a land title. Others received titles but found that some of their neighbors had been using their assigned lot to cut through to a public road. Roads connected the emergent production in the Argentine valleys with Chilean cities. In the aftermath of the border negotiations and through the 1920s, expanding agriculture in Llanquihue impacted the Argentine valleys. More land for wheat and barley meant fewer pastures for cattle, which increased the demand from the eastern side of the Andes. Some companies, such as the Chile-Argentina Trading and Cattle-Breeding Company, purchased land and hired free-riding herdsmen, managers, and laborers. General stores blossomed from southern Neuquén to western Chubut, where people met, traded, drank, and sometimes gambled. State arteries into the northern Patagonian Andes sought to construct a landscape that would make resources available for national and international markets. Governing elites in Chile and Argentina believed that roads and railroads would consolidate these landscapes of progress, as Pedro Navarro Floria calls them, by extracting production and facilitating colonization. The materiality of roads and railroads offers a window into how authorities and residents understood the geographical space of the northern Patagonian Andes and how they experienced it. However, with time, trans-Andean productive orientation in the Argentine valleys irked some and annoyed others at the local and national levels. Soon, an anti-Chilean sentiment would brew in Argentine Patagonia. Movement was at the center of the nationalizing mission of the northern Patagonian Andes. What happened when spatial mobility was also a form of transgression?

Chapter Four

SPATIAL DISCOURSES FOR A HEALTHY NATION

1910–1925

On a warm Sunday in November 1912, Manuel Sales hosted a barbecue in El Bolsón to celebrate the completion of his home. Guests included twenty-four-year-old Vicente Fernández Palacios, the local judge, and Valentín Bustamante, a businessman from Bariloche who simultaneously served as alternate judge and manager of El Bolsón's police precinct. As wine flowed at Sales's home, Fernández Palacios's vociferous complaints about the police filled the conversation. Bustamante, who "did not like" the judge, felt the commentary was aimed at him, a civilian managing a police station, and his business colleagues in El Bolsón, whom the judge accused of cheating law enforcement. Bustamante left but returned a couple of hours later when he heard gunshots. The judge had fired his Colt revolver and perforated a couple of kettles. No one was hurt, but he was still swinging the weapon dangerously. As a figure of authority, Bustamante repeatedly demanded he surrender the revolver, and three times the judge refused. At this time, a police officer lunged at the judge and took his weapon. When the judge composed himself, he attacked Bustamante with a "horse whip, breaking the scalp and sinking [in] the skull," and he fell "unconscious to the ground soaked in blood." At this sight, the judge promptly mounted his horse and sped away.[1] Bustamante filed a report in Bariloche because his managing position at El Bolsón's police station prevented him from reporting the incident there. The police officer in Bariloche tried to cover up the investigation, so Bustamante re-filed the assault and the subsequent cover-up with the Border Police, a new force in western Patagonia. As a result, Judge Fernández Palacios was removed from office in June 1913, probably when the investigation hit the desk of the director of national territories, and only appeared in the written record afterward as a lower-ranking aide in the Ministry of Agriculture.[2]

The Fernández Palacios episode introduces public disorder as a symptom of an unhealthy society and people like Bustamante, a civilian aiding the police, as the truly patriotic cure for the nation. In southern Argentina, hygienic views perpetuated centuries-old spatial discourses of Patagonia as a no-place, constituting a persisting Desert. In the first decades of the twentieth century, local elites in Argentine northern Patagonia used spatial discourses to advance their own vision of the nation. Historians of Argentina have examined the state's efforts to create spaces—from schools to prisons—for regulating the behavior of desirable inhabitants.[3] Here, I shed light on how old and new ideas about Patagonia animated the ways in which landowners, merchants, and professionals construed Patagonia to define their own role in the making of Argentina in a frontier space. At the turn of the century and through 1930, authorities in Buenos Aires increasingly viewed Patagonia as a dangerous place, marked by old tropes of difference in new ways. At Sales's home, at least thirty other men sat at the table (we only know of one woman present, roasting the meat), including nine Chileans, three Argentines, and a Spaniard. The host himself was "Arab," per a police report. Indeed, investigators and witnesses made note of people's ethnic identity, evidencing an awareness of difference that irrigated anti-foreigner sentiments. For national authorities, only strict enforcement of legislation would ensure that nascent communities remained "healthy" while eliminating "bad elements." This certainly points to the contrast examined by Kristin Ruggiero, in which the expansion of individual rights in Argentina overlapped with "measures to ensure the health and security of the social whole."[4] In the national territories, however, this rhetoric fell apart because citizens did not enjoy the same civil rights as in other parts of the country.

I begin by contextualizing hygienic views of Patagonia relative to Buenos Aires. Labor strikes in the capital city directly informed how authorities addressed public disorder in the national territories. Governors of national territories reported to the director of national territories in the Ministry of the Interior, the same ministry that administered the police of the capital city. Hence, law enforcement's responses to protests in Buenos Aires modeled policing throughout the national territories. Additionally, in 1902, a year after the border negotiations were finalized, the national government stretched the reach of policing experiences to Argentina's southernmost frontier by creating a penal colony in Tierra del Fuego. As Lila Caimari has argued, authorities hoped inmates would be "purified" by life away from urban centers and in a specific frontier environment.[5] Against this backdrop, the murder of a Welsh businessman in 1909 catalyzed concerns about criminality in northern Patagonia. Increasingly, local elites blamed foreigners and Indigenous people for violent assaults and public delinquency. Crimes against property and proprietors constituted attacks on the nation. As a result, the Argentine government created an autonomous force, the Border Police, in the 1910s to patrol key routes and keep an eye on local law enforcement. In essence, this cadre replicated decades-old understandings of Patagonia as a space devoid of the rule of law. But simultaneously, the Border Police legitimized the criminalization of suspicious activity, often imposed on foreign individuals. These sentiments did not divide Patagonian society along Argentine/non-Argentine lines. On the contrary, several immigrants, both of European and Chilean origin, claimed to contribute to the building of the Argentine nation, while they accused other foreigners of theft, murder, and harboring criminals. As nationalist tensions brewed in the 1920s, so did anti-immigrant sentiments.

The chapter closes with a look at the government's introduction of physical health as a desirable quality of Argentine male citizens, even if they were foreign born. Scholars have especially examined the intersections between masculinity and sports for consolidating a sense of modern, national self in critical moments of state formation, such as early republican Brazil or postrevolutionary Mexico.[6] Here, I focus on the creation of a local chapter of a national shooting club in the 1910s to explore how male social elites in Bariloche embraced a program that portrayed the Argentine citizen as physically fit and morally adept in order to construct the nation. The club exemplified how local elites—namely, merchants, landowners, authorities, and professionals—sought to embody a masculine ideal, exemplary in behavior and morality.[7] I interpret such concerns with shooting dexterity as an echo of the military columns that advanced more than three decades earlier. Simultaneously, I see the quest for vitality in the 1910s and 1920s as a prelude to the explosion of outdoor activities in the 1930s geared toward (but not exclusively) tourists.

Local elites used hygienic views as dire spatial discourses of the Patagonian landscape to demand more policing and attack immigrants—almost always Chileans—all while asserting their moralizing mission through the press and sports organizations. Public disorder, seen as an expression of the old trope of barbarism, evidenced that the incorporation of Patagonia into the nation was far from complete. Hygienic approaches to social and spatial control in northern Patagonia continued to rest on decades-old ideas of "conquering" or "filling" the Desert beneath the new veil of public health. Building the nation in northern Patagonia amounted to creating a healthy society.

THE HEALTH OF THE NATION

The Fernández Palacios episode condenses hygienic anxieties about social order in northern Patagonia, which were rooted in national authorities' organicist views of nation. Such views construed society as an organism and an interpretation of crises as treatable, even preventable, diseases. This metaphor gained traction after 1871, when an epidemic of yellow fever in Buenos Aires propelled hygienists to the center of the national political stage for decades. The outbreak began with carnival festivities, when the beats of the *comparsas* (companies of street dancers) deafened warnings from physicians. The spread killed about 7.7 percent of the city's population, mostly in crowded tenements of immigrant families.[8] It did not escape contemporaries that the outbreak symbolized the barbarism of an unplanned, growing city combined with the unpredictable spread of the disease. A healthy city could be restored only by investing in sanitation works, imposing hygienic practices, and policing them. By healing Buenos Aires, both in terms of infrastructure and demography, hygienists emerged as tempered leaders that placated anxieties about a common enemy by disseminating information, recommendations, and regulations. Through the 1900s, the spatial transformation of Buenos Aires constituted what Jorge Salessi defines as a refoundational moment that informed policies across the board. The utopia of an organized, modern, hygienic city served as a prism through which to later control the Patagonian space.[9] Hence, the administration of public health and public order in the capital directly informed how authorities applied policies in the national territories.

Argentine hygienists, as others across Latin America, subscribed to neo-Lamarckian principles of environmental adaptation and Benedict Morel's theory of degeneration. The former argued that an organism could develop new traits to adapt to its environment and pass it on to its offspring. The latter stated that acquired traits of one individual could be passed on to their children. For Latin American hygienists, this meant that vices and habits—acquired behaviors—could be inherited. Hence, an individual's "upbringing, social context, and education" directly contributed to the improvement of the population.[10] Indeed, authorities across Latin America applied these theories to public health interventionism in an attempt to perfect an imagined national race. Nancy Stepan notes that this "preventive eugenics" in fact introduced "biologically-governed norms of social behavior... in the name of hereditarian science."[11] Neo-Lamarckism provided a framework to interpret and treat what doctors diagnosed as hereditary degeneracy.[12]

Argentine and Chilean hygienists believed inebriation was among the most degenerate social diseases. They credited heavy drinking as the root of other "ailments," such as vagrancy, poverty, and criminality, leading to the "decadence of the race, the stagnation of the population, overwhelming criminality... idleness, and consequently slowness of production."[13] Social diseases thrived on excessive behaviors. Excesses represented the absence of temperament, measure, and observance of the rule of law, especially in public. Excessive alcohol consumption underpinned unwanted behaviors in public. Morality would be to social diseases what hygiene was to ailment, a decisive deterrent. Temperance signified a well-rounded, modern, civilized individual, but succumbing to the temptation of drink evidenced weakness of character that could lead to brawls, vagrancy, disorderly conduct, and ultimately, violent crimes.[14]

National authorities conflated other forms of social disturbance, such as labor strikes, with excessive behavior that evidenced poor social health. They viewed ideas that challenged the established order as particularly menacing for the health of the nation. In Argentina, civic disruption in the capital city illuminated the ways national authorities introduced policies in Patagonia designed to sanitize society from unwanted behaviors. Beginning in 1890, police forces in Buenos Aires reacted to workers' celebrations of May Day with increasingly violent repression. They particularly targeted anarchists, an ideology tied to the Argentine labor movement of the turn of the century, especially in urban settings.[15] Because of this, authorities often conflated anarchists with immigrants, who composed the majority of urban workers. However, as Ryan Edwards reminds us, "Not every immigrant was an anarchist, and not every anarchist was an immigrant."[16] Legal action accompanied tight policing of bodies and ideas. Congress sanctioned the Law of Residency (1902), which allowed the national government to block entrance to or expel anyone on the basis of national security and public order, and the Law of Social Defense (1910), which forbade group associations that "disseminated anarchist doctrines . . . or instigated attacks against national laws."[17] At its heart, this legislation sowed anti-immigrant sentiments that would germinate in the 1920s, framing anarchist and communist ideologies as foreign born and therefore a clear threat to Argentineness. Hygienists would squarely place the labor strikes at the turn of the century in the category of "excesses" (*excesos*) and treat them with the same urgency as a rapidly spreading disease.

Within the conservative government, a group of Argentine reformers, the Liberal Reformists, sought to address social issues by redesigning national politics, with mixed results.[18] In 1912, the Sáenz Peña Law instituted the right and obligation for all adult men to participate in all state elections. Coupled with the obligatory military service and compulsory elementary education, it spearheaded a legislative front to assimilate immigrants into Argentine civic life. However, the Sáenz Peña Law had very little impact on voting rights in the national territories. Here, governors and other authorities were appointed by the president; only a handful of towns had councils with elected officials. Imbued with hygienic views, the Liberal Reformists believed socialist and anarchist ideas spread quickly, like tuberculosis, in the crowded quarters of Buenos Aires. Hence, minister of agriculture Ezequiel Ramos Mexía designed a plan in 1908 that would attract immigrants away from urban centers (where they could organize their labor) and disperse them in the fertile lands of the national territories. He proposed to "increase the value of public lands" with transportation infrastructure, such as railroads and ports, that would connect potentially productive regions with national and world markets. He stated, "There are no more deserts; what is left are unoccupied fiscal lands."[19] In his view, situating undesirable immigrants in an empty space was a step toward improving that land and regenerating the individual. Patagonia would model good citizens.

ATTACKS ON PUBLIC ORDER

In Patagonia, newspaper editors echoed hygienic theories on the degenerative force of inebriation. Excessive drinking transformed conversations into discussions and jokes into offenses.[20] It not only propelled men to squabble, sometimes fatally, but it also represented "humanity's worst plague."[21] Children engendered "on tavern days," argued an article, were "weak and rickety." Editors believed that drinkers' descendants could not only become alcoholics themselves but also inherit a "predisposition to tuberculosis, nervous and psychic disorders, idiotism, epilepsy and mental degeneration." What is more, cerebral lesions from excessive alcohol ingestion coupled with other believed complications would, in the eyes of local elites, result in moral weakness.[22] Men, including judges, "altered their values and abandoned their [exemplary] place."[23] In the case of Fernández Palacios, accounts differ as to what exactly the judge did, from "demanding a barbecue and seven liters of wine" to shooting a kettle, breaking dishes, punching and then hitting Bustamante with a horse whip, and breaking a guitar.[24] It is hard to imagine that Fernández Palacios was the only person who had too much to drink at that event. Yet by all accounts, he was the only one to cause a public scene as a direct result of alcohol consumption. Inebriation disrupted public order.

Regional newspapers congratulated the "campaigns of social purification" that national authorities discharged on alcohol sales and consumption, as well as other excesses. Especially telling are articles that appeared in *La Nueva Era*, a newspaper published in two cities at the mouth of the Negro River, Carmen de Patagones and Viedma, which circulated across northern Patagonia. In 1914, Enrique Feinmann, a prolific hygienist, wrote that Argentine legislation, "inspired by the spirit and practice of the highest civilization," needed to follow other civilized nations "to ward off the danger of alcohol that threatens humanity ... with the worst of poisons."[25] Another note echoed international calls to "reduce and completely abolish the consumption of alcoholic beverages."[26] In essence, *La Nueva Era* amplified idealist arguments that called for a collective effort for the advancement of society, where "submission to the social duty [did] not diminish individual freedoms." Indeed, much like a body coordinated efforts to fight a disease, aspiring to the common good provided contingency against shared threats.[27] In reprinting such notes, *La Nueva Era* situated cosmopolitan conversations about excessive drinking in the homes of the local elites in northern Patagonia. In doing so, the newspaper helped read the problem of inebriation in Patagonia through the lens of London, New York, or Buenos Aires. Additionally, it provided a classist analysis of the drinking reality in Patagonian towns, insisting that drinking presented "a public danger for the working classes and, especially, for the Indigenous population."[28] If excessive drinking threatened nation-making and civilized society, it certainly pertained to the hardworking settlers of northern Patagonia.

Among such self-proclaimed hardworking, tempered citizens was Bustamante, the man who stood up to Judge Fernández Palacios. He presented himself as an emblem of moderation, the most efficient form of prophylaxis against excesses. In fact, across Latin America, social elites in capital cities founded civic organizations that carried out educational campaigns, boycotted distilleries and saloons, and lobbied for harsher legislation.[29] Under the auspices of international federations against the consumption of alcohol, Argentine women founded the Argentine League against Alcoholism (Liga Argentina Contra el Alcoholismo, 1915) and the National League for Temperance (Liga Nacional de Templanza, 1916).[30] Simultaneously, the Argentine Congress taxed the production, distribution, and consumption of alcoholic beverages to tame drinking urges among the population. It also criminalized behaviors that weakened society in the Rural Code of the National Territories (1894). For instance, the code fined public inebriation that resulted in "yelling, cursing, or scandals" with five pesos the first time and ten pesos thereafter.[31]

During the 1910s, more and more people would openly feel like Bustamante, appalled by excessive behaviors of public disorder, and feel the need to take matters into their own hands. World War I disrupted international trading circuits, severely crippling the Argentine economy, maybe even more so than during the 1930s.[32] In 1919, workers and rural laborers went on successive strikes across the country. Similar to what had happened in the 1900s, demonstrations were met with disproportionate repression. In Buenos Aires, violence reached a high point in what became known as the Tragic Week, when police forces quashed a strike by steelworkers in one factory with deadly brutality. The workers' deaths only summoned solidarity from tanners, millers, builders, and maritime workers, among other industries, bringing the city to a halt. Paramilitary groups assisted police forces, including the Defenders of Order and the National Association of Labor, formed by companies and landowners to coordinate the recruitment of strikebreakers.[33] President Hipólito Yrigoyen, elected as a result of universal male suffrage, deployed the army to assist in disbanding the strikes, which drove workers in other cities to join.

After the Tragic Week ended in mid-January 1919, the Defenders of Order transitioned into a new, national organization, the Argentine Patriotic League (Liga Patriótica Argentina). The league crystalized the anti-leftist (but not necessarily anti-liberal) anxieties that had been brewing since the beginning of the century. During the events of January 1919, many of the league's members had participated in the first pogrom against perceived anarchists and Bolsheviks, whom they ethnically identified as Jewish immigrants from Eastern Europe.[34] The league institutionalized what Sandra McGee Deutsch identifies as a counterrevolutionary movement in the face of turmoil in postwar Europe and the fear of its ripple effects that were already shaking Argentina's streets.[35] This reactionary aspect has been debated by scholars. Fernando Devoto highlights the nationalist character of the league as a form of Argentine protofascism. For Osvaldo Bayer, the league operated as a counterunion, a sort of mutual aid association without state oversight, which Lisandro Galluci describes as "combative conservatism."[36] The league's membership filled up with myriad supporters, from elected officials to landed proprietors and industrialists. Imbued with eugenicist views of society, its first president (1919–46), Manuel Carlés, assessed the "social unrest . . . and unusual impulses" as a "temporary fever," not "chronic illness." The league, which included male and female members, sought to eradicate such disease by spreading their ideas through pamphlets, parades, and signs. They also directed their efforts at breaking labor organizing. They tried to counteract the influence of unions through their own workers' brigades and feeble teachers' associations, while they also resorted to violent tactics such as defamation and harassment.[37]

In Patagonia, the league was very active in great part against a widespread labor strike in 1920–21. The end of World War I interrupted the growth of wool prices, aggravated by the establishment of tariffs with Chile and the opening of the Panama Canal. The strike was, of course, disruptive but nonviolent. Workers were strictly prohibited from drinking, looting, or harming cattle or people.[38] Yet, President Hipólito Yrigoyen, who rode into office on the back of the extension of civil rights, authorized a brutal repression of workers at the hands of the army and the Patriotic League.[39] Arbitrary executions left 400 dead, a high toll for a population of 20,000.[40] Local reports and the national press described the bloodshed as a successful defeat of the anarchists' and bandits' attempts to set up communist brigades to advance toward Buenos Aires.[41] From then on, the league's president, Carlés, made it a central goal to form brigades in the South to counteract the influence of unions and immigrants. The events in Santa Cruz serve as a backdrop to understanding the anxieties around social unrest in Patagonia, a symptom of an unhealthy society, and increased calls for "moralizing action" (*medida moralizadora*) at the hands of law enforcement.[42]

A MURDER RATTLES THE NORTHERN PATAGONIAN ANDES

One evening in late December 1909 (sources disagree on which day exactly), two men entered Llywd Ap Iwan's store in Arroyo Pescado (Chubut) and shot him.⁴³ Ap Iwan managed a branch of the Chubut Trading Company's general stores at the crossroads of the steppe and Esquel.⁴⁴ He was an engineer educated at the University of Cambridge and a member of the first Welsh families that colonized western Chubut in the 1860s, where he enjoyed a comfortable social status. That tragic day in December 1909 had been particularly windy, preventing many customers from coming into the store. By the end of the day, however, three men had come in: an Indian, a man who spoke in Spanish, and a North American, whom Ap Iwan's employee described as an Englishman because he spoke English. Probably because of this, the shopkeeper "did not think this man any worse than anybody else."⁴⁵ Unfortunately, the English-speaking customer was not a client but a robber, and he was not English but US American. His partner also entered the store, and they drew revolvers and demanded that Ap Iwan open the safe. But the safe was nearly empty as it had been a quiet day. The thieves thought the Welshman was hiding money elsewhere. Maybe out of frustration or maybe out of disbelief, they shot and killed him. The bandits fled immediately, heading south and cutting telegraph lines along the way so that police officers could not alert other stations.⁴⁶ Law enforcement began a long chase to find the North American thieves. At first, authorities thought the robbers were Robert LeRoy Parker and Harry Alonzo Longabaugh, also known as Butch Cassidy and the Sundance Kid, who terrorized the Patagonian valleys. However, the investigation revealed that Ap Iwan's murderers were two other US American outlaws, William Wilson from Texas and Robert Evans from Montana, who had ridden with the famous members of the Wild Bunch and adopted their robbing style.⁴⁷ The police suspected that Wilson and Evans were also responsible for breaking into a bank in Río Gallegos (capital of the territory of Santa Cruz).⁴⁸ Although police officers knew who they were, the "North Americanos," as they nicknamed the pair, got away with murder.⁴⁹ They were only caught three years later and were killed while resisting arrest.⁵⁰

Local and national elites grew increasingly alarmed by the ease with which bandits freely roamed. Though probably exaggerated, a very real "fear took hold of the inhabitants" in northern Patagonia, a newspaper reported.⁵¹ Landowners, ranchers, store managers, journalists, and professionals viewed bandits' ease of movement as an attack on their own freedoms. If criminals could prowl the valleys with impunity, property owners were unsafe. Attacks on property and "honest citizens" evidenced "the depredations carried out by the delinquents who swarm there without any authority to stop them."⁵² The figure of the bandit represented immorality, disruption of progress, and social disease, recalling earlier depictions of Patagonia as a monster and as a Desert.

Local elites saw themselves at the frontlines of the nationalizing mission to fill in the Desert and, therefore, deserving of the right to live without fear for their lives or property. In Argentina, these local elites construed their patriotic identity on the basis of a higher morality, which, in turn, was predicated on the primordial tension between civilization and barbarism. By portraying northern Patagonia as a dangerous space, local elites proclaimed their patriotic duty to transform it into a productive, law-abiding landscape. As a result, those who lived and traveled across northern Patagonia often carried weapons.

In Argentina the Rural Code of the National Territories allowed for concealed carry of weapons, but the murder of Ap Iwan showed that this was not enough to keep people safe.[53] Self-identified hardworking citizens channeled their frustrations through the regional press in the hope of drawing the attention of authorities in distant capitals. Newspapers used Ap Iwan's murder to depict a dire situation in northern Patagonia. Ap Iwan himself had also described the serious situation to relatives back home in the weeks leading up to his death, noting "the utter lack of security for life and property in the neighborhood of Colonia 16 de Octubre, and the growing state of lawlessness in that area."[54] Newspaper editors rendered store employees at the mercy of assailants' "Hands up!" order, to which "there was nothing to do but obey."[55] In doing so, they called attention to the risks "hardworking people" took to advance the nation in frontier spaces. According to one article, Ap Iwan "defended the safe with all his might," even though this was not likely, because "it contained but very little money."[56] News articles also used Ap Iwan's murder to reaffirm the tension between civilization and barbarism that drove the colonization of Patagonia. For them, the Welshman epitomized the Western commitment to the protection of the basic (liberal) rights of property and life.[57] Presented by newspapers as symbols of civility, Ap Iwan's heritage, education, and personal wealth contrasted with the pillaging brokenness of his murderers.[58]

Other reports on banditry soon followed Ap Iwan's murder, adding to the narrative of the dangers of the northern Patagonian Andes. In 1911, Julián Gonzalorena, a Chilean living in Ñorquinco, Argentina, stated that cattle rustlers had taken 300 horses and 2,000 cows from "honorable people." Even though some people fought off robbers, they received death threats to prevent them from going to the police. But in Gonzalorena's opinion, the governor needed to not only send more officers but also improve their equipment. *La Nueva Era* amplified the story as evidence of "the fear that has taken hold of the inhabitants of that area in the face of the pillaging carried out by the delinquents."[59] Once again, hardworking people were at the peril of assailants, subtly tapping into the tropes of barbarism and civilization.

News articles also reinforced the idea that bandits were foreigners. Chilean newspapers highlighted the US American nationality of Ap Iwan's murderers. They described the robbers as "Yankee raiders" who "spoke English with a pronounced American accent."[60] As such, the newspapers distinguished between US nationals and other English-speaking immigrants, such as the Welsh or British. In doing so, they built on a prejudice against US Americans shared among some cattle ranchers. Certainly the news of Butch Cassidy and the Sundance Kid's gang pillaging peaceful towns in the Andes contributed to that jaundiced eye toward the US American accent. Argentine reports echoed a version of the anxieties in Buenos Aires around immigration and radical ideologies. In contrast, Ap Iwan's murder, where victim and perpetrators were both foreign born, revealed the contradictions in such a simplistic view. Previously, Argentine national authorities had seen Indigenous peoples as enemies of the national state and contrary to the civilized nation they were trying to build. But by the 1910s and 1920s, the symbol of the "ferocious Indians" was increasingly becoming an image of the past, an origin story of sorts of Patagonian settlement.[61] One newspaper reported, "There are no savage Indians in the vast expanse of the [territory of] Río Negro."[62] Indigenous presence would have meant that the tenets of a modern nation were still absent in northern Patagonia.

Increasingly, thus, local and national elites imagined the quintessential Patagonian bandit not as Indigenous but as Chilean, often conflating one with the other. Mestizo and Indigenous Chileans had migrated to Argentina in the last decades of the nineteenth century, particularly to Neuquén. Chilean legislation on landed policy had pushed many farmers and cattle herders out of their land. Simultaneously, Neuquén offered land available for ranching, especially at the turn of the century when demand increased in Chile. Argentine land distribution was slow, which enabled several herders, many of whom were of Mapuche background, to bring their cattle to graze across the Andes, especially in public lands. Susana Bandieri argues that their presence led *neuquina* population to see them as "intruders," not only in terms of their Indigenous-Chilean background but also in terms of their participation in capitalist production, perpetuating earlier conditions of marginality.[63] Some articles on Ap Iwan's murder conveyed the presence of Chileans and possibly Indians in Evan and Wilson's gang, as if their presence would make it easier for readers to disapprove of the murderers.[64] Residents of Arroyo Pescado traveled to the nearest police station, in Colonia San Martín, and blamed foreigners, particularly Chileans and US Americans, for what they characterized as alarming rates of banditry.[65] One article even included dangerous animals as part of the criminal band: "[Ap Iwan] had been attacked and nearly killed by Indians, Chilenos, and savage Pumas."[66] For newspaper editors and landed elites alike, the unpredictable marauding of criminal gangs recalled nomadic tribes in northern Patagonia.[67]

Argentine authorities conflated crime with national origin and with unpleasant consequences for regular people that had nothing to do with bandits. For instance, after the murder of Ap Iwan, police forces in Chubut and Santa Cruz set out to chase the two delinquents. They were so desperate to find the murderers that at some point officers detained anyone unknown to them (*desconocidos*). For example, the police arrested a traveler passing through a city in Santa Cruz because they did not recognize him. He was released three days later, "tormented by hunger . . . and devoured by insects . . . cursing the North Americanos, the Comisario, and all concerned."[68] On another occasion, a North American man who had been living in Argentina for several years went camping with a friend and a local landowner. A stranger came up to their campfire saying he had heard "the North Americanos are in the neighborhood." Upon scrutinizing the group and identifying two of them as foreigners, the newcomer added, "In fact, I should not be surprised if some of the señores sitting here could enlighten us."[69] The stranger turned out to be a police lieutenant and was openly accusing the two North Americans of being the robbers every police officer in Patagonia was trying to find.[70]

The murder of Ap Iwan revealed an underlying condition of northern Patagonian life. Despite the efforts to subdue the territory, organize it, and distribute its land, northern Patagonia remained a figuratively empty space and in reality a site of conflict. Self-proclaimed hardworking people felt they had no freedoms and lived in fear while bandits rode freely up and down the roads. It was imperative to impose order.

CALLS FOR MORE POLICING

In the aftermath of the murder of Llwyd Ap Iwan, local newspapers called for increased security in the Andes: "We reiterate the need for an armed unit stationed [in Chubut] that efficiently protects the lives and interests of those colonies."[71] Criticism of lax policing in Patagonia gained momentum in 1910. Requests highlighted the need "for a professional outpost that surveys the cordillera to avoid attacks like the one in Arroyo Pescado."[72] A group of residents of Colonia 16 de Octubre, near Arroyo Pescado, wrote to the minister of the interior, Indalecio Gómez, protesting that the lack of weaponry, munitions, and competent staff enabled violent crimes like the murder of Llywd Ap Iwan. In their view, the logical response to violence was a more heavily armed and better-prepared policing force.[73] The demands reached a peak in 1911, when a cattle rancher from Corcovado (Chubut) was kidnapped, allegedly by Robert Evans and William Wilson, the same men who murdered Ap Iwan.[74] Territorial governors amplified these concerns. For the governor of Neuquén, Eduardo Elordi, the scattered population, slow communication between police stations, and few competent and honest personnel delayed any effective action.[75] The minister of the interior attempted to increase the number of police officers in northern Patagonia by recruiting Paraguayans escaping the political instability of their country and crossing over to Chaco and Formosa. However, when the political situation stabilized, many potential recruits returned home.[76]

Partly riding on these demands and partly as a generic response to the 1909 strikes in Buenos Aires, Minister Gómez restructured the police in most national territories. For both the minister of the interior and for the governors there was little doubt that "progress of the Territories is intimately dependent on their security; without good police forces, the axiom 'To govern is to populate' cannot be translated into practical and real facts."[77] In fact, they made police reform a central theme of the first (and only) conference of governors of the national territories.[78] Initially, the expansion of surveillance translated into new police stations in the northeastern territory of Misiones and eight additional officers in the penal colony of Ushuaia (Tierra del Fuego).[79] Within a couple of years, other territories also received more police agents, sixty in La Pampa, twenty in Chaco, and ten in Formosa.[80] In Patagonia, Bariloche received a station in 1911, which "was called for because of the importance of police services in this territory which, due to its richness and vast extension, requires constant vigilance."[81]

A snapshot of the statistics in Neuquén illustrates the surveillance landscape in the northern Patagonian Andes. Neuquén was divided into twelve departments. I used data from a census and from a government report to the Ministry of the Interior to calculate the ratio of police officers to population for each department (table 4.1).[82] At the beginning of the 1910s, the department containing the capital city had the largest number of officers both in absolute numbers and per 100 inhabitants. In addition to regular forces, Neuquén City had a regional prison and the main police precinct (*jefatura*

Table 4.1. Statistics for analyzing the number of police officers in each department of Neuquén relative to its population and area, 1910s.

Department	Population	Density (hab/sq miles)	Police officers (total)	Police officers (per 100 inhabitants)	Sargentos	Cabos	Gendarmes	Size (sq miles)	Police officers (per 100 sq miles)
Confluencia (with prison guards)	2,559	1.11	73	2.93	4	5	64	2,300	3.17
Ñorquín	5,314	1.00	27	0.51	2	3	22	5,339	0.51
Los Lagos	3,272	0.71	33	0.98	3	5	25	4,625	0.71
Chos Malal	3,110	1.31	19	0.61	2	2	15	2,368	0.80
Minas	3,069	1.31	23	0.75	1	2	20	2,340	0.98
Picun Leufu	1,031	0.31	8	0.78	1	1	6	3,302	0.24
Aluminé	2,511	0.72	17	0.68	1	2	14	3,464	0.49
Limay Centro	2,054	0.80	12	0.58	1	1	10	2,575	0.47
Colorado Arriba [Río Colorado]	1,598	0.49	9	0.56	1	0	8	3,247	0.28
Collon Cura	1,468	0.73	8	0.54	1	1	6	2,016	0.40
Las Lajas	1,176	1.51	15	1.28	1	2	12	781	1.92
Aselo [Añelo]	312	0.07	3	0.96	0	1	2	4,162	0.07
Total	**27,474**	**0.75**	**247**	**0.90**	**18**	**25**	**204**	**36,519**	**0.68**
Confluencia (without prison guards)	2,559	1.11	39	1.52	2	2	35	2,300	1.70

Sources: For population, see Dirección General de Territorios Nacionales, Censo de población, 227–30; for number of police officers, see Gobernación del Neuquén, "Expediente 1643N—plano e informe del Territorio de Neuquén solicitado por el ministro del interior," April 18, 1910, Fondo Ministerio del Interior, Expedientes Generales, box 7, Archivo Intermedio, Archivo General de la Nación, Buenos Aires. For the area of each department, I overlaid a map of Neuquén from 1923 (which still had the subdivisions of the early 1910s) onto Google Earth Pro and calculated the area by tracing each district with the polygon tool. See Alfredo Weber, "Mapa geográfico-comercial."

de policía), which required having more officers.[83] The department of Los Lagos, in the southern part of the territory and bordering Lake Nahuel Huapi, was the second most populated department in the territory (3,272), after the largest district, Ñorquin (5,314). Yet Los Lagos had more police officers (35) than any other district besides Confluencia (the one with the capital city) (39, plus 34 prison guards). Similarly, Los Lagos ranked third in the police-officer-to-inhabitant ratio. These numbers might suggest that rising concerns about criminality in the Patagonian Andes were met with an increased police presence. Territorial authorities in Neuquén joined their peers in Río Negro

Figure 4.1. Screenshot of a section of "Mapa geográfico-comercial con la red completa de ferrocarriles de las repúblicas Argentina, Chile, Uruguay y Paraguay," by Alfredo Weber (1923), overlaid onto Google Earth. The image shows the department of Los Lagos in dark gray. The map of Neuquén looks tilted because I adjusted it to match the curvature of the earth. This tilt exemplifies how maps fail to fully grasp the geography of our world.

and Chubut in portraying the environs of Lake Nahuel Huapi as a dangerous place that needed surveillance.

The geographical distribution of these numbers, however, tells a slightly different story. By overlaying a historical map of the departments of Neuquén onto Google Earth, I calculated their surface area using the polygon tool (fig. 4.1). With this data, I could compare the number of police officers per 100 square miles for each department, concluding that Los Lagos was not as heavily policed as I had initially thought (or as the governor had led me to believe). Far from having a relatively high number of officers compared to other departments, Los Lagos was outnumbered by the smaller but more densely populated departments of Chos Malal, Las Lajas, Las Minas, and Confluencia. All of these were on the northern side of Neuquén.

In both Chile and Argentina, national authorities viewed law enforcement in northern Patagonia as one of the most efficient treatments for social degeneration. However, often the cure was as bad as the disease. Governors on both sides of the Andes received complaints of mistreatment, abuse of authority, and poor behavior from law enforcement officers. Governors in Llanquihue received complaints about police officers confiscating animals and returning them "half dead."[84] In 1916 in Neuquén, the police captured a group of escapees from the local prison and massacred them. A former police officer turned journalist, Abel Chaneton, reported on this violence. Unfortunately, his vociferous condemnation of police brutality led to his murder the following year.[85]

In another case, a Chilean woman, Natalia Toledo, left Benito Crespo in charge of her land in Chenqueniyen, near Ñorquinco. She was away in Chile for eighteen months taking care of her late husband's business across the Andes. Crespo was a former judge and a sought-after estate manager, with good connections and vast experience. Upon her return, Toledo found that Crespo had allowed three people to herd sheep and cattle on her land. Her land stretched across three different jurisdictions, Bariloche, Ñorquinco, and Pilcaniyeu, so she filed a report in the three police stations within months. While all three precincts sent officers to kick the intruders out, Toledo complained that the sergeant from Pilcaniyeu protected the squatters instead of removing them because he was friends with Crespo. She accused the men of conspiring to displace her from her own land because she was a woman. It seems her position as a widow provided her more room for garnering sympathy from authorities than her national origin. After a brief investigation, the trespassers were removed from Toledo's land and the officers from their positions. Nothing happened to Benito Crespo.

Often, abuses of authority occurred as plain negligence in the care of prisoners. In 1920, a Chilean citizen, Luis Oyarzun, filed a complaint with his consulate against Benjamin Farrington, a police officer in western Chubut. Oyarzun claimed he had been arrested on February 18 "without just cause" and held in a prison in Leleque for thirty-two days. He argued that Farrington frequently mistreated prisoners, and he included the complaints of four other detainees he met while under arrest. A man called Pedro Velázquez said he was chopping wood when Farrington "detained [him] and whipped [him] in front of several people," adding humiliation to the unjust detention.[86] José

Santos Bascuñar claimed he had been imprisoned for sixteen days and that Farrington had taken the money Bascuñar had for his family. The third man, in jail for twenty-eight days, accused the officer of not providing the minimum medical care for his tertiary syphilis. And the fourth, Manuel Ramos, had been incarcerated for more than a month without anyone taking a statement. The four complaints in Oyarzun's file are telegrams Oyarzun sent on the same day at the same time—March 19, 1920, at 6:00 p.m.—from the station in Cholila, thirty miles southwest of Leleque. More than 100 years later, we do not have any way of knowing if the telegrams were sent together because they arrived together from the jail in Leleque or if, led by Oyarzun, all four prisoners decided to denounce their treatment at the same time. At the request of the Argentine minister of the interior, the governor of Chubut sent his secretary to open an investigation, which resulted in the removal of Farrington and his commanding officer.[87]

The establishment of police stations and increased personnel was probably rushed. The system soon revealed its flaws. It was hard to find new recruits, training was mediocre, and equipment was lacking. These shortcomings, theorized national authorities, resulted in officers' abuse of authority. For governors, "the progress of the Territories [was] intimately dependent on their security." In their view, only with "good police forces," that is, appropriately trained and equipped but also morally sound, would the civilizing of Patagonia be complete.[88] It is safe to assume, as the case of Neuquén showed, that stretching scant resources over vast areas resulted in little oversight, poor embedded accountability, and probably more (underpaid) labor. The minister of the interior frequently received requests for leaves of absence or reappointments to Buenos Aires.[89] Who, then, was willing to police the national interests in the frontier?

THE BORDER POLICE

The creation of the Border Police (Policía Fronteriza) in February 1911 stands out as one of the most influential law enforcement policies in northern Patagonia. This initiative deployed a cadre of relatively autonomous officers who answered to neither local nor provincial authorities.[90] This force operated in the northern Patagonian Andes, stretching from Neuquén to Chubut (it was then extended to Santa Cruz). Independent from territorial governors, the Border Police worked not only to aid local law enforcement but also as a federal tool of surveillance in the Patagonian northwest.[91] Most surprising, perhaps, is the fact that the prime objective of the Border Police was not to police the border. In other border regions of the world, border policing emerged to prevent illegal migration or illegal trade.[92] In contrast, the Border Police sought to restore "the times when it was possible to travel safely to Chile."[93] Its goal was to protect travel and trade, not police the border.

Reports on the effectiveness of the Border Police in rounding up bandits projected a sense of improvement in public safety. In 1911, Adrián del Busto, chief of the Río Negro section of the Border Police, charged more than sixty Argentine men in a relative short period of time. Unfortunately, the indictments used up all the available paper in Bariloche and surrounding areas. Del Busto sent his secretary to purchase paper in Puerto Montt to finish the reports. In 1911, it was easier and faster to travel to Chile than to Neuquén or Viedma, let alone Buenos Aires. This episode did not escape the notice of the editors of *La Nueva Era*, who wondered whether Chilean paper was appropriate for indicting Argentines. Would lawyers use this as a reason to free their clients? They concluded that del Busto's sound decision showed that, under his leadership, the Border Police was "the immovable pedestal where the tranquility of the population of the cordillera is secured."[94] Unfortunately, these residents of the cordillera had no paper on which to write their praises.

The Border Police also "closely oversaw the behavior of police personnel" in places "isolated from the main authorities."[95] Minister Gómez appointed Mateo Gebhard to command the Border Police in western Chubut. He spoke in broken Spanish with a pronounced German accent, which led some sources to believe he was Austrian.[96] Following the same logic they used when appointing the leader of the Capital Police in Buenos Aires, federal authorities thought that someone with military experience would be an effective leader of the Border Police in Patagonia. Gebhard received autonomy for operations, distribution of personnel, and discipline and reported directly to the director of national territories, not to the territorial governors.[97] Gebhard informed the governors about the corrupt behaviors of police officers under their control. Such was the case of *Subcomisario* Maximiliano Montero, an officer in the regular force in western Chubut, who was frequently intoxicated. While on the road on official business, he tried to force two subordinates to commit theft. When they objected, he himself stole a saddle at one stop and "hens, cigarette boxes, and silk scarves" at another.[98] Gebhard found out about these abuses because the two junior officers reported Montero to his supervisor, who did not suspend him. To Gebhard, this was as much a violation of public duty as Montero's misdemeanors.

Numerous complaints targeted the Border Police for abuse of authority and violence. With little oversight, "energetic and zealous" officers in Bariloche captured "near a hundred citizens" on suspicion of banditry in the force's first eight months. They kept their prisoners "poorly housed, underfed and brutally treated." *La Nueva Era* denounced these practices as unequivocally barbaric, as if they evoked a more violent, less cultured time.[99] As the discourse about criminals shifted toward Chileans, it is not surprising that this force was also accused of profiling. Gebhard earned a reputation for his bias against Chileans. He unapologetically denounced the Chilean consul in Chubut to the director of national territories. On paper, the consul had granted some permits to Chileans to exploit the Andean forests on their side of the border. In reality, he had apparently turned a blind eye to Chileans crossing over to the Argentine side to collect timber. What added insult to injury was the fact that the consul also drank heavily, especially absinthe, and quarreled frequently under the influence. In various gambling dens, he squabbled "with Juan Fardon, Chilean; with someone named Fasiasi, Italian; with Guillermo Davies, Welsh; [and] with Alberto Caso Rosendi, senior and junior, both Argentines." Gebhard had been reticent to file a report, "since [he was] gratuitously attributed with a certain grudge against the Chileans; an assumption that completely undermines [his] temperate behavior." Instead, Gebhard's statement tapped into the dichotomy of inebriation as an image of degeneration and his own moderation as moral virtue. By echoing hygienic ideas about inebriation within the Patagonian context, the chief of the Border Police—an immigrant—positioned himself as an agent of moderation and thus of nation-making in the still barbaric Patagonian frontier.[100]

Chilean newspapers denounced the mistreatment of their fellow countrymen in Argentina. With the creation of the Border Police, Argentine forces allegedly "apprehended numerous Chilean citizens innocent of recent robberies, under the pretext of pursuing banditry."[101] Newspaper editors described family life, farming, and steady work as markers of civilized society. They framed Chileans living in Argentina as active contributors to the growth of the country. For instance, Chilean Froilán Muñoz had lived in Epuyen (Chubut) for more than twenty years "with his mother, wife, and children" and owned about 1,500 animals. Although Chileans living in western Chubut and Río Negro "were wealthy people," concluded *El Llanquihue*, chasing bandits was "a pretext for dispossessing Chilean citizens who have been living in the Argentinean countryside for years."[102] The minister of the interior admitted that "several merchants and working and well-regarded people" were detained by the Border Police. However, they "were involved by covering for the bandits."[103] Editors in southern Chile dexterously portrayed bandits in the cordillera as escapees from the overwhelmed Argentina police, implying that they only made it to the West as a result of another force's ineptitude. Other times, Chilean newspapers stressed the foreign origin of rustlers, usually US Americans. With this, they replicated the idea that, like a disease infests an organism, a foreign element threatened progress in Llanquihue.[104]

In 1913, two years after the creation of the Border Police, the minister of the interior, Indalecio Gómez, proudly reported that "the plague of banditry has been suffocated in Río Negro and Chubut." With that, he not only called to maintain the police force in those two territories but also to expand it to Santa Cruz, farther south, where "[bandits] had sought refuge."[105] Regional newspapers were a little wary of the minister's assessment. *La Nueva Era* reported in 1913 that the Border Police had caught the murderer of a businessman from Pilcaniyeu but set him free. A succession of six crimes that followed, "where victims were mostly women and children," prompted the newspaper to suggest they were committed by the same person who had killed the businessman. The editors denounced the liberty that such a perpetrator enjoyed and blamed the police force for the fact that "life in the settlements of the cordillera [is] less and less appealing."[106] Despite the minister's assertion of the Border Police's success, a shaky budget and denunciations against it forced him to disband the force in 1914.

However, the Border Police was reinstalled four years later to tackle rustling. Indeed, stealing livestock undermined trans-Andean trading networks. As one newspaper acknowledged, "Contraband does not exist in the Andes. ... What exists and needs to be dealt with is cattle-raiding."[107] In other words, the Andean region needed more policing, regardless of what it was called: "It seems [the Andean region] does not need a Border Police, it simply needs a police force."[108] Some residents of the area acknowledged that more policing of the border region had overall improved the situation. José Vereertbrugghen, a resident of Bariloche, asserted that "when ... the Border Police arrived here ... the situation was worse. Within a short time, everything changed radically and thanks to their campaign, we have enjoyed some calm years. Murderers got a good scare."[109] In 1921, a gang of bandits attacked some ranches in western Río Negro, emboldened by the laborer uprisings in rural Santa Cruz. The Border Police, however, reacted promptly and chased them across the valleys to the north, "as they fled to Neuquén."[110] During the second stint of the Border Police, *La Nueva Era* amplified the voices of some concerned residents of the northern Patagonian Andes (such as Vereertbrugghen) demanding increased police presence, including a permanent military garrison. Whether reporting the Border Police's effectiveness in chasing down bandits or the lack of it, the editors of the Viedma-based paper made the case that "the [Border] police cannot fulfill its role.... It is necessary to establish military units in strategic points."[111]

The creation of the Border Police disrupted the narrative of progress that authorities had pushed in northern Patagonia. For them, advancing capitalist interests would bring prosperity in a space that was viewed as being empty. Land distribution would lead to agricultural production, which would activate trade and demand infrastructure. More people would want to settle there. However, as the minister of the interior reported in 1913, as the population increased in the Andean region, "the bad social elements settled there [too]." Such newcomers, however, did not establish themselves with the goal of advancing the economic and cultural values set forth in much of the legislation. Instead, argued the minister, they settled "either as intruders ... in state-owned fields as in a domain conquered in good law, or resolutely throwing the mask to let the eyes of the frightened populations see the sinister silhouette of the bandit." The problem was not rooted in the border per se but in who settled where and whether land occupation was legal or not. "Bad elements," as authorities frequently referred to people who undermined a desirable order, hampered progress and degenerated society. It was imperative to dispense "once and for all with the mountain banditry" by eliminating the unwanted population with more police forces and introducing settlers that would propagate good behavior across the northern Patagonian Andes.[112] This solution certainly echoed decades-earlier portrayals of Indigenous communities as the enemy of the nation. If in the 1870s, the presence of Indigenous peoples in northern Patagonia evidenced the absence of the national state, the perceived high crime rates in the 1910s continued to point to a weak state. Increasingly, authorities and local elites turned to blame immigrants for the situation in Patagonia.

ANTI-IMMIGRANT SENTIMENTS IN ARGENTINA

"Things are going bad in the Andean region," began an article on December 21, 1919, in a newspaper published in the Atlantic towns of Carmen de Patagones and Viedma. The article continued: "It looks like guarantees for life and livestock are an illusion.... All accounts that we receive ... [call for] authorities to severely and energetically suppress banditry." The article made its case for more policing in the Andean region by including a letter from José Vereertbrugghen, a Belgian physician who resided in Bariloche. The letter denounced the murder of Argentine Juan Torrontegui presumably at the hands of a Chilean man named Montero, who also shot the arresting officer. Besides outrage at the crime against an Argentine and a police officer, Vereertbrugghen condemned the fact that Montero wandered freely around Lake Nahuel Huapi with blatant impunity. Montero's freedom was a slap in the face to the efforts that people like Vereertbrugghen made to "develop the Argentineness in those places today abandoned to foreigners that harass our honest and hard-working settlers." A foreigner himself, what pushed a Belgian doctor to claim to be more Argentine than a Chilean bandit?[113]

Vereertbrugghen's complaint exemplifies how local elites increasingly conflated ethnicity with criminality. Specifically, they considered Chileans a threat to nation-making in similar ways that banditry was an obstacle for the economic progress of "hardworking people." Landowners, businesspeople, professionals, and local authorities positioned themselves as nation-makers in the northern Patagonian Andes, regardless of their national origin. As Vereertbrugghen's letter suggests, people of specific nationalities, such as Chileans or, in some cases, North Americans, were increasingly vilified. For instance, an observer who traveled to Neuquén blamed Chilean immigrants for importing alarming quantities of alcohol. In his view, drunkenness made "every household, every shack [*rancho*] . . . a clandestine business where immoral orgies, crimes in an array of disgusting images, and even murder [were ubiquitous]."[114]

Anti-Chilean sentiments grew among Argentine authorities. The governor of the National Territory of Chubut, Julio Lezama, accused Chileans of bringing "an inconvenience and a danger" to Argentine valleys. The geography of the border, "located in the previously disputed zone," enabled Chileans to cross into Argentina. Lezama suggested that the Palena and Futaleufú River passes, which had sparked numerous debates during the border negotiations, attracted "the worst [people] and settle in the enviable Argentine lands located one step away, as if it were a 'res nullius.'"[115] The governor, who resided on the Atlantic coast, perceived the Andean valleys as no-man's-lands precisely because of the presence of Chileans. The governor's anti-Chilean attitude was reflected in law enforcement. Police officers grew increasingly worried about the number of Chileans among the troops' ranks. In 1907, for example, some prisoners escaped from the jail in Bariloche and fled toward El Manso Pass, which connected the eastern Argentine slopes with the Chilean village of Cochamó. The police commissioner led a futile chase across the Andes. He accused the troops of tipping off the outlaws: "The soldiers, who are Chilean, warn the bandits or, if they are captured, they help them escape."[116] The perception of Chilean disloyalty to the Argentine state surfaced vividly in the way police officers treated settlers born on the western side of the Andes. Froilán Muñoz denounced the local police in 1911, saying they were "harassing cattle-owning Chilean settlers to emigrate back to Chile and [wanting to] take their estate," even though he had lived in the Epuyén valley since 1890.[117] Upon settlers' refusal to leave their land, the police imprisoned them "in Bariloche, Jeleque [Leleque], and Súnica" without indictment.[118] The targeting of foreign-born settlers did not wane, despite some protests in Puerto Montt and a timid intervention by the Chilean consul in Neuquén.

The high Chilean percentage of the population in Neuquén worried law enforcement. In 1922, the deputy sergeant of the Neuquén police accused Capt. Adam Giménez of the national guard (*gendarmería*) of insubordination to the governor, Francisco Denis. These were two very distinct forces. The police were the regular law enforcement troop that stretched throughout the territory. The national guard was a temporary force established specifically in the capital city. This small body was created in response to a massive escape from the Neuquén prison in 1916 and (historically unrelated) uprisings in Santa Cruz in 1920.[119] The national guard fell under the umbrella of the Ministry of the Interior, so its officers did not report to the governor or any other local authority. The national guard's leader, Giménez, accused the governor of attempting to arrest him in front of his 300 foot soldiers, challenging the force's mission. With a droplet of melodrama, Giménez reminded his superior that the national guard's purpose was to cooperate "effectively with the surveillance and security of the Territory [of Neuquén] ... [and] to establish a patriotic [sentiment of] Argentineness in these extensive and far-off regions where population, commerce, ranching, and even monetary circulation are Chilenized almost completely."[120] In Giménez's view, the widespread presence of Chileans in Neuquén undermined the nationalizing mission of the national guard (or any police force, for that matter). The governor responded to this challenge to his administration by accusing the national guard of being "composed mostly by Chileans, whose antecedents leave a lot to be desired," implying that they had criminal records.[121] With these words, Governor Denis sought to undermine the authority the national guard claimed under the umbrella of "nationalization." If a force was majority foreigners with criminal records, how could it monitor the making of Argentina in the frontier?

Certain members of the local elite in northern Patagonia believed that Chileans were undermining Argentine efforts to nationalize the frontier, despite the fact that many Chilean businessmen were themselves part of this elite. José Vereertbrugghen, the Belgian physician from Bariloche, protested that an alleged murderer, a Chilean, had not been arrested and wandered freely around Nahuel Huapi: "It seems that to reside here we need permission from Chilean bandits and not from the Argentine government."[122] Although Vereertbrugghen wrote these words in a letter to a friend, they were published in *La Nueva Era*, which took him completely by surprise.

Chileans living in Argentina pushed back against skeptical portrayals of their presence on the frontier. Arturo Ríos, the Chilean consul in Bariloche, accused the physician of manipulating the facts of an alleged murder to "unjustifiably and categorically attack Chileans living in the [Andean] region."[123] Ríos refuted point by point Vereertbrugghen's exaggerations, including the claim that the supposed murderer was Chilean when, in fact, he was a Spaniard. He concluded: "The truth is ... we travel along the roads of Bariloche with tranquility, day or night, making any weapon unnecessary." In addition, while Vereertbrugghen frowned upon the high number of Chilean officers in the local police force, Ríos argued that "Lieutenant Ávila [from Bariloche] is satisfied with these officers ... and has recalled two other Chileans because they excelled."[124] Luis González, member of a branch of the Chilean Center (Centro Chileno), also reacted to Vereertbrugghen's assertion that high crime statistics were directly connected to Chilean immigration in Argentina. Chilean Centers were social clubs for Chileans residing in Argentina to gather and celebrate their culture, usually hosting national festivities.[125] Accusing the Belgian physician of conflating nationality with crime, González argued that "ninety percent of the [Andean] region's progress rests on the effort of Chilean workers. Neither a house, nor a fence, nor clearings, streets, crops, roads, or anything that indicates the progress, or the work of man has been done by anyone other than the Chilean."[126] In addition, he contended that by having their children in Argentina, Chilean immigrants would be responsible for 80 percent of "the real Argentine population in the region." By "real" González referred to a generation of Argentines born in Patagonia as opposed to being transplanted from somewhere else and becoming Argentine by choice, like Vereertbrugghen's son. At heart, anxieties about Chilean presence in northern Patagonia reveal more about the ways in which Patagonia posed a problem to Argentine authorities and the ways, albeit very clumsily, they tried to solve them.

HEALTHY CITIZENS ARE ARMED CITIZENS

Building the nation in northern Patagonia amounted to creating a healthy society, physically and morally. For the business owners and landed elites of Bariloche, advancing the civilizing frontier entailed the creation of social spaces for supporting and expanding healthy practices. To that end, on April

30, 1915, forty men gathered at the Perito Moreno Hotel in Bariloche. The presence of Pedro Serrano, the governor of the territory of Río Negro, gave the meeting an official status. Attendance by Otto Huber, commander of the territory's military district, and Juan Carlos Pérez Colma, chief of the local police precinct, also provided auspicious support. The rest of the attendees were members of the local elite: a judge, a physician, businesspeople, farmers, and estate managers. The meeting produced the founding constitution of Bariloche's chapter of Tiro Federal, a national shooting club.

Founding a local chapter of a national shooting club was embedded in the hygienic efforts to mold a "national," healthy population across the country. For Argentine political elites, as for Latin American authorities elsewhere, a healthy citizen embodied and projected a modern nation.[127] Indeed, the practice of physical activities, including sports, contributed to civilizing the Argentine masses, including immigrants.[128] Physician Enrique Romero Brest, founder of physical education in Argentina, developed a physical education program for schools that correlated healthy bodies with moral strength. Physical education would determine "the subsequent biological and social fate of a nation" and thus lay "the inescapable foundations of the moral and intellectual life."[129] As a result of his efforts, school curricula incorporated a revised physical education program, which included cestoball, a sport of Romero Brest's creation.[130]

Authorities extended their support of physical education beyond schools to sports organizations, notably the Boy Scouts based in Buenos Aires. The organization aimed to "create citizens of character, healthy in body and mind, self-sacrificing, resourceful, self-reliant, and capable of exalting the country [*patria*] and its institutions." The scouts soon earned official support as a "school of civility and democracy" and enjoyed state funding, free rail and sea tickets outside Buenos Aires, a reduced military service, and preference for entry-level jobs in public service.[131] Francisco Moreno, a key figure in the Chilean-Argentine border negotiations, was one of the Boy Scouts' earliest supporters, particularly after founder Robert Baden-Powell's visited Argentina in 1909. Learning to dominate nature and withstand its challenges reinforced heteronormative ideals of masculinity that advanced the myths of dominance over the environment. Moreno became president of the Boy Scouts, and that combined with his membership in the National Education Council (Consejo Nacional de Educación), which oversaw schools in the capital and the national territories, consolidated the cross-pollination of formal and informal education in the nationalizing mission of liberal elites. Not coincidentally, Moreno befriended Theodore Roosevelt, another Boy Scout enthusiast, during Roosevelt's journey across Chile and Argentina in 1913. Hiking in areas outside of the urban centers equated a sovereign act of knowing the nation and instilled in young recruits the value of physical effort, self-discipline, and solidarity. In a country still anxious about using the taming of nature as a metaphor for constructing the nation, the Boy Scouts provided a breeding ground for the next generation of voting Argentines.[132]

For adults, Argentine authorities endorsed the foundation of sports clubs as sites of male socialization and discipline. Since the late nineteenth century, clubs formalized the practice of (British) sports, such as football (soccer), rowing, and cricket, which acted as cultural markers of (European) modernity, as Nortbert Elias argues.[133] The funneling of foreign investment, especially British, into the productive pampas and Northwest also propelled the emergence of clubs in the shadows of factories, train stations, and sugar mills.[134] But other clubs, such as Gymnastics and Fencing (Gimnasia y Esgrima), instituted military discipline within civic society, forging bodies ready to defend the nation. Like military service, sports clubs invited people from different ways of life, including from diverse nationalities, to abandon their factions to join a common, albeit recreational, cause, a sentiment that could be easily transferred to the nation.[135]

The forty men who gathered in Bariloche founded Tiro Federal under different circumstances but with the same nationalizing mission as sports clubs elsewhere. Tiro Federal Bariloche emerged under the tutelage of local and regional elites, as evidenced by the attendees. However, it was born far from Argentine capitalist markets, which were oriented toward the port of Buenos Aires. The Perito Moreno Hotel, named after the Argentine explorer, was an apt location for such a momentous event. Founded in 1902, the same year the national government signed the act that created Bariloche, the hotel frequently lodged ranchers and merchants and soon became a gathering site for local store owners, authorities, and professionals. In this social venue in April 1915, forty men created a club that would allow them to project their ideas about the nation and perform their role in it. In their view, they constituted a third wave of nationalizing forces in the imagined Desert that was northern Patagonia, right after the military columns of 1879–81 and the surveying explorers of the border negotiations. As a result, authorities hoped the club would develop into "an indicator of the ability and skill of ... all the inhabitants of the region." To foster "such noble purposes," Governor Serrano pledged the support of his office as well as that of the Ministry of War with weapons, ammunition, and property for the shooting range.[136] Inscribed into the constitution was the requirement, or hope, of owning a national flag with the name of the club sewn in gold as a symbol of the commitment to advancing the nation's interests with strictly regulated shooting competitions and social functions. In 1915, forty men saw themselves as legitimate guardians of the nation.

The material and symbolic endorsement from the government of Río Negro and the military consolidated the nationalist mission of Tiro Federal. Its president, Emilio Frey, requested a subsidy that would alleviate the shipping costs of munitions. Such a modest endowment (M$N100) would serve "as a stimulus for us Argentines in this border region . . . that we will not leave, as a reward to the country to enlarge the already long list of good shooters." Frey viewed the nascent club as a metaphor for frontier life, where hardworking people would "Argentinize" "this frontier region, so distant [from other centers]," with "good shooters."[137] Indeed, other shooting clubs in northern Patagonia viewed their mission as that of training brave, armed citizens, ready to defend the national territory in the frontier they inhabited.[138] The virtuous citizens-in-arms that inspired the foundation of the Tiro Federal also nourished the ranks of the local Patriotic League brigades. For instance, the League's chapter in Viedma elected to its board two former Bariloche authorities, Adrián del Busto (who was *comisario*) and Benito Crespo, former judge and estate manager and member of Bariloche's shooting club.[139] Across Argentina, a chapter of the Tiro Federal would mold a citizen "who, knowledgeable and skilled with weapons, would guarantee the defense of individual and collective interests."[140] Born in a borrowed locale—a hotel—and with no equipment of its own, the club admitted its outlier status in terms of material resources.

Tiro Federal inserted its local rhetoric into the national agenda to mold Argentine citizens. It dovetailed with the pedagogical mission of national commemoration that Yrigoyen's government introduced to school curricula. Commemorations of historical events sought to bolster a nationalistic sentiment among Argentines and non-Argentines. Celebrations of national holidays, including Chilean Independence Day, and religious festivities all took place at the same place in Bariloche, the Historical Cypress. This was the tree where, allegedly, Indigenous *lonko* Valentín Sayhueque held Francisco Moreno prisoner in 1880 in the military campaign to Río Negro. When former president Theodore Roosevelt disembarked in Bariloche in November 1913, Moreno took him to the historic tree before anywhere else. By then, the local council had raised funds to build a fence in an attempt to permanently distinguish the site from the rest of town. In doing so, they separated a piece of nature—a tree—with historical significance from the business of everyday life. In doing so, the local council cemented the national sentiments the tree evoked in a specific history of territorial appropriation. This history became the nation's history in the locality, with all its implications. The tree represented a stitch in the fabric of civilization, upholding liberal notions of democracy, property, and hegemony and excluding undesired, "barbaric" populations. Somewhat ironically, the municipality of Bariloche felled the Historic Cypress in 1958 to facilitate traffic.[141]

CONCLUSION

Throughout the 1920s, two factions emerged in Bariloche, one that aligned more with the work of the local council (often composed of several immigrants) and the other with an opposition emboldened by the actions of the Argentine Patriotic League. The local economy, crippled by the tariffs on the border, never fully recovered from the postwar crisis. In 1927, the only power plant caught fire and was not replaced for years. Work on the railroad to Viedma was delayed and then halted in 1928. In a context of literal darkness, by the end of the decade a feud had erupted between a member of the league, physician Luis Pastor, and local Italian businessman Primo Capraro. Accusing Capraro of embezzlement, Pastor ridiculed his heavyset build and apparent loud voice, stating, "It is not possible that there is anyone so noisy, so thunderous, so boisterous, so tumultuous."[142] At the same time, Pastor's home was plastered with pamphlets that claimed Italians brought "civilization" to Patagonia and therefore were equally righteous participants in its nationalizing efforts. To support Pastor, Pedro Alcoba Pitt, a local teacher and news correspondent for *La Nueva Era*, called attention to the events as evidence that tempered behavior, at least in public, was a desirable, masculine trait for the frontlines of the nation. He used the tropes of "clean air" and "clear skies" to call for purification of the local demeanor, an image that would persist into the following decade.[143] In the election of 1930, authorities initially refused to register foreigners to vote (though they were allowed to vote), ignoring "the sacrifices that foreigners have made for years for the progress and good of this people." The effects of the economic crisis of 1930, coupled with an electoral defeat and some business setbacks, drove Capraro to commit suicide on October 4, 1932.

Local confrontations in Bariloche encapsulated bits and pieces of a hygienic rhetoric that had been circulating for decades. At the national and local levels, elites construed the Patagonian space as a site marked by danger. Their discourse increasingly incorporated hygienic elements that diagnosed the backwardness of the region as an ailment that affected the whole nation.

For instance, these elites saw the rising crime rates (which were rising only because there was more reporting) as an unequivocal symptom that prevented northern Patagonia from becoming a productive space. At least two incidents shaped anxieties about progress in northern Patagonia: the social unrest in the port of Buenos Aires in the 1900s and late 1910s and the labor strikes in Santa Cruz in 1921–22. These events fed into authorities' imaginaries about an anarchist/Bolshevik conspiracy to overthrow the government, which, as Ernesto Bohoslavsky has argued, is probably why the massacre of labor strikers in Santa Cruz was forgotten until the 1970s.[144] Such fears arose in the form of police brutality in Buenos Aires and Patagonia; paramilitary far-right organizations that multiplied across the country, especially the Patriotic League; and legislation that allowed the government to expel crime suspects from the country. Together with these concerns stemming from the capital, the specific experience of living in the northern Patagonian Andes rendered nationalist discourse particularly attractive, even for immigrants.

In the first decades of the twentieth century, national authorities' portrayals of Patagonia reinforced the spatial discourses of difference contained in earlier depictions of the region as containing monsters, barbarism, and emptiness. Hygienism provided a new vocabulary to understand (or diagnose) the rapidly changing Argentine reality. In northern Patagonia, local elites of various national and ethnic origins appropriated this framework to assert their influence under the guise of the always appealing nationalizing mission. Like Valentín Bustamante, these elites saw with horror the excessive behaviors that obstructed their work. Inebriation, corruption, and overall lack of temperament undermined the health of the nation. Economic and social elites saw northern Patagonia as a dangerous place, a site where the nation was at its most vulnerable, and requested more state presence in numbers of police officers. When this failed, civic organizations sought to fill that vacuum with their own nationalist agendas, including the Tiro Federal Bariloche and the Patriotic League. Even though such associations might have been animated by the same national sentiments, such feelings were deeply personal and highly evolving. However, from the perspective of national authorities, northern Patagonia continued to remain unincorporated, presenting a vulnerability to alternative (non-Argentine) identities. This, in turn, posed a risk to the integrity of the nation. How would authorities resolve this tension?

Chapter Five

NATIONAL AESTHETICS IN THE ARGENTINE LOCALITY

1905–1945

Since 1940, a complex of four buildings arranged in a U shape constitutes the Bariloche Civic Center (fig. 5.1). Combining local tuff stone with cypress and larch situated the picturesque complex in conversation with its surrounding landscape. Yet this architectural style is strikingly reminiscent of a French medieval village. Together with the Andean backdrop, this eclectic style reinterpreted the medieval charm of southeastern France through a modern lens: it cemented a way in which landscape and built environment should constitute, together, a symbol of Argentinness.[1] In a town founded and fueled by Chileans of German descent, evoking the charm of southern France constituted the most anti-German,

Figure 5.1. Partial view of the Bariloche Civic Center, ca. 1950, from the archways facing west. Photograph by Bruno Ricardo Sálamon. Courtesy of Alcoba Pitt Collection, Archivo Visual Patagónico.

and therefore anti-Chilean, form of nationalist transformation of the built environment.

The Civic Center was part of a constellation of new buildings that the new National Parks Bureau (Dirección de Parques Nacionales, DPN) built in Bariloche and its environs.[2] From hotels to public offices, authorities mobilized a coherent aesthetic in support of a distinct nation-building strategy, a process we continue to observe in other postcolonial nations from Mexico to the Arabian Peninsula. In Bariloche, this strategy sought to create a spatial discourse that locals and visitors could recognize in the context of a border region as inherently Argentine and, eventually, appreciate it differently. They would recognize in the built environment a place imbued with specific national values that inspired onlookers to embrace, love, or defend the fatherland. Place-making rested on a constellation of meanings bestowed onto space that was often fragmented, as I have examined in previous chapters. In the 1930s, the DPN sought to change this by introducing a cohesive array of cultural values that resulted in what Thomas Lekan calls the "nationalization of landscape."[3] For instance, the DPN defended a historical narrative that praised the civilizing force of the nation-state, represented by a monument to Julio A. Roca, and the conservation of certain landscapes for their economic exploitation, particularly tourism and logging. Taken together, these meanings of space generated differentiated value in distinct landscapes, as if some were worth protecting and visiting more than others.[4] The material intervention in the environs of Lake Nahuel Huapi sought to create a place where visitors and locals could contemplate the nation. The architectural aesthetic chosen by the DPN in the northern Patagonian Andes marked the persistence of a national state by asserting its presence in the region while simultaneously punctuating the landscape with cultural markers of the nation. Through buildings, roads, and materials, architects in the DPN's technical office reacted to a regional past that barely included Buenos Aires and sought to forge a vision for the future looking toward the Atlantic Ocean. This nationalist aesthetic, soon known as "Bariloche style," only made sense in the context of the northern Patagonian Andes. It was not an aesthetic that authorities reproduced elsewhere. It represented a coordinated attempt to consolidate "the national character" (*lo nacional*) in local styles. In fact, the DPN also promoted, directly or indirectly, tourism to other parts of Argentina, including Iguazú Falls, Córdoba, and Mar del Plata.[5] To flip Alon Confino's take, locality was a metaphor for the nation.[6]

This chapter traces the aesthetic arc in Bariloche and its environs to illustrate how different actors used the built environment to instill meanings (and recast them) in the geographical space. It begins by analyzing the built environment of the early twentieth century, when the influence of Chileans in Lake Nahuel Huapi materialized in homes and warehouses. In these early decades, the Chile-Argentina Trading and Cattle-Breeding Company consolidated its trans-Andean operations. It owned land, warehouses, stores, and docks on both sides of the Andes and monopolized the Pérez Rosales Pass. Its reach could be seen, quite literally, in the unifying architectural style of its distant properties. Even after the company's downfall in the 1910s, travelers, locals, and authorities synthetized this sense of trans-Andeanness of sorts by labeling the northern Patagonian Andes as Chilean or Argentine (or sometimes Chilean Argentine) Switzerland. Such abstraction condensed a desirable characteristic of the Alpine country into the environs of Lakes Llanquihue and Nahuel Huapi, supported, undoubtedly, by the similar scenery. In Argentina, the label particularly conferred value to specific landscapes, the snowcapped mountains, the evergreen forests in the wide valleys, and the deep-blue and emerald-green lakes.[7]

In the 1930s, however, the Argentine DPN re-signified this discriminated appreciation for the Andean regions from a site of production to a site of conservation. It harnessed emergent portrayals of Lake Nahuel Huapi as a potential tourist destination among the region's more prominent voices and packaged these depictions as the "awakening of Bariloche."[8] In doing so, the DPN sought to distance the history of Bariloche and its environs from its Chilean heritage by introducing myriad public works—from hotels, resorts, and churches to roads, hospitals, and schools—through a unifying aesthetic. Additionally, it recycled earlier versions of a desirable inhabitant by promoting travel among the wealthy urban classes, who could afford train tickets, car rides, hotel fees, and excursions.[9] At the center of this policy was the creation of the first two Argentine national parks, Nahuel Huapi and Iguazú, in 1934. This move gave the bureau's first director, Exequiel Bustillo (1934–44), unprecedented autonomy to alter the built environment. Bustillo found himself uniquely positioned to advance nationalist interests, especially in northern Patagonia, a space he viewed to be threatened by Chilean presence. In this context, conserving nature constituted a more patriotic endeavor than exploiting it for agriculture. The picturesque and Alpine stylistic choices, as represented in the Bariloche Civic Center, repurposed an earlier understanding of the environment from an agrarian village to a place to admire and enjoy. Its productivity still served the nationalist aspirations of the national government, although it looked different.

This and the following chapter make similar arguments from different angles. They examine how governments created national landscapes that could elicit love of country among locals and visitors, serving old objectives of territorial control through new methods. In doing so, states forged a collection of sites (natural or built) to re-signify agrarian spaces as pristine vistas, even among locals. This way of seeing specific places as extraordinary and, therefore, attractive is what John Urry calls "the tourist gaze."[10] Argentine and Chilean authorities developed tourist destinations that curated sensory and emotional experiences of the nation. In this chapter, I focus on how the Argentine DPN created a national landscape in the built environment of Bariloche and its environs, an aesthetic that condensed the nation into a frontier space as the interventionist presence of the state. This aesthetic program reacted directly to Chilean presence in Nahuel Huapi, which national authorities exaggerated and viewed as a threat. In Chile, the creation of a national landscape did not react to a perceived danger. Instead, authorities sought to consolidate a national character rooted in the modern practice of travel. The next chapter shows how the outdoors offered a way of seeing and experiencing the nation that was rooted in a gendered view of nature. Thus, both chapters examine the creation of a national landscape through two distinct environmental aspects, built interventions and outdoor enjoyment. State agencies in both countries produced and reproduced sites of attraction in as many publications as they could, including guidebooks, magazines, reports, and correspondence. These written and visual discourses molded travelers' expectations in terms of views, activities, and facilities, highlighting or erasing certain aspects of the destinations.[11] In doing so, they bolstered locals' and visitors' "sense of identity; they encouraged awareness of and loyalty to place."[12] The nation extended into the local landscape.

A COMPANY IN STYLE

The political program of the National Parks Bureau, condensed in but not limited to the Civic Center, reacted to a perceived danger from foreign influence. Especially in the northern Patagonian Andes, authorities viewed the Chilean families that lived there, the work they had, and the businesses they ran as a threat to Argentine sovereignty. The most emblematic of such businesses was the Chile-Argentina Trading and Cattle-Breeding Company, with properties on both sides of the Andes and a monopoly in the Pérez Rosales Pass. In 1905, the Chile-Argentina Company went public, constituting a commercial society valued at £275,000.[13] By then, the holding owned six properties in the environs of Lake Nahuel Huapi and several warehouses in strategic ports in Puerto Moreno, Puerto Bueno, and Puerto Blest on Lake Nahuel Huapi, where it stored goods and docked its fleet. Additionally, it managed the frequent international mail. In 1907, more telegrams went to and from Puerto Montt along this route than any other, without taking into account official correspondence.[14] The freight was so voluminous that managers designed a cable car that would go from Peulla, Chile, to Puerto Frías, Argentina. The Chile-Argentina Company bought all the needed materials (cables, iron, and machinery) in Germany and, as it had done with boats, assembled the cable car's towers on-site. This required the felling of large extensions of millenary forests, which has left a visible mark on the vegetation even today. The company purchased two cargo vessels, the *Cordillera* and the *Condorsol*, for transatlantic journeys and regional passenger travel.[15] In the Pérez Rosales Pass, the Chile-Argentina Company built an ox-pulled wagon that went along wooden rails and made transportation easier. Rumor has it that Theodore Roosevelt rode it on his crossing from Chile to Argentina in 1913.[16] The former US president observed the impact of the Chile-Argentina Company in Bariloche, noting that most conversations gravitated toward the company's infrastructure—the docks, the sawmill, the flour mill, the buildings—and its capital, most of German origin.[17] Over the years, the Chile-Argentina Company acquired estates from other landholding firms and expanded its ranching operations from southern

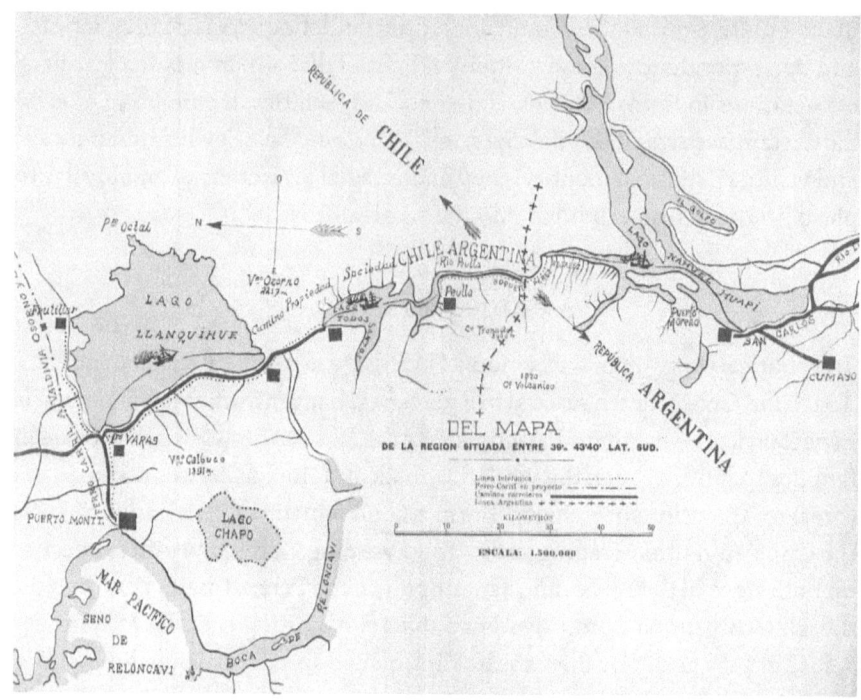

Figure 5.2. Map of the Pérez Rosales Pass, which united the Chile-Argentina Company's properties in the two countries, 1904. Some of the company's branches are marked by a black square. In *Chile y Arjentina*.

Neuquén to the Maullín River valley.[18] The business relied on the real estate of eighteen warehouses and stores, four steamboat routes, and a cable car (we know now it was never finished) and roads, trails, and telegraph stations. In the absence of the state, private sovereignty thrived (fig. 5.2).[19]

Architectural aesthetics threaded together distant properties of the Chile-Argentina Company, from the Patagonian steppe in Argentina to the Chilean port of Puerto Montt. German architecture, a long-standing tradition among German immigrants in southern Chile, heavily informed the visual style of the Chile-Argentina Company.[20] Transverse gables, frame or massive timber construction, and wooden shingles on exterior walls typified the look of most of the company's properties. The first building for La Alemana—Carlos Wiederhold's first store in Nahuel Huapi, which seeded the town of Bariloche—reflected an early stage of *blockhaus*, or massive timber style: a barnlike building with no gable, a wooden frame, and exterior larch tiling (fig. 5.3).[21] The main building of the Chile-Argentina Company improved upon the previous typology (fig. 5.4). Its larger dimensions allowed for a more complex distribution of interior space. Along the Pérez Rosales Pass, where the Chile-Argentina Company had exclusive rights, warehouses and stations displayed the same rectangular base and larch-shingled facade, except in Casa Pangue. Here, builders used vertical wooden beams. The environmental conditions along the pass required specific adaptations of the *blockhaus* template. A store-hotel in Ensenada, a hotel in Peulla, and a warehouse in Casa Pangue had a covered porch along one side, probably added to protect travelers from the heavy precipitation that typifies that section of the cordillera. The main house in Estancia San Ramón, the company's estate closest to Bariloche, had a similar sheltered porch, designed to echo Spanish colonial construction.[22] San Ramón sat to the east of Bariloche, where rain was less frequent than in the cordillera. More likely, managers had designed the porch to offer some shade in the Patagonian steppe.

The aesthetics of the Chile-Argentina Company's buildings homogenized a transnational space. But despite its unifying force, a shared aesthetic left room to hierarchize the multiple components of the company. Taking up two-thirds of the block, a two-story corner building in Puerto Montt headquartered the business as of 1901. Its lime-and-brick facade showcased a neoclassical style that trended in urban spaces in Latin America at the turn of the twentieth century. A balustrade and two columns framed a balcony on the upper floor. Unlike the first floor or other company buildings, a semicircular casement adorned the windows on the second floor, crowned by an ornamental keystone. A cyma recta and a parapet wall emphasized the building's flat roof, a rare sight in rainy Puerto Montt. The chamfered corner allowed for a small balcony with iron banisters and crowned by a pediment with the letters *H* and *A*, for the managers' names. Inside, the building awed clients with its spacious shopping area and elegant taste. The first floor offered diverse merchandise, from garments and textiles to groceries and decorative objects. The wholesale shop was located on the second floor, together with machinery and haberdashery equipment. Visitors agreed that the Chile-Argentina Company's elegant and modern architecture "attract[ed] and prepar[ed] the buyer's spirit to purchase there."[23] The company's store in Puerto Varas, on the western shore of Lake Llanquihue, mimicked the sophistication of the one in Puerto Montt in a simpler construction. The main floor also had rectangular windows, which the other stores had adorned with rectangular encasements. The gable had been expanded to a full-height second floor. A small pediment arched the windows, and color-blocked frames added detail. The store in Frutillar, seventeen miles to the north, exhibited the same characteristics at a smaller scale and with no gable. These were the company's

Figure 5.3. Carlos Wiederhold's store, La Alemana, ca. 1902. Courtesy of Roth Collection, Archivo Visual Patagónico.

only two stores that mirrored the exterior of the headquarters building in Puerto Montt, made of bricks and fine masonry, not larch shingles like all the other constructions.

After making a series of poor decisions, the Chile-Argentina Company began to decline in the early 1910s. Investors left, forcing the board to liquidate assets, mostly among its employees, vendors, and neighbors.[24] But after its final demise in 1918, the spatial influence of the company persisted in the environs of Lake Nahuel Huapi, at least partly because Chilean architecture of German heritage symbolized a desirable nation south of the Bío-Bío. Perhaps those representations resonated with settlers on the Argentine side; perhaps it was simply the technique and style that most people were familiar with. In any case, visitors and residents recognized a shared stylistic choice that transcended the border and reminded them of the Alps.

A SWISS LANDSCAPE IN THE ANDES

Changes in Llanquihue and, by extension, in the Argentine Andes resulted in a new way of seeing space. Far from being a Desert, the Andean valleys were fulfilling the promise of prosperity that explorers had foretold only years earlier. In the first decades of the twentieth century, travelers and locals alike began using "Chilean Switzerland" and "Argentine Switzerland" to describe the environs of Lakes Llanquihue and Nahuel Huapi. The Switzerland trope

Figure 5.4. Main building of the Chile-Argentina Trading and Cattle-Breeding Company in Bariloche, ca. 1910. Courtesy of Cherubini Collection, Archivo Visual Patagónico.

evoked not only an attractive landscape but also a productive space. Hence, it implied a history of colonization and a promise of a bright future. In Chile and Argentina, this reference has powerful reverberations even beyond the chronological boundaries of this book and into the present day.[25]

Since at least the 1880s, travelers have used the terms "Argentine Switzerland" and "Chilean Switzerland" to refer to the environs of Lakes Nahuel Huapi and Llanquihue. This image synthesized both an incomplete reality and a permanent aspiration. On the one hand, national authorities, travelers, and local elites applauded immigrant settlers who brought progress to an "empty" region. With the support of a feeble national state, newcomers were able to build ports, open roads, grow crops and cattle, and open businesses. On the other hand, much needed to be done: schools functioned in rented properties, police officers were undersupplied, and communication was slow. Early references to the Argentine Switzerland evoked a site valuable for its natural landscape—but not for the Indigenous peoples living there. Such depictions aimed to undermine the worth of Mapuche societies and assert Argentine sovereignty in the Andean valleys.[26] Francisco Moreno probably cemented the use of "Argentine Switzerland" to describe the northern Patagonian Andes, especially the environs of Lake Nahuel Huapi.[27] The national newspaper *La Nación* reported on one of Moreno's talks in 1905, inadvertently introducing the label into the media.[28] In Chile, writers began using "Switzerland" to describe the southern provinces in the late 1910s.[29] The inauguration of the railroad in Puerto Montt (1912) facilitated travel to Llanquihue, which established southern Chile as an attractive landscape.[30]

Local businesspeople appropriated the "Switzerland" trope to portray the northern Patagonian Andes as an attractive place for visiting and for investment. Unlike national authorities, some local elites adopted the image of Switzerland as a descriptor for a transnational space. As a result, they talked about the "Chilean Argentine Switzerland." For instance, when Federico Hube and Adolfo Achelis scouted for investors for the Chile-Argentina Trading Company, they described the location of the company as attractive because of the beauty of the terrain. They told their prospective shareholders, "Travelers and tourists that wish to visit the impressive and colorful landscapes of a part of the Andean region of Argentina and Chile, rightfully called the Chilean and Argentine Switzerland, will find in this brochure the views and maps of that lovely journey full of volcanoes, glaciers, rainforests, lakes, as well as labor and trade."[31] Infrastructure would transform a beautiful space into a fecund landscape: "The Chile-Argentina Trading and Cattle-Breeding Company will build steamboats in Lakes Llanquihue and Nahuelhuapi; it will erect a cable car to cross the highest peaks, and [it will introduce] other improvements in the short sections of the land pass."[32] "Switzerland" represented both beauty and production, an attractive image for investors.

Despite the geographical particularity of the "Argentine Switzerland," its meaning sometimes stretched across the Andes. For instance, in 1913, *La Nación* commented on President Roosevelt's trip from Chile to Argentina across the Pérez Rosales Pass. This pass connected the two trading hubs of Bariloche and Puerto Montt along a route that passed through four lakes, two in Chile and two in Argentina. In this specific instance, when the former president of the United States visited Bariloche, the newspaper initially described "Argentine Switzerland" as the area "between the international landmark, Lake Frías, and Lakes Nahuel Huapi and Gutiérrez."[33] The article described the multiday journey from Puerto Varas to Bariloche along the Pérez Rosales Pass, which required several stops. In each instance, travelers could admire "vast lakes ... surrounded by temperate forests."[34]

The "Chilean Argentine Switzerland" descriptor for the northern Patagonian Andes persisted in a few, albeit widely spread, publications. A guidebook published in 1920 highlighted the trans-Andean experience as Chilean Switzerland. The authors of *Turismo en las provincias australes de Chile* opened with advice on how to travel from Buenos Aires to Puerto Montt. In another chapter, "The Chilean Argentine Switzerland," they focused on the mesmerizing journey between Puerto Varas and Bariloche. The narrative revolved around natural landscapes, with repeated mentions of lakes, mountains, and rivers, and very little human presence.[35] The Andean landscape evoked a pristine attraction for travelers. Another guidebook, with at least two editions, was Germán Wiederhold's *Turismo en la provincia de Llanquihue a través de la Suiza chilena y argentina*. Germán was Carlos Wiederhold's youngest brother. He had participated in the family business but later focused on his passion, photography. The guidebook zoomed in on the history and geography of southern Chile, traveling options to the environs of Lake Llanquihue and parts of Chiloé Island, and information about crossing into Argentina. For Germán Wiederhold, Lake Nahuel Huapi and Bariloche represented an undeniable extension of any trip to Lake Llanquihue, in part since Chilean initiative had grown a storefront into a trading hub in Bariloche. Additionally, Bariloche had survived the breakdown of the Chile-Argentina Company, because of several entrepreneurial men "who devote[d] their energy and capital to increase the industrial and commercial drive of the region."[36] Soon, advertisements and newspapers began to use the shortened "Chilean Switzerland" to refer to the environs of Lake Llanquihue.[37] For Wiederhold, Chilean Switzerland represented a transnational space built on the work of immigrants (or people of immigrant descent) and the endorsement of the national state.

In Argentina, journalist Ada María Elflein reproduced Germán Wiederhold's original photographs in her own book *Paisajes cordilleranos*. In it, she described her impressions of her trip to the "Chilean and Argentine Switzerland." Accompanied by two teachers, Elflein traveled from Buenos Aires to Zapala (Neuquén) by train, then to Valdivia across the Andes, where she took a second train to Osorno, where she noted how the burning of wood and coal thickened the air, and later to Puerto Varas and Puerto Montt.[38] She left the environs of Lake Llanquihue to cross back into Argentina through the Pérez Rosales Pass, and after spending some time in Bariloche, Elflein took a car to Neuquén City and from there to Buenos Aires.[39] Elflein located "Argentine Switzerland" in the "lakes region" that orbited Lake Nahuel Huapi. Although she owed this description to the geographic resemblance to the European country, the writer also incorporated economic, historical, and cultural tropes. Like Germán Wiederhold, Elflein minimized Indigenous presence to a distant episode of "bloody encounters" and maximized efforts to colonize the region, from Spaniards to contemporary Argentines, Chileans, and Europeans living in Bariloche. Argentine Switzerland was a promise of prosperity for newcomers. It was a "traveler laughing at the desert" as "a steam force" carried her across Lake Nahuel Huapi, something that would have contented Federico Hube. It was the possibility of a national epic, like the Alps had been for European nations, engendered in the "legitimated patriotism" of Bariloche's multinational residents and supported by the advancement of the national state (preferably in the form of schools and banks).[40]

People from Bariloche, many of whom were immigrants, consistently recognized the Chile-Argentina Company's role in founding their town. On the thirtieth anniversary of Carlos Wiederhold's inauguration of his store in Bariloche, inhabitants from around Nahuel Huapi honored him with the status of "first resident," acknowledging in his persona and his country the origin story of their town.[41] This honor included a before-and-after diptych that illustrated how Bariloche had changed in thirty years. Where before there had been empty hills and a lake, there now stood an array of houses and a dock. The town owed its character to Wiederhold and other Chileans who had participated in its making. Indeed, the landscape of the northern Patagonian Andes changed abruptly between the 1890s and 1920s. An expanding agriculture in Llanquihue impacted the Argentine valleys. Companies like the Chile-Argentina acquired land and employed herdsmen, managers, and laborers. General stores sprang up from southern Neuquén to western Chubut, serving as hubs where people gathered to trade, socialize, and drink. These hubs attracted more people from the steppe and from abroad, including Chileans. The Chile-Argentina Company imprinted on Bariloche an architectural style tied to Chile. But such trans-Andean productive spatial discourse in the Argentine valleys irked some and annoyed others at the local and national level.

A PARK FOR THE NATION

Chilean and Argentine efforts to create national landscapes emerged from renewed anxieties about the nation. Nationalist sentiments began to brew in the 1920s and gained force throughout the 1930s and 1940s across the continent. As a state ideology, nationalism emerged in Latin America mostly as a result of international wars or domestic conflict, perhaps best theorized by José Itzigsohn and Matthias vom Hau. The drive to legitimate state authority moved public officials to instill national cultural points of reference, from popular art in Mexico to samba in Brazil.[42] Thus, for example, new public buildings in Buenos Aires used monumentality to project the strength of an administration that remained in power despite electoral fraud.[43] Since the 1920s, a rising cohort of Argentine military intellectuals had developed an interest in national politics as a centerpiece of national defense. Like other militaristic nationalisms of the time, these men supported policies that would bring Argentina closer to economic autarchy.[44] Such factions within the military joined far-right civil associations in criticizing "the vices of democracy" to support Gen. José Félix Uriburu's coup against President Hipólito Yrigoyen on September 6, 1930.[45]

Far-right nationalism offered an ideological response to immigration. Nationalist groups believed newcomers undermined the core values of the country because they had no clear loyalty and imported radical ideas, such as communism, that could undermine the Argentine nation. While Uriburu distanced himself from nationalists, their anticommunist, antiliberal, xenophobic, authoritarian views permeated political alliances around the next president, Gen. Agustín P. Justo (1932–38), and his vice president, Julio A. Roca Jr. Justo dominated the political sphere for the following ten years through a coalition of conservatives, Radicals, and independent socialists.[46] During this period, the military developed a nationalist consciousness anchored on concerns about industrialization and geopolitics, which directly informed how ruling elites, military and civil, saw Patagonia. The sudden death of several political leaders (President Roberto Ortiz and former president Marcelo Alvear in 1942 and Agustín Justo and Julio Roca Jr. in 1943) weakened the fragile balance between the conservative alliance and the Radicals, precipitating a second coup on June 4, 1943. Among the military members of the new de facto government was Col. Juan Domingo Perón, who won the 1946 election after famously garnering popular support on October 17, 1945. Before that, the first director of Argentina's DPN, Exequiel Bustillo (1934–44), had viewed Perón's rise to power with disgust as a feeble continuation of the progressive administrations of the 1920s and as a clear threat to his work.

The creation of national parks in northern Patagonia and northeastern Argentina responded to rising concerns about national sovereignty in frontier regions.[47] The first two national parks in South America, Nahuel Huapi in northern Patagonia and Iguazú in northeastern Argentina, were located in border regions of immense natural beauty and significant geopolitical value. Across the Americas, national parks represented the "ongoing presence" of central governments in natural landscapes.[48] Indeed, Nahuel Huapi and Iguazú were two of fifty-five national parks created across Latin America between 1930 and 1945. Additionally, the creation of protected areas (which I use here as synonym for national parks) responded to local circumstances, as Emily Wakild and Frederico Freitas have examined in Mexico and Brazil, respectively.[49]

In Argentina, Bustillo was perhaps more influenced by the United States, where the establishment of Yellowstone (1872), Yosemite (1890), and Grand Canyon (1919) National Parks "appealed to citizens' desire to preserve what was grand or unique about [North] America, or to preserve historically meaningful landscapes."[50] In part, this vision incorporated Francisco Moreno's ultimate goal for the region of Nahuel Huapi. On November 6, 1903, Moreno donated to the state 92.6 square kilometers (35.75 square miles) of land he had received as a reward for his role in the Chilean-Argentine border negotiations. This lot sat on the boundary and included the Argentine side of the Pérez Rosales Pass, which, in the early 1900s, was managed by the Chile-Argentina Company. In a letter he included with his donation, Moreno argued that in the "quiet magnificence" of the Patagonian Andes, "inhabitants of both sides of the Andes" would find ways to "solve problems" that diplomacy could not. "Rest and relaxation" created community, and community sparked the will to work together.[51] If for Moreno national parks would bring people together, for Bustillo they served not only to protect a part of the territory but also to "uphold . . . our territorial sovereignty."[52] *National* parks (as opposed to federal, wild, or natural parks) clearly encapsulated the underlying objective of defining a national landscape.[53]

A longer history of conservation in the Americas seeded the civilizing mission the DPN designed for itself. The United States had led the way in separating out federal lands for the enjoyment of nature, heightening the valuation of "unspoiled nature as an increasingly scarce commodity," especially in industrialized societies.[54] National parks like Yosemite and Yellowstone offered spaces for urban elites to export modern democratic values (as limited as democracy was in the late nineteenth-century United States) to the wilderness, which excluded Indigenous presence. Bustillo echoed such aspirations in the DPN's mission to "conserve the natural state of . . . nature," as it constituted a source of "science, history, culture, art, [and] curiosity." In his last report as director, he drew a direct line between this and the American conservation movement, describing former president Theodore Roosevelt's meeting with Francisco Moreno in Bariloche to allegedly discuss the vital importance of national parks for any "great country."[55] Authorities in the United States as much as Argentina found in the taming of nature an avenue for creating "a vision of nature that satisfied Euro-American . . . parameters."[56] The natural pristineness of national parks offered a backdrop to disentangle the vicissitudes of everyday life from an atemporal vision of the nation. In this context, Nahuel Huapi National Park created the myth of a separate time and a separate space, as if it were divorced from the ecological context and historical flows that surrounded it. The DPN's task, then, was to create a stand-alone understanding of space that served as a lens for interpreting the history of the nation through tourist experiences in the park.[57]

In 1934, two events changed Bariloche's orientation toward the Atlantic Ocean: the arrival of the first locomotive in May and the creation of the DPN and the Nahuel Huapi National Park in late September. The new railroad connected the town directly with Viedma and Buenos Aires. This rail line had been under construction since 1908, coming to a halt in 1928 in Pilcaniyeu, just fifty miles west of Bariloche. Work resumed in 1931 and finished three years later. Then, Congress created the DPN, an autonomous and far-reaching state agency under the umbrella of the Ministry of Agriculture.[58] Transforming the built environment to create an attractive, national landscape was at the center of the DPN's purpose. The bureau would "execute works of infrastructure" and "mobilize tourists" to visit "distant sites of the [national] parks." Beyond its obvious scope, the DPN also built churches, schools, and hospitals, erected monuments, paved streets, appointed teachers, and even attempted to remove residents from their land.[59] In short, it provided a "civilizing, and therefore patriotic," push, which authorities deemed necessary for developing a border region like northern Patagonia.[60]

Exequiel Bustillo was part of an administration that saw itself as restoring the conservative values that had dominated politics prior to 1916. He shared his peers' opinions that the Argentine journey to progress required the firm hand of government. This authoritative stance was hardly innovative as right-wing regimes emerged across Europe and the Americas in the 1930s, and it was a particularly unoriginal posture toward Patagonia, where national governments had consistently prevented the expansion of civil liberties. But the 1930s also saw the armed forces position themselves as protectors of the constitution and guardians of democracy. For members of the Conservative Party, including Bustillo, the economic development of Patagonia would not only address the immediate financial crisis but, more important, tackle concerns of national security. Nationalist ideas about Patagonia heavily rested, as Bohoslavsky argues, on numerous conspiracy theories about Chilean, Jewish, British, and communist territorial claims in the region.[61] But Bustillo did not fear, at least on paper, Bolivian or Nazi "infiltration" in Patagonia as some of his colleagues did. Yet he particularly worried that Chile, "squeezed between the sea and the mountains," posed a real threat to Argentina's "vital spaces."[62] Once, he tapped into those fears to request a supply of freshwater for the town of San Martín de los Andes, claiming he could probably get it done if he asked the Chilean government, which the Argentine minister of water works categorically protested.[63] Bustillo acted on his suspicions that Chileans were infiltrating the Andean valleys by soliciting recommendations from former military officers to become park rangers.[64] After all, protecting the national park, which included surveilling worker crews (made up of detainees) and taming horses, was akin to protecting the country.[65] He also barred foreigners from filling the low ranks of the DPN, especially Chileans. Bustillo believed the demographic pressure from across the Andes "was already beginning to stifle or dilute the weak national sentiment."[66] Yet military reports seldom mentioned Nahuel Huapi as a site for concern. Rather, they feared Chileans undermining far more crucial places, such as oil fields in Plaza Huincul (Neuquén) and Comodoro Rivadavia (Chubut).[67]

WHO MADE THE NATIONAL PARK?

Bustillo was in Paris when the military coup of September 6, 1930, ousted President Yrigoyen and reinstalled conservative rule in Argentina. Bustillo promptly returned to Buenos Aires, pulled by his "patriotic yearnings" in service of the "nation," which meant in support of the military government.[68] In Paris, he had befriended Luis Ortiz Basualdo, who co-owned a property on the Huemul Peninsula on Lake Nahuel Huapi. Sealed over coffee at the Parisian Ritz hotel, this friendship resulted in a 1931 trip to Bariloche that changed its history forever. From then on, Bustillo would describe the environs of Lake Nahuel Huapi in a before-and-after manner, referring to how his work, and later the DPN's, deeply transformed it. The journey to Bariloche, "back then, was very long, tiresome and full of discomfort."[69] The group traveled from Buenos Aires to Carmen de Patagones, crossed the Negro River by raft, and hailed another train from Viedma to the then railhead town of Pilcaniyeu across "an inhospitable, barren desert, covered only with bushy vegetation."[70] Only fifty miles of rutted paths separated Pilcaniyeu from Bariloche, mostly across a monotonous landscape that disappointed Bustillo. Where was "the beauty he had heard so much about?"[71] But then, the topography changes (it "raises its hierarchy," says Bustillo) and, in that critical moment, "an admiration is born and will never leave us."[72]

After his first trip in 1931, Bustillo purchased two lots of 625 hectares (1,500 acres) on the northern shore of Lake Nahuel Huapi to establish a summer estate. He called it Cumelén, a Mapudungun word he found, somewhat ironically, doing research in the library of the Jockey Club, an exclusive club for the *porteño* men of the social elite. Bustillo liked both its meaning, "carefree," and the sound of it. More importantly, Cumelén "properly expressed the colonizing purpose" of settling in Patagonia, even if it were only for summers. In many ways, Cumelén worked as small-scale experiment, and many of Bustillo's policies in Nahuel Huapi can be traced back to his attitudes about

Figure 5.5. "Lo que va de ayer a hoy" (From yesterday to today), a visual account of the Argentine DPN's infrastructural impact on travel, 1938. Dirección de Parques Nacionales, *Parque Nacional Nahuel Huapi: Historia*.

his own estate. The sentiment that mobilized him was that Nahuel Huapi was a region marked by backwardness and that the national park's core mission was to develop it.[73]

Bustillo used media articles, official publications, and even private correspondence to amplify the messianic role of the agency he presided over to improve the material substance of Nahuel Huapi. A DPN guidebook published in 1938, for instance, includes a visual portrayal of the transitions between an Indigenous time, a not-so-distant past, and the present (fig. 5.5). It depicts three elements where the DPN had developed infrastructure and that would be of interest to visitors: crossing the Limay River, ground transportation, and lodging. The "before" frames portray Indigenous populations, implying a primordial time of primitive practices. Two of these frames are illustrations and the third is an image provided by the La Plata Museum, indicating that the only place where anyone could find vestiges of Indigenous culture was in exhibit boxes. The second set of images, between "before" and "now," recognizes settlement prior to the 1930s but conveys it as still insufficient for a desirable future. The ox-pulled carts, rafts across the rivers, and precarious hotels were redeemed, in the eyes of the DPN, by the sturdiness of bridges, the accessibility of roads and motor vehicles, and the impressive Llao Llao Grand Hotel, which I discuss in the following pages.

In this spatial transformation, Bustillo positioned himself simultaneously as one of many "regular settlers" and as indisputable architect of the nationalizing strategy of the DPN. He attached the work of the DPN to the conservative agenda but also to his own persona, tapping into his network at dinner parties in Buenos Aires. Bustillo saw himself as "a kind of unofficial representative" of Bariloche to whom local residents came when they needed something done. He also took on the task of organizing the agency and instituting the national parks as his personal legacy. He relied almost exclusively on his personal contacts within and outside the administration to advance his agenda. Convinced of his central role, he got upset when things did not go his way, accusing colleagues and workers of lacking "good faith and interest in works of progress."[74] By extending the reach of this network through invitations, appointments, and donations, Bustillo symbolically linked Nahuel Huapi and his own persona to a "prestigious genealogy" of historical actors, like Moreno and Roca, and to the cultural values of a social class he tried to replicate.[75] In doing so, he consolidated a narrative centered on the Argentinizing mission of the conservative elite as a second act of the oligarchic pre-1916 government.[76] As a former resident of Bariloche put it then, "Bustillo and the park are the same thing."[77]

Bustillo was not alone in envisioning a transformation for Patagonia. In 1935, only a year after the creation of the DPN, Col. José María Sarobe analyzed the economic, social, political, and geographical challenges that Patagonia posed for the national government in an award-winning monograph, *La Patagonia y sus problemas*.[78] With a prologue by an elderly Ezequiel Ramos Mexía, the book outlined major points of continuity between the Liberal Reformists of the early twentieth century and the new generation of Argentine politicians. Sarobe agreed with Bustillo about the central roles played by Roca and Moreno in incorporating Patagonia into the Argentine nation. His view synthesized a collective assessment of violence during and after the military raid of 1879–81 as the first of many steps the Argentine government took to bring progress to Patagonia. Unlike Bustillo, however, Sarobe viewed the period of 1895–1912 as a golden age in Patagonia, when the lack of customs encouraged free trade and when field research, such as that of the Hydrological Studies Commission, informed legislation. However, a disproportionate dependency on wool prices, which was far from desirable, had displaced small landowners, "true settlers," in favor of latifundia.[79] Like others before him, Sarobe believed the trans-Andean valleys facilitated the incursion of a "nomadic and undesirable population."[80] At the very least, land accumulation was a symptom of good-intentioned legislation with negligible execution. Akin to the nationalist views of the 1930s, the colonel proposed effective legislation and a strong state presence in the form of police forces, judges, and teachers.[81] Additionally, topographic and climatological differences in Patagonia pointed to multiple natural resources that could be exploited, such as gold in Neuquén, oil in Chubut, and tourism in Nahuel Huapi. But overdependency on one product or mild regulation of landed property were hardly the only Patagonian problems.

Sarobe joined his contemporaries in picking a bone with the seemingly overwhelming foreign population south of the Colorado River. Although there had not been a census since 1914 (and there would not be another one until 1947), Sarobe estimated the Patagonian population at 200,000 people, a low number vis-à-vis the 1.6 million Chileans across the Andes.[82] A classified document prepared for army senior officers synthesized Sarobe's conclusions as a warning against a possible war with Chile and Brazil.[83] Given Patagonia's vast resources and dispersed Argentine population, "the enemy will try to take the best advantage of this preponderant situation in the border region and in the far South, undertaking an offensive action that will allow him to quickly reach the *possession of Patagonia*."[84] In addition to military preparedness, the document proposed two nonmilitaristic actions: a plan to exploit Patagonia's natural resources, which would include improving communication, and "the expansion of Argentine influence" through national population and selective immigration.[85] The wording certainly borrowed from Ramos Mexía's ideas about progress in Patagonia. Yet the new generation frowned upon immigrants who "did not assimilate our customs," such as the Welsh in Chubut, who had kept their language for at least six decades.[86]

THE NATION IN THE BUILT ENVIRONMENT

The work carried out by the DPN in Bariloche included opening and paving roads; building hotels, campsites, public offices, and even housing; erecting monuments; and opening tourist sites. This chapter focuses on the built environment; the next chapter will discuss outdoor activities. Both the spatial transformation and tourist experiences supported a vision of the northern Patagonian Andes as a recreational site where modern citizens could find their love of country. At the center of and as an entry point to the civilizing mission of the Nahuel Huapi National Park sat the town of Bariloche. Its strategic location in the border region situated Bariloche at the center of the DPN's development policy, as it "could not be left to a slow and spontaneous growth."[87] Under Bustillo's leadership, the DPN sought to transform what "looked like a Far West town" into a modern city that embodied a nationalist program, albeit through an Alpine, French aesthetic.[88]

The nationalist aesthetic that the DPN developed (as did other state agencies) had its roots in the early twentieth century, when questions about the nation gained renewed impulse. The centennial of the Revolution of May 25, 1810, considered the beginning of the Argentine independence movement, sparked reflections on the first 100 years of Argentine history, present fears of national disintegration by way of immigration, and avenues for the future. Artists, writers, and thinkers produced a loose cultural movement that sought to address the tension between past and future in an array of materials that celebrated the nation, commemorated its (mythical) foundation, and outlined a vision for it.[89] The national government used this artistic agenda to cement its own hegemonic role through a series of cultural symbols. For example, monuments and buildings sprang up across the city of Buenos Aires in the years leading up to the centennial as an "amusement park" of a sanitized national identity. They moved social conflict out of center stage and instilled instead the everlasting aspiration of progress, embodied, for instance, in cattle exhibits (reflecting productivity) and monuments (creating sites of memorialization). All in all, the first decade of the twentieth century provided fertile ground to create a cultural inventory of spatial references, consolidating a sense of shared national, albeit contradictory, values.[90] This cultural movement, far from homogenous, did animate a new reaction to old questions about the nation that persisted through the 1930s, when they reached the Patagonian Andes.[91]

The Bariloche Civic Center exemplified an aesthetic geared to evoke Argentina in a border region. It was inaugurated in 1940 and housed a police station, the customs office, the municipality (created in 1930), a museum, a library, and a post office. Clustered together, these sites offered residents spaces to become citizens, even though those living in the national territories could not elect officials beyond the local level. The Catholic chapel, a staple in Spanish colonial main squares and, traditionally, a marker of local time, was not part of the Civic Center. One was built later about 600 meters (0.37 miles) away from the complex, signaling that Catholicism, though relevant, was not constitutive of the town's future growth.[92] Instead, a clock tower replaced the old sundial, marking the mechanization of time as one of the most visual facets of modernity in the city.[93] The center also reserved a spot for a coffee shop, but it seems that "nobody wanted it."[94] The Civic Center's location was no accident. The DPN had bought the lot from Primo Capraro's heirs, who had made Bariloche their home. Capraro had bought the land, including the dock, from the Chile-Argentina Company and installed a sawmill, flour mill, woodshop, and personnel quarters.[95] The center embodied the before-then-now progression that appeared in the guidebooks but with an adjusted timeline. The base layer had been Carlos Wiederhold's properties, a moment in time that perhaps the DPN was eager to erase. Then Capraro had improved the facilities, and now the DPN had erected a long-lasting complex out of stone. Two archways on the eastern side allowed traffic to flow from the square into the main artery, Mitre Street. The Civic Center condensed in its style and location what Adrián Gorelik and Graciela Silvestri have called a "mythical origin and a mythical destiny."[96] For example, the archways symbolized two arcs of triumph of sorts, a passage between the old village and the new city. A boulevard on the south side connected the complex to the local offices of the DPN, built in a matching architectural style. The boulevard emulated the wide avenues of Haussmann's Paris while symbolizing a balance of modernity and beauty.[97] In the 1930s, a 300-foot boulevard in a frontier town encapsulated the Argentine fin de siècle, materializing the restoration of conservative government that Bustillo supported and embodied. If, as Liliana Lolich has pointed out, the DPN's spatial interventions in Bariloche worked on the urban matrix as a tabula rasa, the boulevard between the Civic Center and the DPN office created a symbolic past to venerate while situating an aspirational future as a modern, recreational city.

The Bariloche style was conceived on the drawing board of the DPN's Technical Office, led by Bustillo's brother Alejandro.[98] Bustillo tasked the design of the Civic Center to a young architect and one of his brother's protégés, Ernesto de Estrada. Estrada's training in Buenos Aires and Paris built on a decades-old inspiration in the French Beaux-Arts movement among Argentine architects, both aesthetically and technically. The imprint of the Belle Époque on Argentine architectural and landscape design evoked the country's Conservative Period prior to 1916. The 1930s, however, brought a fresh, nationalist perspective, and architecture was no exception. During his studies at the Free School of Social Sciences in Paris (today, Sciences Po), Estrada witnessed the French modernist movement gain force. Somewhat institutionalized via yearly conferences known as the International Congresses of Modern Architecture (Congrès Internationaux d'Architecture Moderne; CIAM), modernism in architecture was all about efficiency in the city. CIAM became the most relevant international forum for city planning and architecture through the 1960s, with a strong imprint in Latin America, from housing projects in Mexico City to Buenos Aires's beltway and Brazil's new capital, Brasília.[99] Yet the argument for functionality over aesthetics that emerged from CIAM IV (1933) and was laid out in its most famous document, the Charter of Athens (1942), did not completely translate into a well-rounded modernist style for DPN's works in Nahuel Huapi.[100] Estrada,

Alejandro Bustillo, and the DPN Technical Office agreed that a functional city should satisfy the needs of a modern society, calling for methodical harmony between landscape and architecture. For them, Bariloche was to provide recreation for tourists from other parts of the country who were performing the modern act of traveling for pleasure. To that end, the architects splintered from modernists' lack of interest in ornamentation and wanted the Civic Center and other constructions to evoke "art and grace" in a relatively homogenous style.[101]

In the Argentine context, the built environment provided space for the conservative administration of the 1930s to materialize a national identity (*lo nacional*) as viewed through a modernist lens. In Buenos Aires, for example, the authorities exploited the tensions between origin and destiny around the 1936 celebration of the 400th anniversary of the first founding of the city of Buenos Aires. An ad hoc committee of historians decreed the date and place of the founding, even though there were no documents to prove either. In doing so, they portrayed the founders "as part of a well-ordered . . . and scientifically resolved conquest" that constituted the "essence" of the city in its origins.[102] A monument to this mythical moment, the Obelisk, was erected at the intersection of Avenida 9 de Julio and Avenida Corrientes, two main arteries that were also widened in the 1930s, fusing the city's origins with its aspirational modernity.[103] In Nahuel Huapi, too, the DPN reinforced the intersection between mythical origin and aspirational future as an image of the nation. For example, the 1938 guidebooks recognized Carlos Wiederhold as the "first settler" and his house as the "first house" in the region. Although the guides included Indigenous histories, they did not recognize them as part of the nationalization effort in Nahuel Huapi. Thus, the DPN did not see them as first settlers or their homes as first homes. Yet the guides made no mention of Wiederhold's Chilean origins or that he had established a lucrative trans-Andean business. Instead, the story quickly moved on to the surveys of the border subcommissions and the decree of incorporation of the city of San Carlos de Bariloche on May 3, 1902. Similar to what happened in Buenos Aires, doubling down on the specific date gave Bariloche a clear civilizing mission in the border region.[104]

The Bariloche Civic Center illustrates the contradiction between the ideas it embodied (that the Argentine nation is present in a frontier space) and the language to transmit them (French eclecticism). The Alpine-inspired style in Nahuel Huapi National Park contrasted with the Spanish Mission aesthetics chosen for the other park created in 1934, Iguazú National Park in northeastern Argentina. Authorities did not choose French or Spanish styles because of what they were but because of what they represented. In Nahuel Huapi and Iguazú National Parks, these styles embodied an Argentine presence vis-à-vis the borderline with Chile and Brazil, respectively. Thick walls, arcades, an internal courtyard, and exposed rafters at the roof overhangs endowed the DPN's building in Iguazú with Spanish heritage, in contrast to the Portuguese on the other side of the river.[105] Similarly, the French inspiration for Bariloche's Civic Center contrasted with the German Chilean style that had flooded Nahuel Huapi.

THE ARGENTINE TOURISTSCAPE IN BARILOCHE AND BEYOND

The Civic Center spearheaded a constellation of buildings developed by the DPN in Bariloche. In 1930, Bariloche had graduated its local council to a municipality. The creation of the Nahuel Huapi National Park immediately generated all sorts of jurisdictional disputes, which the DPN tried to quash with a vigorous and well-funded urban development plan. Standing at the doors of the DPN building and facing north, one could see the boulevard opening into the complex, which encircled a square. At the center of that square stood a statue of Julio A. Roca on his horse, also facing the lake (fig. 5.6). The coup of 1930 and the subsequent presidential election of an army general, Agustín P. Justo (1932–38), situated the military at the center of Argentine state-building. While statues and monuments had long symbolized a nation-centered narrative across the country, it was during the 1930s that this memory incorporated into its pantheon Julio A. Roca, former president and leader of the military campaign of 1879. The governing elites of the 1930s, of which Bustillo was part, found in Roca's dissolution of the internal frontier (through the military campaign in Patagonia) and demarcation of the external frontier (through his negotiations with Chile) an archetypal historical reference to state-building as a patriotic endeavor.[106] In Bariloche, the Civic Center would pay homage, as Bustillo put it, "to he who had freed [Patagonia] from the Indian who plagued it."[107] Bustillo commissioned a statue of Roca to sit in the middle of the square surrounded by the buildings of the Civic Center, quite literally at the heart of the city. Bustillo also pushed for the name of the square to honor the soldiers of the military campaign, a symbol that persists to this day as the Plaza Expedicionarios del Desierto.[108] Such spatial intervention inserted the DPN's work into a sequence of violent attempts to incorporate Patagonia into the nation. In doing so, it established a public narrative about waves of nationalizing efforts that situated state efforts not

Figure 5.6. Bariloche Civic Center seen from the National Parks Bureau office and facing north, ca. 1955. Photograph by Bruno Ricardo Sálamon. Courtesy of Capraro Collection, Archivo Visual Patagónico.

only at the center but as the only path for progress. Since its origins, the Civic Center constituted the epicenter of an *Argentine* touristscape.

Probably the most understated piece of infrastructure in Bariloche was the Avenida Costanera (coastal parkway). Coastal boulevards beautified Latin American cities in the first half of the twentieth century. They offered infrastructure for enjoying sights (*paisajes*), which increasingly became objects of consumption and profit. In Acapulco, Rio de Janeiro, and Lima, visitors could enjoy natural and urban views from the comfort of a modern pathway.[109] In Bariloche, the DPN designed the Avenida Costanera to go from the train station to the main dock and then to the Llao Llao Grand Hotel. The dock was where Carlos Wiederhold had established his fleet at the turn of the century and where Primo Capraro had expanded his business to build the local sawmill. The DPN acquired the dock from the Capraro family in 1938, right across from the lots where it would build the Civic Center. The coastal parkway threaded together public buildings, like the Transportation and Workshop Building of the DPN (1936) and the cathedral (1946), built in the same style as templates for future construction.[110]

Building the Costanera required the expropriation of lots along the shoreline of Lake Nahuel Huapi. This was possible because national parks legislation gave the DPN jurisdiction over 35 meters (0.02 miles) of shoreline, overriding the municipal authority. Of course, some local proprietors, like José de García and Pablo Mange, opposed relinquishing their land to the whims of the DPN, especially since they had homes or businesses there. This was not a simple feud between local interests and national priorities.

Julio Comezaña, a local businessman, was one member of Bariloche's elite who supported the expropriation of lots in favor of the Costanera. He was frustrated by those "insensitive to all noble sentiments and contrary to the spirit of progress in all its forms."[111] Although some families, like the Capraros and Lahusens, had leverage because of the improvements (*mejoras*) they had made to their lots, other people simply farmed their lands without much protection against the force of the DPN. If they refused to sell, Bustillo would not hesitate to order an expropriation.[112] The expropriation of lots to build the Costanera is one of the few debates about the DPN's work that made it into the historical record. Local businesspeople certainly embraced the transformation their town was undergoing and sought to play a role in it. The presence of the state was visible in the local branches of the national bank, post office, telegraph office, and radio station. A hospital cured the body, and a museum uplifted the soul. The presence of the army secured the border, and the Civic Center instilled style, which several businesses on the adjacent Mitre Street appropriated for their facades.[113] Yet, some long-standing residents who lived in and around the national park complained that the DPN interrupted their way of life, levying excessive taxes and charging fees for campsites locals had opened and had been using since before the park was established (without acknowledging that they had also appropriated land from previous inhabitants).[114] Hotel owners were particularly hostile toward those constructing the DPN's grand hotel, Llao Llao.[115] Local employees of the DPN also complained about the toxic, often unethical work environment and delayed funding, which impacted performance and, overall, limited the efficiency of public works.[116]

The nationalist views of space that radiated from the DPN were heavily grounded in a curated historical narrative. In northern Patagonia, this chronicle began with Julio A. Roca and the military campaign of 1879–81, followed by the border negotiations, the extension of the railroad, and the creation of the national park. The DPN carved this narrative into Bariloche with monuments, street names, and a museum. Housed in the Civic Center, the Francisco Moreno Museum of Patagonia joined the pantheon of institutions, from the police to the customs office, that heralded the nationalizing mission of the 1930s and early 1940s. While including the museum in the center's design was laudable, Bustillo worried about how to fill its collection. How could "a secular desert like Patagonia," which "lacked a rich history to obtain certain representative objects," attract the curiosity of travelers and scholars?[117] In other words, the museum was a small-scale representation of the challenges authorities perceived in Patagonia. Like the Patagonian emptiness that persisted in the minds of national authorities, the museum also provided a new space to frame the Argentine nation relative to its frontier history. The museum purchased its initial collection from private collector Enrique Artayeta, who was then appointed the museum's first director (1939–50). The collection amounted to a variety of Indigenous artifacts, including valuable textiles that Artayeta claimed had belonged to the last *lonkos*. A section of regional flora and fauna complemented the first exhibit, depicting northern Patagonia—and the region of Nahuel Huapi in particular—as an ahistorical site. The museumification of nature portrayed it as part of the passive space of the showcase, waiting to be visited, observed, and admired.

Bustillo and Artayeta reinforced the portrayal of northern Patagonia as a no-place by introducing a narrative that explained the history of Bariloche solely as a story of state advancement. Like the La Plata Museum decades earlier, the Museum of Patagonia established the military campaign of 1879–81 as the focal point of an ideological program, echoing earlier linear narratives about civilizing an empty space.[118] Around it, the DPN constructed a teleological narrative of nation-making that began with the failed Catholic missions to Nahuel Huapi, continued with the scientific explorations of the nineteenth century, grew with the Welsh settlements in Chubut, and finally sealed Patagonia's destiny with Roca's military columns crossing the Negro River. Artayeta used his network to supplement the collection of taxidermy animals and Indigenous artifacts with relics from the military campaign, such as weapons and Argentine symbols. Including these objects in the permanent exhibit reaffirmed the patriotic mission of the campaign and the pedagogical role of the museum.[119] Other forms of memorialization, such as street names (Villegas, Roca, San Martín) and monuments, continued to consolidate the legacy of the 1879 campaign beyond the museum.

Naming a main street and the museum after Francisco Moreno and placing Roca's statue right across from it brought the goals of the campaign full circle. Such a pairing, while not historically inaccurate, reinforced the role military and scientific violence played in the making of Patagonia, a partnership that authorities considered alive and kicking in the 1930s.[120] With the focus on such a violent event, the museum folded Indigenous history into natural history as a means to delegitimize Indigenous participation in the making of the nation. Within this Argentine-centered narrative, Artayeta sought to remove the term "Araucano" from the exhibit, a term used by non-Indigenous Chileans to refer to the Mapuche people. As he put it, "It is high time ... we differentiate [our Indians] from the Chilean race."[121] This differentiation spoke more to the urgency to delineate the historical boundaries of Indigenous peoples as

"Argentine," even though they crossed back and forth across the cordillera, than to the accuracy of the label.

The colonizing mission of the DPN would have been incomplete without the building of Catholic temples. Bustillo, the architect, designed at least two buildings, a main temple in the town of Bariloche and a chapel on the grounds of the Llao Llao hotel. For the DPN, it was crucial that the temples "kept the harmony" of other constructions, giving them a sense of belonging to the urban landscape and not of dissonance. In their view, chapels represented "a tribute to the first Jesuit missions that populated and civilized the region."[122] Bustillo also sought support to give the cathedral the status of national monument, as if this category would bring full circle the colonization efforts begun in the late sixteenth century. It was ironic, however, that he neglected to acknowledge that the missions came from across the Andes. For Bustillo, building temples was part of the moralizing agenda of the DPN. For him, the cross symbolized "the colonization of the desert" and the "reign" of "Christian civilization."[123] He hoped Pope Pius XI would support his enterprise with the creation of a diocese in the cordillera separate from Viedma (which managed all of Patagonia). Only in 1993 did the Pope create a separate diocese in Bariloche. Despite these views, the funding and location for the DPN's chapels show that "the building of temples was one of the most irrelevant to the purposes of our institution."[124]

The DPN situated both the chapel in Bariloche (Our Lady of Nahuel Huapi) and the one in Llao Llao (St. Edward's) at an outlying location from the main focus, the Civic Center and the hotel, respectively, signaling the secular flare of modern cities. Bustillo supported the idea of building temples but didn't want to fund their construction through the DPN. Instead, he solicited support from Josefina González-Devoto, whose donation made possible the construction of St. Edward's and its inauguration in January 1938.[125] The size of the temple in Bariloche was disproportionate to the size of the town, anticipating that the town would grow. Again, Bustillo used his personal connections to request that wives of his friends create a ladies committee to fundraise for the temple in Bariloche.[126] Years later he claimed the committee "was just a facade," because he was the one really pulling the weight to get anything done.[127] Unfortunately, the committee only raised enough to pay for the stained-glass windows. These featured scenes from the Bible and relayed the history of Catholicism in Patagonia, including the martyrdom of Nicolás Mascardi and Beatus Ceferino Namuncurá, who originated from Río Negro.[128] Alejandro Bustillo donated the neo-Gothic designs that were used to build the chapel. As in his other projects, the style evoked French medieval architecture. The chapel was named after an image of the Virgin Mary that had accompanied missionary Nicolás Mascardi in the late seventeenth century, inserting the work of the DPN into a history of colonization and consolidation of the nation. At the end of the day, the federal government had to step in and donate funds, allowing the church's inauguration to take place in 1946, even though, ironically, the building was never completely finished.

A CONSTELLATION OF TOURIST VILLAS

In addition to the functional city ideal borrowed from French modernism, the garden city movement informed city planning for the DPN. The garden city movement, which originated in Britain at the turn of the twentieth century, sought to address the urban consequences of industrial capitalism, including overcrowding, poor ventilation, inadequate drainage systems, and gas emissions. Its defenders proposed decompressing cities into carefully planned networks of garden towns. Each town would be economically independent from the next, each surrounded by a green belt of arable land, but connected to one another. The garden city movement was especially concerned with the urban setting's relationship to its natural context. Latin American urbanists adapted different facets of the movement from Britain, France, the United States, and Spain to create local versions of the garden city.[129] Beyond providing an immediate solution to urban problems, the garden city movement and its derivatives served as an ideological "tool for reconstruction ... resource extraction, population resettlement, and territorial dominion."[130] The DPN did not conceal its goals to restructure the social fabric of Nahuel Huapi and Iguazú National Parks, prioritize access for desirable settlers and travelers, and consolidate Argentina's sovereignty in a border region through material discourse.[131]

Particularly in northern Patagonia, the DPN sought to erase any vestige of Chilean and Indigenous presence by extending the "colonizing mission" of Bariloche to a series of satellite towns (*villas*).[132] The DPN took up the role of the Office of Lands and Colonies at least partly to approve land sales in the environs of Lake Nahuel Huapi. After 1935, applications for land titles from Chileans were rejected, and Chileans were barred from participating in auctions for lots.[133] Simultaneously, some residents used their connections with Bustillo to request the land titles of lots where they had lived for years. Such was the case of Rodolfo Koessler, a German physician living in San Martín de los Andes who "had invested [M$N]30,000" in his property, "only guided by the desire of progress for this town."[134] Similarly, a Swiss hotelier applied for financial aid from the DPN to establish a hotel in Lake Traful, and an Argentine polyglot requested a job with the road repair crew.[135]

Conservation in Nahuel Huapi National Park also involved landscaping the environment to meet the DPN's vision of the park. In 1937, the agency established a botanical garden and greenhouse on the south part of Victoria Island, the largest island in Lake Nahuel Huapi. Forest engineers imported plant species from all corners of the world to evaluate which ones would create better landscapes. These were the species that the DPN would plant in sites ravaged by fires and along the Avenida Costanera or use for timber. If foreign species were deemed productive, the tourist gaze on local vegetation considered Patagonian species as "pretty." Yet, in some instances, native flora was seen as uninteresting vis-à-vis foreign species.[136] On the northern section of the island, the DPN created a zootechnical station to observe local fauna and introduce new species. Technicians brought to this "zoo of sorts" the spotted deer (*Axis axis*) and wild boar (*Sus scrofa*), which reproduced quickly, severely affecting native huemul and pudú deer populations, as well as the autochthonous flora.[137]

Anxieties about undesirable populations and absence of "progress" were hardly new. However, the aggressive efforts to create a national landscape in Nahuel Huapi certainly reproduced practices of exclusion.[138] For instance, back in 1907, Antonio Buenuleo, an Indigenous farmer, applied for the land title of the lot where he and his family had lived for six years, breeding cattle and horses. However, he was pressured by neighbors such as Primo Capraro, who coveted the lot, to drop the application. Capraro offered Buenuleo a job as manager of the land in exchange for his surrender, which he took. Buenuleo then received a title to lot 127, a lot to the south of his previous land but with no access to Lake Gutiérrez.[139] As time went by, the area between the two lots became known as Pampa de Buenuleo, a beautiful valley that never made it to the pantheon of touristscapes in Nahuel Huapi National Park. The map in the DPN's guidebooks reduced this Indigenous presence in the valley to a small label that read "Buenuelo's Post," on the road from Bariloche to Lake Gutiérrez.[140] What this map does not tell is the grazing fees the DPN charged the Buenuleos, which sank them in debt, or the number of times state agencies (including the DPN) asked them to move their homes in the second half of the twentieth century.[141] The DPN categorized these erasures of Native population as the de-Chilenization of Nahuel Huapi, a policy rooted in Argentine prejudice against residents of the neighboring country, as I have examined in chapter 4. Argentine authorities ascribed Chilean nationality to

Indigenous peoples, including the Buenuleos, even though their ancestors had inhabited the Andean valleys since before either country existed.[142] By labeling unwanted populations as Chilean, Argentine authorities reproduced justifications to violently remove people from their lands.[143] Removal from their lands did not prevent people with Indigenous heritage from receiving permits for animal grazing and farming or from working for the DPN. For instance, Juana María Antimil, an illiterate woman, applied for and was granted a permit to have eight grazing animals, four mares and four cows, even though her family had lived near Nahuel Huapi since before 1902.[144]

In the eyes of the DPN, visitors represented a desirable population. Thus, it was their experiences more than that of the locals that were prioritized in the spatial planning of the park. Particularly, a constellation of tourist towns around Bariloche would facilitate visitors' enjoyment of the natural environs of Lake Nahuel Huapi. The DPN's Technical Division, headed by Alejandro Bustillo, designed these villages in the same Alpine style as the new constructions in Bariloche. Immersed in a national park, Bariloche became a city garden par excellence, both as an entry point to the surrounding landscape and as a template of style and design for the subsequent tourist villages. Orbiting Bariloche, Villa La Angostura, Villa Llao Llao, and Villa Catedral perhaps materialized best Bustillo's interpretation of the garden city movement.

Estrada conceived Villa La Angostura, located on the northwestern shore of Lake Nahuel Huapi, near Bustillo's estate, Cumelén. Villa La Angostura embodied Exequiel Bustillo's vision of a tourist village "at the heart of the cordillera." It symbolized "the placement of a milestone marking the influence of our nationality in those latitudes."[145] Bustillo claimed that, when he established Cumelén in 1931, "there was nothing in Villa La Angostura." However, even his own descriptions of the site contradicted this assessment. Consistent with before-and-after narratives that illustrate the work of the DPN, Bustillo described that the "absolutely nothing" he had found in the small plain that would become Villa La Angostura was indeed "a primitive cemetery or rather a burial ground fenced with logs."[146] In fact, the decree that officially founded Bariloche in 1902 included lots on the opposite side of the lake. The first recipients of one of these lots were two *lonkos*, Ignacio Andriau and José María Paisil.[147] Their presence hindered the DPN's expansionist plans. As a result, the Andriau and Paisil families saw their taxes increase and their documents disappear from official records. When the situation became unbearable, they sold their lands to the DPN at below-market value.[148]

The DPN developed winter sports in Mount Catedral. At the foot of the mountain, it built a hotel, Catedral Hotel (1944), and a homonymous tourist

Figure 5.7. View of the Llao Llao Grand Hotel and part of its golf course against the backdrop of Mount López, 1942. Photograph by Bruno Ricardo Sálamon. Courtesy of Caspani Collection, Archivo Visual Patagónico.

village. The modernist aspirations designed for Nahuel Huapi became reality in Villa Catedral, where the works of the DPN offered new residents running water and electricity. For Bustillo, Villa Catedral's charming character evoked the Alpine hamlets in Switzerland or Tyrol. In that vein, he used his network connections to reach out to Austrian skier Hans Nöbl, who began spending his summers in the South American winters and getting acquainted with Bariloche's skiing community. Nöbl proposed Mount Catedral as the best location to establish the DPN's ski resort, in opposition to locals skiing in Mount Otto, closer to town.[149] He, like others, insisted on the importance of a cable car to climb up the mountain, which was completed only in 1950.[150] In order to enjoy nature, it needed to be altered.

Among all the tourist villages around Bariloche, Villa Llao Llao offered the most luxurious and exclusive experience to visitors. This experience revolved around the Llao Llao Grand Hotel (fig. 5.7), located between Lake Moreno and Lake Nahuel Huapi, fifteen miles west of Bariloche. The hotel opened its doors on December 31, 1938, and was officially inaugurated a week later. Ten months after the inauguration, a fire destroyed it almost entirely. The reconstruction took less than a year, and it was fully operating just before Christmas 1940.[151] The hotel could accommodate about 400 visitors in 169 rooms, and it offered up-to-date amenities, including a reading room, a small movie theater, a ballroom, a hair salon, a post office, a bank, and a retail wing.[152] To support its services, the hotel also included a bakery, a sparkling water factory, a laundry facility, its own electric power plant, a garage, stables, and a small printing press. The DPN purchased the natural harbor, known as Puerto Pañuelo, located across the road from the hotel.[153] Like the Civic Center, the hotel incorporated the lake as part of its landscape. However, the lake did not symbolize a connecting avenue between productive hubs but rather became a silent recipient of the tourist gaze.

The timber used to build the hotel opened space for a nine-hole golf course in the same way that the DPN cleared slopes for skiing on Mount Catedral. To uphold the expectations of affluent visitors, the DPN entrusted the course to Luther Kootz, an American engineer brought to Argentina by famous golf course designer Alister Mackenzie. Additionally, Harrods Gath y Chaves, a British department store in Buenos Aires, clothed the caddies.[154] Alejandro Bustillo had also built a chapel on the grounds of the hotel in a matching style so that not even on Sundays did visitors need to travel to Bariloche. The self-sufficiency of the Llao Llao Grand Hotel echoed the Bustillo brothers' ideological and architectural aspirations to create a functional garden city. The chosen site, atop a hill in the isthmus of the Llao Llao Peninsula, gave the hotel the perfect backdrop of Mount López's rocky profile, bringing the garden city vision to life for synchronization between the built and natural environments (fig. 5.7). Soon, the hotel became a "finishing touch" in the work of a political generation born with the military coup of 1930, an embodiment of "good will, progress, and love for country."[155] Alejandro Bustillo replicated the Bariloche style in most of the private residences he designed nearby, cementing the Alpine undertones in an Argentine touristscape.

CONCLUSION

In 1937, Exequiel Bustillo, director of the Argentine National Parks Bureau, received a letter protesting plans to erect a statue honoring Julio A. Roca in Bariloche. Bustillo saw Roca as a clear reference for his work. For the anonymous author of the letter, who signed it, "A soldier who knows the truth," Roca had been a temporary political appointee. The author of the letter all but accused Bustillo of historical amnesia, especially as Bustillo's father had participated in the military campaign against the Indigenous polities of northern Patagonia in 1879–81. "The annual reports of the Ministry of War must exist in your father's library," he wrote and asked Bustillo to "enlighten" himself "so as not to perpetrate the injustice of placing General Roca in Bariloche, a place that only corresponds to General Conrado Villegas."[156]

For this veteran, the statue was honoring the wrong man. The monuments, buildings, roads, public offices, and other initiatives of the DPN remained beyond scrutiny, as did its professed nationalizing mission in the northern Patagonian Andes.

At the time of writing this conclusion, debates have reignited in Bariloche on whether to remove the statue of Julio A. Roca. For some people, the monument honors the violence of the military campaign of 1879–81; for others, it is in itself history, so moving it would amount to erasing history. Like most monuments, the statue tells us more about the political agenda of the time it was erected than about the history of Patagonia. It condenses the nationalizing mission that the DPN believed it inherited from the pre-1916 generation of political figures, from Roca to Moreno. It also works as a focal point of that mission, sitting in a square that honors the soldiers of the military campaign, bounded by streets named after national heroes (which replaced old street names like Chile and O'Higgins), and surrounded by buildings that eradicated any remainders of Chilean influence. In Bustillo's inception, Roca's statue represented a chronological origin for the advancement of the state in Patagonia and the disarticulation of Mapuche *lofs*. In his view, the monument condensed the onset of Chilean and Argentine nation-making, which he construed as in confrontation with each other.

The history of the built environment in Bariloche offers a window into how people gave meanings about "the nation" to space and how these meanings changed over time within the frontier space of the northern Patagonian Andes. At the turn of the twentieth century, national authorities bestowed on the northern Patagonia Andes an agrarian mission. The fertile valleys would yield grains, fruits, and cattle under the unwavering care of settlers, transforming wild vistas into productive farmlands. The Chile-Argentina Company embodied the entrepreneurial success of that vision, even though the company came from across the Andes. The rural features of the nascent economy, anchored in the trading hub of Bariloche and the export port of Puerto Montt, were branded as the Chilean and Argentine Switzerland. Yet, nationalist concerns brewed in Buenos Aires, where authorities perceived some people and ideas as threats to a national sense of self. In the Patagonian Andes, these increasingly nationalist views inspired anti-Chilean sentiments, as discussed in chapter 4. In the 1930s, these worries morphed into policy. The conservative administration that came to power with a coup and stayed in power due to electoral fraud sought to reaffirm an idea of the nation that eliminated Chilean influence in the Andes and reoriented trading circuits toward the Atlantic Ocean. A substantial part of this policy, both economically and symbolically, consisted in removing Chilean and agrarian character from the northern Patagonian Andes and replacing it with a nationalist aesthetic centered on the enjoyment of nature. The architectural style and urban design for the villages in Nahuel Huapi National Park sought to condense a shared sense of national self. In visiting Bariloche, tourists were knowing Argentina. In enjoying the outdoors, tourists and locals were building the nation.

Chapter Six

THE OUTDOOR DESTINATION

1905–1945

In 1937, journalist Carlota Andrée published an article in *En Viaje*, the travel magazine of the Chilean State Railway Company (Empresa de Ferrocarriles del Estado; EFE). In it, she called for Latin American countries to "open ... their doors in the frontier" and allow tourists to "assertively penetrate the kingdom of panoramic beauty[,] of multiple and diverse beaches, mountains, and forests." For her, visiting "the panoramic views ... hidden from men's gaze" would elicit from travelers a "triumphant praise of this unappreciated and fine country."[1] Scholars would describe Andrée's portrayal of nature, beautiful and passive to the visitor's eye, as an example of the male tourist gaze. John Urry defined the tourist gaze as seeing specific places as particularly attractive because of how different they are from one's own environment. But this attraction is inscribed "in circles of anticipation, performance, and memories."[2] In the 1930s, marketing materials such as pamphlets, guidebooks, and advertisements for tourist destinations across the Americas used language attractive to white, upper-class, heterosexual men to generate that anticipation among would-be travelers from urban centers.³ What was appealing to the male tourist gaze was appealing to all.

In this particular article, Andrée echoed state authorities' well-established way of seeing southern Chile, where its perceived backwardness could only be remedied by outsiders. Now, however, the civilizing mission rested not on immigrants, as it had before, but on tourists. Yet Andrée's article also epitomizes an important shift. EFE, the state agency responsible for tourism policies, curated a series of tourist destinations in southern Chile that recast nonagrarian southern landscapes as visibly attractive to visitors. It was as if national authorities were trying to rediscover nature after decades of promoting agricultural expansion. To do this, EFE used a feminized language to revamp the southern landscape as a "she-land."⁴ Feminizing nature was hardly new. Accounts written by intrepid Spaniards first described Patagonia to Europeans in the feminine—as a she-land—an approach replicated by explorers of the nineteenth century. Feminizing nature reaffirmed a power imbalance between a colonist, an explorer, or a tourist and the landscape before them. In the 1930s, the Chilean government incorporated this discourse into state policy. It publicized this policy as central to the spatial configuration of the national territory through EFE's two main publications, *En Viaje*, a travel magazine, and *Guía del Veraneante*, a yearly guidebook. Such configuration confirmed that the south of the country was peripheral in relation to the Central Valley but in a different way. Rather than focusing on the fertility of the land and the importance of working it to bring about progress, the government now highlighted the pristine quality of nature (not land) to attract visitors. To enhance the attractiveness of untouched spaces, *En Viaje* and *Guía del Veraneante* used images of seemingly unmarried women in their childbearing years. Seeing space as fertile and virgin marked a discursive shift within state policy, but at heart it perpetuated the same way of imagining the Chilean nation.⁵

EFE established a feminine visual reference for the landscapes of southern Chile to reassert the notion of pristine, beautiful nature. This visual grouped together three elements: a mountain or volcano, a body of water, and a forest. This peak-water-forest triad became a shorthand in the pages of *En Viaje* and *Guía del Veraneante* for the provinces in the South. Indeed, "snowcapped volcanoes," "lakes as soon asleep as bristling," and "immense forests of crowded trees," as Andrée put it, "astonish and overwhelm the spirit," prompting "triumphant praise" from the tourists admiring their "beautiful country."⁶ The agrarian character of Araucanía, Valdivia, and Llanquihue was moved to the background. Similar to what happened in Nahuel Huapi National Park across the Andes, the government used a visual composite to convey nationalist values. Traveling to southern Chile became "a quest of intense patriotism," because there visitors could "find the original source of national history."[7] Admiring specific landscapes meant admiring one's country as a whole. This corresponded with what the Argentine National Parks Bureau (DPN) did in Argentina. Perhaps the main difference was that EFE construed outdoor landscapes as inherently Chilean, while the DPN especially (but not solely) focused on the built environment to construct a shared aesthetic across Nahuel Huapi National Park.

Chile's military victory in the War of the Pacific (1879–84) had underpinned a sense of a homogenous, racially superior, national self based on steady economic expansion.[8] However, World War I abruptly cut off the demand for Chilean nitrates, used for gunpowder, which never recovered, unable to compete with synthetic nitrates. This crisis rippled across Chilean society, which questioned its identity and demanded action from a slow-moving parliamentary government.[9] Nationalist ideologies gained momentum in Chile, where president Arturo Alessandri Palma succumbed to their pressure and resigned in late 1924. During the following eight years, political power alternated between Alessandri and Carlos Ibáñez del Campo, a colonel who rallied the armed forces (and who had served as Alessandri's minister of war). A new constitution in 1925 attempted to articulate Chile's updated sense of self. Specifically, it reverted authority from Congress to the president, a move that centralized government in strong federal offices, like the General Controlling Office, the National Police (*carabineros*), or EFE. Alessandri's second term (1932–38) rested on an alliance against leftist parties. But the rise of fascism in Europe called for actions against the Far Right and pushed centrists to turn to socialists and communists for the 1938 elections. This fragile but successful alliance, the Popular Front, installed Pedro Aguirre Cerda as president.[10] Aguirre Cerda reinforced state economic planning, which had proved a useful tool to counteract the effects of the Depression and of a devastating earthquake that hit Chillán (Ñuble) in 1939. This policy represented a dramatic shift from the export-oriented economy dependent on foreign investment. In this context, Aguirre Cerda created the Production Development Corporation, an agency responsible for accelerating industrialization by creating state companies. While EFE was established in 1884 with the initial purpose of overseeing the expansion of rail networks to the west and south of Santiago, it was not until 1927 that it took on the role of developing a tourism policy aimed at promoting travel along its railway routes.[11]

Chileans did not escape the urge to locate the idea of the nation in cultural creations in the 1920s and 1930s. This quest was rooted in turn-of-the century repudiations of the foreign influence of modernism and validation of local traditions. These explorations resulted in a new literary movement inspired by rural life, *criollismo*. Intellectuals and artists across Latin America celebrated regional culture, language, and identity, especially in the face of the rising hemispheric influence of the United States. In the 1930s, the Chilean monthly magazine *En Viaje* provided a platform for second-generation *criollista* writers, who described the nation through a spatial lens. *En Viaje* frequently published articles and short stories by Mariano Latorre, Luis Durand, and Sady Zañartu, all paradigmatic *criollista* writers. Pamphlets, guidebooks, and periodicals show that the state appropriated *criollista* language to construct a nationalist understanding of the Chilean space.[12] EFE's propaganda office was in charge of publishing and distributing both the monthly travel magazine and the tourist guide. The *Guía del Veraneante* compiled information about major attractions, hotels, and routes and had no clear editorial team. *En Viaje*, on the other hand, was founded by EFE's printing press director, Wenceslao Landaeta Sepúlveda, who directed the magazine until September 1939. Washington Espejo succeeded him in October, and he was replaced by a senior editor, Carlos Barella, in June 1943.[13]

This chapter begins by examining how the male tourist gaze of national authorities and tourists created a sense of pristine, beautiful nature in southern Chile. This understanding of space masked earlier portrayals of those same landscapes as productive farmland. Then, based on the male tourist gaze, the next section analyzes how EFE constructed parallelisms between nature and women's bodies to affirm the allure of the South. In doing so, it situated the South, in the form of the peak-water-forest triad, within the pantheon of landscapes that evoked a sense of national identity for people in other parts of the country.[14] The third section studies how state agencies constructed traveling for leisure as a modern experience almost exclusive to the urban upper-middle classes. Akin to what Jacob Dlamini has examined, political elites saw "the right to seek leisure where and when they desired ... linked to their self-regard as modern subjects."[15] To that end, EFE compared traveling to southern Chile to taking promenades in the city, especially by publishing nearly as many centerfold images of the South as of the Central Valley. The fourth section shows how locals' and tourists' enjoyment of the outdoors in Chile and Argentina cemented understandings of the northern Patagonian Andes as national landscapes. This is not to say that enjoying the outdoors elsewhere did not result in pride of country, but here it constituted a distinct goal of state agencies. Overall, the re-signification of the Andean landscape from a site of agricultural production to a site of admiration remained incomplete, as I examine in the last section.

THE MALE TOURIST GAZE IN SOUTHERN CHILE

The male tourist gaze upon the Chilean-Argentine Patagonian Andes resulted in an image of virgin nature. Seemingly untouched patches of forest and unexplored valleys were alluring to travelers, evoking the way a man fell for a woman. Such a view reinforced Western, heteronormative understandings of nature, refusing to acknowledge the role of Indigenous communities in sustaining those ecologies. Travel publications like *En Viaje* and *Guía del Veraneante* increasingly drew attention to the "virgin forests" near Concepción, Cañete, and Lake Lleu-Lleu, in Araucanía, silencing any Mapuche presence.[16] Virgin forests spread along lakeshores, riverbanks, and mountain slopes, making the train journey from Santiago to Castro visually appealing.[17] Writers also highlighted the feminine tropes of southern Chile by ascribing masculine symbolism to the mining North: "The South calls the men of the North and invites them to enjoy its majestic views," for example.[18] The male tourist gaze encompassed a way of seeing space, regardless of the gender of the traveler, because it mapped out how different regions of a territory should relate to one another. In the 1930s, Chileans and Argentines in the capital cities saw the Patagonian southern districts as peripheral spaces, and the male tourist gaze validated this view as modern and desirable.

Visual and literary elements in travel publications characterized the relationships between visitors and touristscapes as heteronormative. The allegedly untouched nature of southern Chile made it inexplicably attractive; hence writers compared visiting the South with falling in love. As one writer put it, "What a joy to fall prisoner to the wonderful traps that Chilean landscapes laid for us, there where we would want to root our feet! We have never felt our bodies and souls so imprisoned by the magical and deep voice of a virgin forest, which, with the blow of a breeze, gives us its most pure accents."[19] Travel publications bolstered the relationship between visitor and landscape by running pieces that often portrayed women in relation to their husbands, fathers, and lovers. Husbands were loyal and loving, and wives "adored [them] like an idol."[20] Columns described men as brave, renowned, intelligent, honest, and hardworking.[21] Likewise, Chile "treasured ... endless beauties ... [in] the southern region."[22] Somewhat like having a romantic crush, "we from [the Central Valley] are attracted by the big southern cities

Figure 6.1. Cover of *En Viaje* illustrating a virgin forest near Puntiagudo volcano, February 1938. The same image was used on the interior back cover of *Guía del Veraneante*, 1938. Courtesy of Memoria Chilena, Biblioteca Nacional Digital de Chile.

and the lake district, so beautiful that those who have seen it once 'can never forget.'"²³ Tourism, then, represented the patriotic-masculine endeavor of discovering, seeing, and possessing the beautiful-feminine landscapes of southern Chile.

The editors of *En Viaje* reinforced the male tourist gaze by supplementing such descriptions with depictions of nameless landscapes, usually in the form of a composite that would become a shorthand for southern Chile, the peak-water-forest triad. This composite condensed "the South" into three elements: a snowcapped mountain range or a volcano, a lake or a river, and a forest. A low snow line signified the Andes south of the thirty-ninth parallel, even though snowcapped mountains did exist farther north. Conic mountains promptly situated readers in southern Chile, a land with thirty-nine active volcanoes.²⁴ Streams or lakes clearly differentiated the surging landscapes of the South from the dryer areas of northern Chile. Forests evoked virgin nature waiting to be discovered by urban eyes. The overwhelming anonymity of each element in the peak-water-forest touristscape meant they could have been from anywhere and everywhere south of the Bío-Bío River. Magazine covers and images accompanying articles portrayed lustrous forests with very little human presence or none at all, a nod to earlier understandings of the South as empty (figs. 6.1 and 6.2).²⁵ The appeal of seemingly virgin landscape invited tourists to reconquer it. It was EFE's mission that visitors admire "the

Figure 6.2. Cover of *En Viaje* showing "virgin" nature interrupted only by a moving train, January 1944. Courtesy of Memoria Chilena, Biblioteca Nacional Digital de Chile.

impenetrable rainforest, majestic in its millenary grandiosity."[26] The triad represented a postcard picture of virgin nature awaiting the tourist gaze, analogous to the Argentine built environment I discussed in the previous chapter.

Yet not all of southern Chile was an Edenic landscape. The peak-water-forest triad excluded agriculture, a force that had spearheaded Chilean colonization in Araucanía, Valdivia, and Llanquihue Provinces. References to rural life in EFE publications focused on the Central Valley, in the form of advertisements, farming advice, and appeals to farmers to travel in the winter months.[27] Additionally, during this season, the capital "turn[ed] on all the lights in its halls and burst into sensual and evocative music," inviting southern residents to spend the cold season in warmer weather.[28] As a way to foreground the modern experience of visiting southern Chile, EFE publications featured the modern urban landscapes of the coastal cities of Puerto Montt and Valdivia.[29] In Pucón, on Lake Villarrica, EFE built a grand hotel whose paradigmatic architecture also became a reference point for southern Chile. Over the 1930s and early 1940s, *En Viaje* and *Guía del Veraneante* increasingly transitioned from depictions of virgin landscapes to descriptions of sites where visitors could appropriate them through an activity. The feminization of the Patagonian Andes situated landscapes and their peoples in a showcase of sorts, where they stayed passive while visitors—outsiders—carried out the active roles of enjoying them through the male tourist gaze.

Figure 6.3. Illustration of a woman with the Chilean touristscape. Back cover of *Guía del Veraneante*, 1942. Courtesy of Memoria Chilena, Biblioteca Nacional Digital de Chile.

THE FEMINIZED CHILEAN TOURISTSCAPE

Chilean guidebooks and travel magazines used female imagery to represent the attractive landscapes of the South. Take, for instance, the back cover and interior back cover of the 1942 *Guía del Veraneante* (figs. 6.3 and 6.4). In them, two women, one drawn, one photographed, pose for the viewer as they balance gracefully, on a boat and a railing, wearing summer attire that reveals their pale skin. As they try not to lose their balance, both women playfully hold on to a line and a pole; their actions do not mess up their hairdos. In figure 6.4, the grand hotel at the back flanked by a volcano situates the woman in Pucón, on Lake Villarrica. Indeed, the image draws parallels between a seemingly untouched landscape and the innocence of the young woman. But the landscape surrounding her, synthesized in the hotel, shows a geographical space marked by the presence of the state. It is because of EFE that this woman can enjoy this touristscape. In figure 6.3, this is less obvious because the landscape surrounding the woman shows no human-made infrastructure. It may not be readily apparent where she is sailing, but the peak at the top left is remarkably similar to the Puntiagudo volcano, near Lake Todos los Santos.

Figure 6.4. Woman with the Pucón Grand Hotel and the Villarrica volcano in the background. Inside back cover of *Guía del Veraneante*, 1942. Courtesy of Memoria Chilena, Biblioteca Nacional Digital de Chile.

The composition points to the role of the state in facilitating her enjoyment through the Chilean flag on the vessel. In both images, the peak-water-forest touristscape serves both as a backdrop for tourist activities and as the main attraction that brought visitors there in the first place.

Editors coupled images of beauty with descriptors of fertile abundance. In previous decades, national authorities and local elites had associated the trope of fertility in southern Chile with agricultural development. Working the land had been a masculine form of nationalizing the South. In the pages of *En Viaje*, however, fertility had more to do with feminine nature and what it had to offer as is. An editorial note, for example, praised the month of February as the best time to visit southern Chile because "the miraculous southern rivers emerge with Nature's overwhelming beauty, abundant in fish and in sunny beaches, where golden sands mix with the polychromous array of swimsuits in restful warmth."[30] EFE insisted on nature's reproductive lure, which brought men outdoors: "It's the resurrection of men drawn by this irresistible force that co-participates him in the joy of nature."[31] Some pieces even compared the popularity of touristscapes with women's fertile age. One editorial, for example, described the lakes in southern Chile as being in "their prime... in full power of beauty... in dazzling fullness of grace." Like "society women," some of these lakes were "trendy" (like Villarrica or Llanquihue), but another one, like a debutante, may be "waiting in line... yet presented to society, but it is getting ready for its touristic debut."[32] In general, *En Viaje* and *Guía del Veraneante* used images of abled women in their childbearing years as a reproductive ideal, a nationalist concern Chile shared with other nations in the first half of the twentieth century.[33]

Articles often drew parallels between landscapes and women's bodies. In 1942, *En Viaje* covered the inauguration of a new road between Puerto Varas and Ensenada in a ten-part photo essay. Only one image showed the skyline of the city, captioned "A part of the beautiful Puerto Varas, seen from the Hotel of the Empresa de Ferrocarriles del Estado." The caption for the only image with women sounds familiar: "Beauties of the beautiful city of Puerto Varas pose for *En Viaje*'s camera, making the stay of tourists more attractive."[34] This caption used three iterations of "beautiful," once to describe the city, once to describe the women, and once to describe how they affected the viewer. A city and women offered the same visual pleasure for travelers. The focus on beauty reinforced the male tourist gaze on attractive, feminized landscapes. Similarly, the editors of *En Viaje* sometimes used photographic layouts to illustrate the correspondence between landscapes and women's bodies. In each issue, the magazine included a photographic centerfold, sometimes several pages long, featuring various landmarks from around the country, from waterfalls to statues. In some of these, the editors juxtaposed images of southern Chile with images of female movie stars, a parallel reinforced by the captions. This was particularly true for Hollywood stars, whose movies had made US American cinema in Chile "an irresistible commodity."[35] Similarly appealing was a display of photographs of actors Elisa Landi, Jean Harlow, and Dorothy Jordan, the three "most beautiful [stars] in Hollywood," next to several images of sites around Lake Llanquihue.[36] Women's bodies commodified nature.

Women's bodies were used more than any other images on the covers of *En Viaje*. Between its first issue in November 1933 and December 1945, women appeared on the covers of forty issues of *En Viaje* out of a total of 132 (table 6.1). I noted all the visual elements on each cover and then collapsed singular and plural labels (man/men, tree/trees, woman/women), though I did not merge "tree/s" with "forest" because some covers showed trees in urban settings. The data shows that *En Viaje* used female bodies more than any other image. This was followed by mountains (36), tree/s (33), and men (27). Grouped together, the mountains, volcano, forest, tree/s, lake, and river categories

Table 6.1. Number of occurrences of each visual component on the available covers of *En Viaje*, 1933–1945.

Visual element	Number of occurrences
Woman/women	40
Mountain	36
Tree/s	33
Man/men	27
Lake	20
Snow	16
Boat/s	15
Forest	15
Beach	12
Train	12
Ocean	9
River	9
City	8
Volcano	8

Source: Author's calculation with the help of Asa Ackerly.

amounted to a total of 101 appearances, followed by what I identified as human components (man/men, woman/women, children, couple, and people) in 79 instances. Far behind came the ocean, beach, boat/s, and harbors set, with 42 appearances. Notice also how transportation services—markers of modern travel—were not as frequent as images of nature or people.

Women appeared mostly unaccompanied on the covers of *En Viaje*, signaling the vitality of youth in the ability to throw on vacation gear and the flexibility of singledom to leave the city. This might explain the high frequency of images of women on their own or with other women (28) compared to images of women with men (8) or children (2). Images of young, able-bodied women also reflected the aspirations of the white upper-middle class in Santiago and Valparaíso. Many of the women in the magazine were blond, slim, and dressed in attire that revealed or accentuated their bodies. They were photographed from suggestive angles or nearly naked, highlighting the attractive parallelism between women's bodies and nature (figs. 6.5 and 6.6). The covers implied the carefreeness of traveling for pleasure and the attractiveness of a beautiful place. On the January 1937 cover of *En Viaje*, a woman sits on the beach with Lake Villarrica and the unmistakable volcano in the background, coupled with the Pucón Grand Hotel (fig. 6.7). She is waving at someone, as if inviting readers to come with her. In contrast, the pages inside the magazine boxed women into expectations of domestic responsibility, quite

The Outdoor Destination

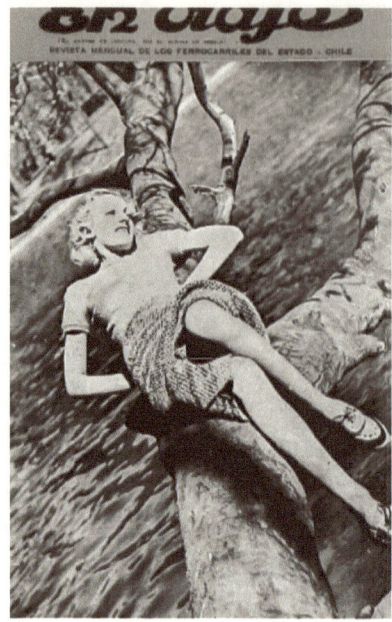

Figure 6.5. Woman posing on a tree. Cover of *En Viaje*, November 1941. Courtesy of Memoria Chilena, Biblioteca Nacional Digital de Chile.

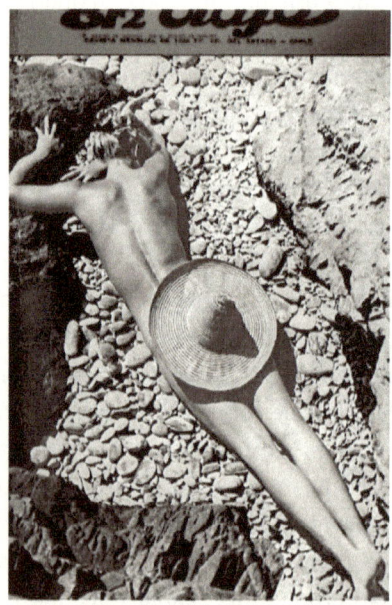

Figure 6.6. Woman sunbathing, covering herself only with a sun hat. Cover of *En Viaje*, March 1942. Courtesy of Memoria Chilena, Biblioteca Nacional Digital de Chile.

Figure 6.7. Illustration of a woman at the beach with the Pucón Grand Hotel and the Villarrica volcano in the background. Cover of *En Viaje*, January 1937. Courtesy of Memoria Chilena, Biblioteca Nacional Digital de Chile.

distant from the carefree affect of the women on the covers. The images on the covers symbolized a way of enjoying the national landscape, the articles delineated fashion trends, household chores, and family responsibilities as the desirable aspirations of modern, upper-middle-class Chilean women. Women's bodies in the fashion pages of *En Viaje* or in photo arrays of socialites showed less skin and more modesty. Features portrayed women with the latest hairstyle, including victory rolls, pompadours, and curls, and wearing utilitarian trends popularized at the beginning of World War II.[37] It was not only women's bodies that made the landscapes of southern Chile attractive. It was also their youthful, modern style.

Social, racial, and aesthetic elements carefully constructed an ideal vacationer: Who did not want to be a privileged woman having fun on the shore of Lake Villarrica? Who did not want to be the man hanging out with her? The train ride mirrored the anticipation of a man meeting a woman, as "the spirit rejoices seeing the succession of landscapes while looking out the train window."[38] And that excitement continued at the destination, where tourists "contemplate the superb views of Nature [with] emotional enjoyment, which comforts the body and soul."[39] Despite the portrayals of carefree women on its covers, *En Viaje*'s articles mostly discussed women in relation to men, as wives, mothers, and daughters. Article topics ranged from household management to marriage advice.[40] *En Viaje*'s notes on women's fashion or tips for happy married life might seem out of place in a travel magazine. Geared toward female travelers, these pages stressed the role society expected from them. Jokes mocked physical or sexual violence against women, while articles taught women to make their husbands happy with a clean house, a slender body, and passive behavior. In many ways, *En Viaje* also taught its readers to maintain such expectations of tourist destinations.

TRAVELING TO THE SOUTH

Chilean travel publications promoted leisure travel as an activity proper of a modern society. They highlighted amenities and activities that complemented the untouched allure of the Andes. The summer months offered benign weather for admiring "the magnificence of the panoramas in which mountains and volcanoes are eternally covered with snow, rivers of clean and torrential flow, [and] the exuberance of a vegetation in many parts still untouched by man" (notice the peak-water-forest triad coming to life).[41] Traveling to faraway destinations could be cumbersome for the urban elites of Santiago and Buenos Aires. State agencies highlighted modern amenities available during the journey and at the destinations that made traveling more comfortable. EFE sought to get city dwellers, especially from Santiago and Valparaíso, to vacation in the south of the country. One strategy consisted of highlighting the up-to-date services available when traveling to and staying in southern Chile. For instance, the company advertised the new sleeper cars available on the journey from Santiago to Puerto Montt as a key feature of the popular route. Given that the trip could last about a day and half with at least one transfer, the convenience of beds and dining cars was not exaggerated.[42] In a handful of destinations, the Chilean agency built "modern and elegant" hotels, such as the one in Puerto Varas and another, larger one, in Pucón.[43] Not only could their facilities host hundreds of guests, but they also offered guest services, a hair salon, a game room, a smoking room, and casinos. Telephones and private bathrooms "contributed to the relaxation" of vacationers.[44] Urbanites wanted to escape the rush of city life but not at the expense of conveniences.

EFE also used the visual appeal of *En Viaje*'s centerfold to emphasize the modern experience of traveling to southern Chile. Sites in southern Chile appeared as frequently as sites in the Central Valley, including promenades in the capital city, implying that vacationing in the South was no less enjoyable an experience than a stroll downtown. The centerfold mostly featured photographs from travel destinations along EFE routes. I extracted each site featured in a centerfold, whether it be a city (like Santiago) or a geographical site (like the Llaima volcano). I then georeferenced all sites and aggregated

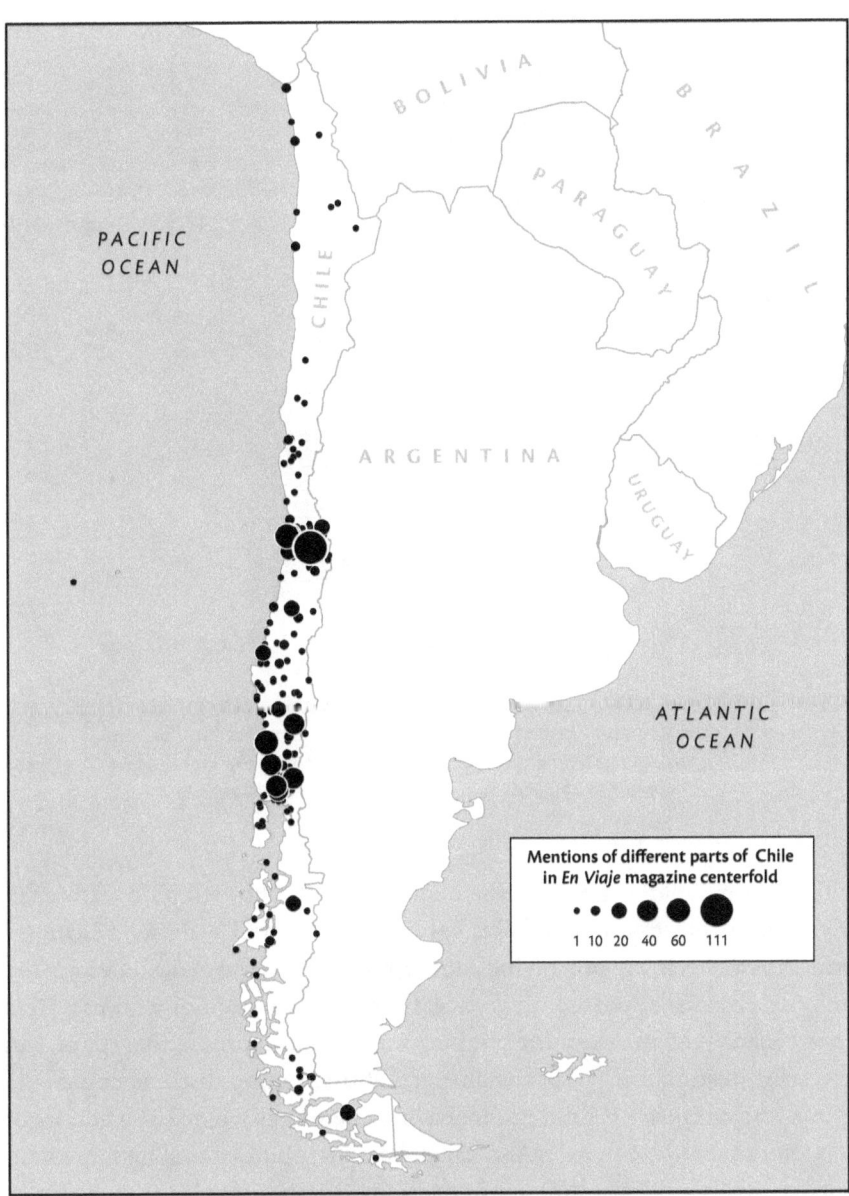

Map 6.1. Aggregate frequency of Chilean places featured in the photographic centerfold of *En Viaje* between 1933 and 1945. Author's rendition using Tableau Public, with the assistance of Matthew Naglak, Yaqing (Allison) Xu, and Asa Ackerly.

The Outdoor Destination

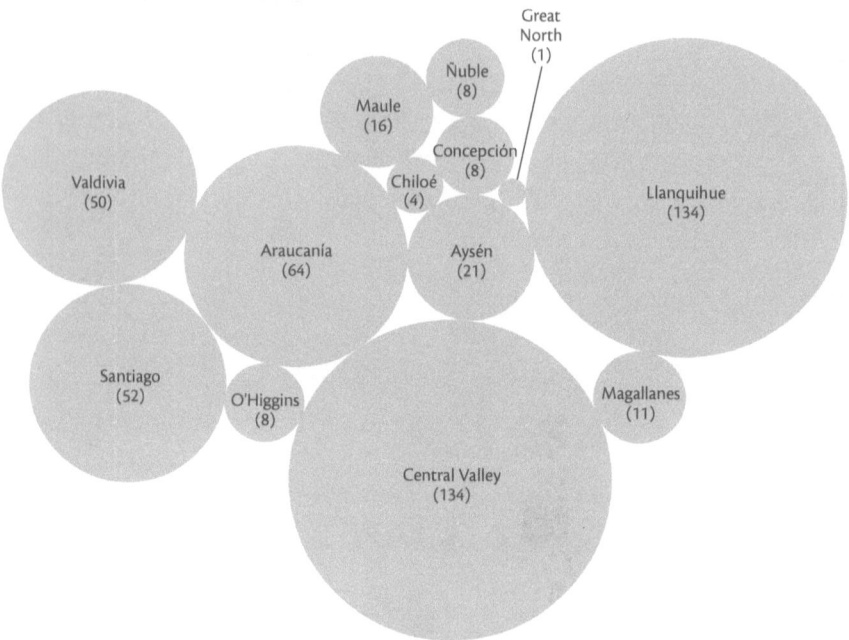

Figure 6.8. Aggregate frequency of Chilean places featured in the photographic centerfold of *En Viaje* between 1933 and 1945, by region. Author's rendition using Tableau Public, with the assistance of Matthew Naglak, Yaqing (Allison) Xu, and Asa Ackerly.

the data on a map (map 6.1). The majority of the photographic features in *En Viaje*'s centerfold were of sites between the Central Valley and Llanquihue Province, clearly portraying southern Chile as a travel destination. The mining districts appeared only once, represented by Iquique. In part, this was because the north of the country was seen as a productive space, not a tourist destination.[45] But the absence of the mining districts could also simply be because EFE did not manage those routes. Hence, the editors of *En Viaje* focused on areas in Araucanía, Valdivia, and Llanquihue more than anywhere else. Yet, we should not overlook the spotlight given to beaches, parks, promenades, and ski resorts in the Central Valley. Images of Santiago populated the centerfold of *En Viaje* just as frequently as views of Llanquihue Province (fig. 6.8). The attractions that the capital offered, such as the botanical garden or the zoo, pointed to Santiago's bustling character. By insisting on tourist destinations farther south, the editors of *En Viaje* portrayed the provinces of Valdivia and Llanquihue as being as modern and desirable as any promenade in the city. Editorial notes complemented these portrayals by describing southern cities as tourist destinations. Particularly, *En Viaje* highlighted the "modern architecture," "affordable amenities," and "ease of travel" by train to Puerto Varas and Pucón, where the state agency had built its hotels in 1935 and 1936, respectively.[46]

The possibility of travel in Chile also symbolized modern democracy, especially in the context of World War II. As Peruvian journalist Alberto Carrasco Hermoza put it in 1942, how better could democratic practices elsewhere in the Americas be improved than by traveling to a country "averse to dictatorial regimes and that is [a] land of the free?"[47] For him, traveling for pleasure was possible in a politically and economically stable environment like Chile, and tourism, then, was the quintessential activity in a functioning democracy. For editors of *En Viaje*, traveling made Chileans better citizens. It exposed them to "our national history and the customs of different regions." Enjoying natural landscapes, in turn, instilled in them "healthy, optimistic views of what our country is."[48]

In Argentina, the national government expanded the road network from 2,000 kilometers in 1931 to more than 30,000 in 1944 (from 1,242.7 to more than 18,641 miles).[49] The expansion of this network, however, focused on the center and north of the country, barely scratching the surface in Patagonia, where the train remained a sought-after means of transportation.[50] Despite this disjuncture, the combination of "roads-cars-tourism" situated traveling across the country as a major drive for national unity.[51] By the 1930s, vehicular transportation was comfortably installed among the Argentine growing middle class. This was possible because of an uptick in national oil production and subsequent regulation of foreign gas and because of lobbying by two major clubs, the Argentine Automobile Club (ACA) and the Argentine Touring Club, in promoting auto driving and racing.[52] In that same decade, the government-owned State Oilfields Company (Yacimientos Petrolíferos Fiscales, YPF) expanded into marketing operations anchored in a curated image of the company as "modern, efficient, and national."[53] In the context of a growing road network, ACA sponsored touring car rallies across the country, where competitors departed from Buenos Aires and over thousands of kilometers "discovered" new regions, such as Patagonia.[54] On two occasions, in 1936 and 1939, the route looped into Chile, went across Mendoza, and came back through Neuquén, turning south to Bariloche and Esquel.[55] The races spotlighted lesser-known routes for drivers and onlookers, who then used this information to recommend speed limits on certain segments, such as Neuquén–Bariloche.[56] YPF and ACA joined forces in the late 1920s and reaffirmed this agreement in 1936 to advance the notion that traveling in one's own country using national gas (though in imported cars) was a patriotic endeavor.[57] The YPF-ACA pact prompted the construction of an impressive network of YPF gas stations and ACA branches along the growing national road system. The architectural styles of these stations were far from coherent—they varied greatly from modernist to regional—but the stations offered recognizable emblems of both institutions, bringing together a national space that still felt disjointed to the urban middle classes of Buenos Aires. Even in the northern Patagonian Andes, drivers could find YPF pumps labeled with its rounded logo, signaling the advancement of the future onto Argentine soil.

YPF collaborated with the DPN in their mission to make Nahuel Huapi National Park a tourist destination. To that end, it created road signs that pointed to tourist attractions such as lakes, hotels, bridges, or other tourist towns in the park.[58] An ACA office in Bariloche and seven gas pumps in northern Patagonia were planned in 1940 to serve the "road to the southern lakes."[59] Drivers could find in DPN guidebooks road maps created by ACA for tourists traveling by car to Nahuel Huapi. These maps also located mechanic's shops, gas stations, types of roads, train stations, and ACA branches. For entertainment, they located tennis courts, golf courses, swimming pools, hotels, and panoramic views.[60] Exequiel Bustillo was adamant that a gas pump be installed in the DPN's forthcoming Llao Llao Grand Hotel as part of its many services offered to elite visitors.[61] The DPN's guidebook also advertised an old YPF camp in the Ñirihuau River valley, where geologist José María Sobral had led exploratory work in 1932–34.[62] In an attempt to cement the shared civilizing mission of YPF (drilling, distilling, and distributing national gas) and of the DPN (constructing national landscapes), the Museum of Patagonia, located in the Bariloche Civic Center, tried to get ahold of the equipment that first drilled Argentine oil in Comodoro Rivadavia (on the Atlantic coast of Chubut).[63] On every sign, in every road, and in every attraction, the DPN sought to inspire among visitors love of country and elicit a patriotic commitment to the nation.

THE OUTDOORS

On November 11, 1931, just as summer approached in the Southern Hemisphere, thirteen people, including at least three women, went on a hike to Mounts Carmen de Villegas and Leones, a little over twenty-two kilometers (fourteen miles) east of Bariloche. It took them about a little over an hour to get to the starting point, "3 kilometers [1.8 miles] from Estancia San Ramón's first gate." Everyone knew Estancia San Ramón, one of the oldest properties in the immediate surroundings of Bariloche and once managed

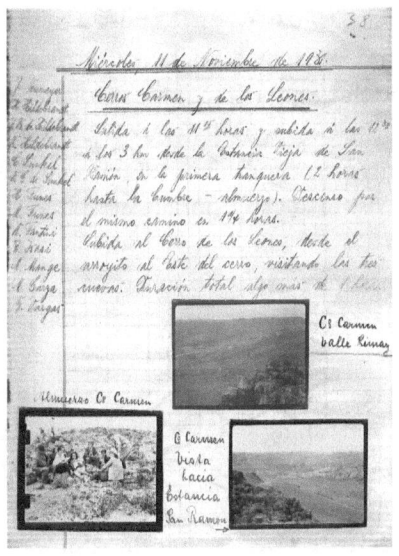

Figure 6.9. Page from Club Andino Bariloche's excursion logbook showing participants, hiking times, routes, and photographs taken. Club Andino Bariloche, "Libro de excursiones no 1," 1931–32, Biblioteca del Club Andino Bariloche, Argentina. Courtesy of Paula Núñez.

by the Chile-Argentina Trading and Cattle-Breeding Company. From the starting point, the group climbed up Mount Carmen de Villegas, had lunch at the 4,714-foot summit, and came back. Then, they hiked Mount Leones, which was lower, and visited three caves in just over an hour. The only trace of the outing ever taking place is an entry in the excursions logbook of Club Andino Bariloche (fig. 6.9).[64]

Hiking, fishing, and skiing constituted a modern form of participating in the nation for both visitors and locals. Residents on both sides of the Andes established clubs that organized outings and fostered mountain sports, like the Club Andino Osorno (CAO, 1935) and the Club Andino Bariloche (CAB, 1931). Club histories recognized geologists and naturalists that preceded them as the first mountaineers in the region. In a brief history of mountaineering in Bariloche, author José F. Finó draws a direct line between "the Jesuit missionaries, [and] the first explorers Dr. [sic] Francisco P. Moreno and Hans Steffen," including "trips by [Fernando] Hess, [Francisco] Fonck, and Bailey Willis."[65] These narratives often presented andinism as a second-generation alpinism in an attempt to reinforce its European roots and stress the settler undertones of mountaineering, something that Mark Carey also notices for the Peruvian Andes.[66] Finó credits a series of German expeditions to the Patagonian Andes in the 1910s as the "first accounts of real ascents."[67] In logbooks, reference points alluded to toponymy that honored either national heroes or first settlers, like Mount López, a name chose by Francisco Moreno to honor Vicente López y Planes, author of the Argentine anthem, or like Mount Otto,

The Outdoor Destination

named for German settler Otto Goedecke, who lived on the northern side of this mountain.[68] Coincidentally, another German immigrant, Otto Meiling, one of the founders of the CAB, moved there after Goedecke's violent death in 1930. Clubs' excursions reaffirmed the colonizing mission of settlers as persistent reverberations of explorations of the cordillera. Hiking and skiing represented active ways for locals and visitors to appropriate what they construed as inert, submissive space.

Unlike explorers, tourists and hikers were expected to have a personal, emotional response that contributed to a collective appreciation of their country. Mountain sports clubs in the environs of Lakes Llanquihue and Nahuel Huapi embraced the messaging from state agencies that exploring nature—by hiking, skiing, sight-seeing, and fishing—was critical to grounding a national sentiment.[69] The views that Club Andino hikers took in "heightened emotions ... at the sight of the huge peaks, at the sublime beauty of the mountains of our homeland."[70] Such activities became meaningful ways of participating in nation-making, especially since Bariloche residents could not take part in national and provincial elections.[71] Across the Americas, including Chile and Argentina, national governments had favored specific recreational activities, like baseball or horse racing, to elicit a sense of belonging within a larger community.[72] As I examined in chapter 4, political elites in Chile and Argentina believed that promoting a healthy body created a stronger society, both physically and morally. Additionally, sports organizations in the Chilean Argentine Patagonian Andes aligned their goals with those of clubs elsewhere, from Vienna to Shanghai, where local elites found in a shared activity a means to reproduce their distinctiveness and consolidate their group identity.[73] In doing so, they distinguished "between work and play," a central facet of capitalist economy and social status.[74]

State agencies like EFE and the DPN favored recreational activities across the northern Patagonian Andes to strengthen national sentiments among visitors. The fresh air of the cordillera and the active lifestyle it offered aided in the "physical improvement of [our] race."[75] Modern life could be tiresome. Traveling "constituted a vital need," like oxygen, to counteract the afflictions of city life. Such sentiments echoed movements across the Western world that strived for a return to nature. The pages of *En Viaje* and the guides of the DPN portrayed visiting natural landscapes as a way to complement modern city life, not oppose it.[76] As an *En Viaje* editorial note put it, "This is the social and medicinal function of tourism."[77] In this vein, state agencies worked with local clubs to expand mountain sports to tourists, who represented an even more transitory type of visitor. It was not uncommon for state agencies to collaborate with local organizations to advance their goals. In 1935, for instance, the Chilean government funded the purchase of boats and construction of a dock for the Lake District Tourist Association based in Puerto Varas.[78] Despite sometimes being at odds with the CAB, the DPN provided funding for the club to purchase ski equipment and build its headquarters.[79]

Sports clubs offered locals a space to organize outings in pursuit of collective experiences of nature and recognize in it a national landscape. Enjoying hikes or slopes allowed them to be tourists in their own backyards while identifying specific activities as attractive for out-of-towners. Moreover, the clubs functioned as social spaces for local political and economic elites to come together outside the constraints of everyday life.[80] For these groups, nature offered a space separate from the social tensions of life. In Argentina, Emilio Frey, a surveyor and long-established figure in town; Otto Meiling, a German immigrant and ski manufacturer; Reynaldo Knapp, a transportation businessman; and Juan Javier Neumeyer, a local physician, founded the Club Andino Bariloche (CAB) in August 1931. For context, this came only a few months after Exequiel Bustillo's first trip to Nahuel Huapi and three years before the creation of the DPN. Within its first year, the CAB received the remaining funds from the "defunct Club Sportverein Bariloche," which were immediately designated for opening trails and building refuges.[81] The club's Chilean counterpart, the CAO, was founded in 1935 with the same objective: to encourage mountain sports among its members. In Puerto Varas, mountain aficionados established the Teski Club in late 1940, which promoted tennis and skiing (hence "Teski"). As with the CAB, members of local social elites filled the boards of the clubs in Osorno and Puerto Varas. For instance, Walter Niklitschek, a businessman from Puerto Varas, participated in the club's foundation and architect Carlos Buschmann Zwanzger designed one of the CAO's refuges and also served as its president.[82]

Activities such as hiking or skiing offered only momentary enjoyment of the northern Patagonian Andes. Ski rides lasted only a couple of minutes before one had to head back up the slope without a lift. When hiking, visitors entered a forest, climbed a mountain, enjoyed the views, and headed back home. The CAB prioritized opening mountain trails (*picadas*) and building mountain refuges (*refugios*) to support its activities. Opening trails involved minimally marking trees with paint to signal the way, felling trees where necessary, and creating rope lines to serve as simple banisters in dangerous places. Such markers punctuated "virgin" nature with human presence, enabling visitors to enjoy the environment without feeling that they were intruding on it. Within its first year and a half of existence, the CAB opened four trails on Mount López and built a mountain refuge for hikers at 1,620 meters (5,315 feet) on a peak of 2,075 meters (6,808 feet).[83] The CAO, conversely, built its first refuge on the eastern slope of the Osorno volcano to facilitate skiing in the winter. In the early 1940s, the CAO built another refuge on the Casablanca volcano, near a new international road under construction. In 1950, it dismantled all of that refuge's structures and moved them to its present-day location in Antillanca, just a few miles south of Casablanca.[84] The refuge on Mount López collapsed under heavy snow in 1957 and was later rebuilt. Such facilities contrasted with tourists' expectations of hotels in Bariloche, where permanence and comfort shaped their vacation.[85] The atemporality that hikers experienced in the natural environment stemmed from an idea of nature as detached from social and political conflict.[86] The ephemeral structure of these constructions and their rudimentary design corresponded with the transitory experience of hiking and skiing, allowing for a break from the chores of everyday life.

In Bariloche, CAB founders tried to find ski slopes right away to promote this sport among a growing membership.[87] Although more people made donations for summer hikes, the executive committee was invested in making skiing accessible to new enthusiasts and purchased some equipment.[88] The club also organized cross-country skiing and slalom races for beginners and advanced skiers, some on the slopes of Mount Otto. Mount Otto was not only closer to Bariloche but also where the access road was more frequently cleared.[89] Competitions were followed by "traditional criollo barbecue" for lunch and an evening party at Suizo Hotel, featuring a local jazz band, dancing (especially the upbeat "typical Bavarian Schuhplattler dance" led by German alpinists and ski manufacturers Otto Meiling and Herbert Tutzauer), raffles, awards, and the election of the party's queen, an antecedent of the present-day National Snow Queen.[90] Over the following years, day races and night parties became a staple at the club and an opportunity to spread the enjoyment of winter sports, with ripple effects in the regional press.[91]

While the CAB and the DPN shared the goals of promoting mountain sports, their relationship was not always smooth. Club members saw hiking and skiing as advancing the civilizing mission of the national government, in no way different from opening roads or building bridges. When heavy snowfall isolated people "in distant places of the cordillera," CAB skiers assisted the national guard in their rescue missions.[92] The club believed it was doing just fine advancing that mission. In fact, many of its members worked with the DPN in some capacity. Emilio Frey was the intendant of the Nahuel Huapi National Park, a point person to execute Bustillo's decisions. Juan Neumeyer served as a physician in the DPN's new local hospital until 1942 and then headed the forestry office.⁹³ Bustillo's archive holds an astounding number of job requests from people in Buenos Aires, Bariloche, and its environs.⁹⁴ Hence, it is reasonable to think that members of the CAB would not want to antagonize the DPN. However, they did fend off proposals that undermined their own work. As late as 1937, the CAB rejected Bustillo's pitch to hire Norwegian ski instructors, because "experts currently teaching [here] satisfy the sports and tourism interests of this institution."⁹⁵ This did not prevent the DPN from hiring Hans Nöbl, an Austrian skier who designed the facilities and instruction in Sestriere, in the Italian Alps. Bustillo agreed with the CAB's general sentiment about mountain sports, but to him mountain sports were a way for the state to assert national sovereignty. To that end, he offered Nöbl's services to the military camp in Bariloche so that the troops could master a skill that could be useful in case of attack.⁹⁶ It was Nöbl's idea to develop a ski resort on Mount Catedral, a present-day staple of Argentine tourism. He scouted sites to build a cable car, installed a ski school, and stocked rescue equipment, all with DPN funding.⁹⁷ Understandably, Nöbl received the cold shoulder from CAB members, which Bustillo saw as a petty attitude that hindered the general interest.⁹⁸ As a result, two ski schools developed in Bariloche. Instruction offered by the CAB and led by Otto Meiling understood skiing as part of everyday life and was geared toward local residents. In contrast, ski lessons given by Hans Nöbl under the scope of the DPN were pitched as a paid activity to a visiting elite.⁹⁹

Enjoying the outdoors was a central aspect of nation-making in northern Patagonia. Visitors and local elites sought to participate in this endeavor through hiking, skiing, and fishing. Through these experiences, they engaged emotionally with the environment to develop love of country while cementing an understanding of nature separate from and subject to society. This way of seeing and using space was very much at odds with earlier agrarian portrayals of the Andean valleys. These aspirations, however, remained incomplete.

INCOMPLETE RE-SIGNIFICATIONS

The spatial re-signification of northern Patagonia did not take place in an empty space. Authorities in both countries worked hard to curate national touristscapes and erase previous aspirations of agro-industrial productivity. Yet old habits die hard. Farming had mapped out the use of space on both sides of the Andes, and it continued to do so in the 1930s and 1940s. Far from being vestigial remnants of earlier periods, these spatial overlaps signal continuations in the actual uses of space. The DPN could build a nice road from

Figure 6.10. Annotated photograph of an excursion by Emilio Frey, Reynaldo Knapp, and Juan Javier Neumeyer, with sheep grazing in the valley. Club Andino Bariloche, "Libro de excursiones no 1," 1931–32, 5, Biblioteca del Club Andino Bariloche, Argentina. Courtesy of Paula Núñez.

the train station all the way to Llao Llao, but visitors would still observe cattle grazing on the farms on the way.[100] Yet as part of the Ministry of Agriculture, the DPN assisted in some of its programs. For instance, it offered scholarships for agronomy graduates from the University of Buenos Aires to do a specialization in forestry abroad.[101] Ranching and sheep farming remained vital in Patagonia; even Bustillo himself owned livestock in Cumelén.[102] At one point, the DPN negotiated the construction of a local slaughterhouse and adjacent meat-packing house in Bariloche, an "area of influence of Nahuel Huapi National Park with meat and meat products and other perishable goods in the best conditions of hygiene, quality and price."[103] CAB excursions sometimes followed "cow trails," paths created by the frequent trudging of cattle. In their search for good ski slopes, CAB members also encountered grazing sheep (fig. 6.10).[104] In 1937 the Ministry of Agriculture established veterinary offices across the region in Zapala (Neuquén), Viedma (Río Negro), Esquel, Trelew (Chubut), Río Gallegos, and Colonia Las Heras (Santa Cruz). Within two years, one of the veterinarians in Río Negro urged Bustillo to talk to the minister about transferring his office from Viedma to Bariloche, signaling the demand for his services in the environs of Lake Nahuel Huapi as well as the influence of the president of the DPN. Finally, part of the DPN's extensive authority included granting permits for grazing and farming. As far as I can tell, men and women receiving such permits had been residents of the Nahuel Huapi region since before the national park. This included people with Indigenous backgrounds, like Pedro Cayun, and people whose family had settled there, like José Domingo Barbagelata. Hence, not only did the DPN fail at re-signifying the agrarian character of the northern Patagonian Andes, but it also accepted it and tried to regulate it.[105]

In Chile, EFE included agrarian districts as part of tourist circuits to remind visitors of the development in this region. For instance, it branded Temuco, a transportation hub in Araucanía, as an attractive place due to its "agricultural, logging, and cattle-raising" character.[106] Similarly, it marketed the rural area west of Puerto Varas, which witnessed land conflicts, as a "charming" example of "fully cultivated fields . . . excellent subdivision of land, and the excellent progress achieved by agriculture."[107] Very subtly, EFE publications pointed visitors toward farming districts where European migrants had settled and where now their descendants enjoyed land titles, social clubs, and financial wealth. At the end of the day, national authorities in both countries understood that tourism was simply another link in the chain of policies that, for decades now, governments had been deploying in northern Patagonia in order to incorporate it into the nation.

Trans-Andean understandings of space were not completely phased out, even though Chilean and Argentine authorities portrayed the northern Patagonia Andes as national touristscapes. In 1938, the same year that the DPN printed a series of guidebooks on Nahuel Huapi, Adrián Patroni published his own travel account of "the Argentine Chilean lakes." Half of the guide offers an account of Lake Nahuel Huapi and the other half of Lake Llanquihue. Patroni had traveled to Bariloche in 1918, so some portions of his report fall into the before-and-after narrative that the DPN instilled in its own publications. However, his transnational approach also contests the notion that Nahuel Huapi, as a touristscape, was separate from its Chilean counterparts.[108] *En Viaje* also incorporated the proximity to Argentine tourist destinations as a way to attract travelers to southern Chile. This was a frequent strategy when advertising, for example, the new hotel in Puyehue, located on a natural hot spring in the cordillera. The hotel was also located on the international road to Argentina, with easy access from Osorno.[109] The *Guía del Veraneante* offered guidance on how to travel from Chile to Buenos Aires via Bariloche and, in slightly less detail, by train across the Mendoza.[110] The guidebook included information on Nahuel Huapi National Park and the work of the DPN, a rarity in a publication about traveling in Chile.[111] Its maps included details

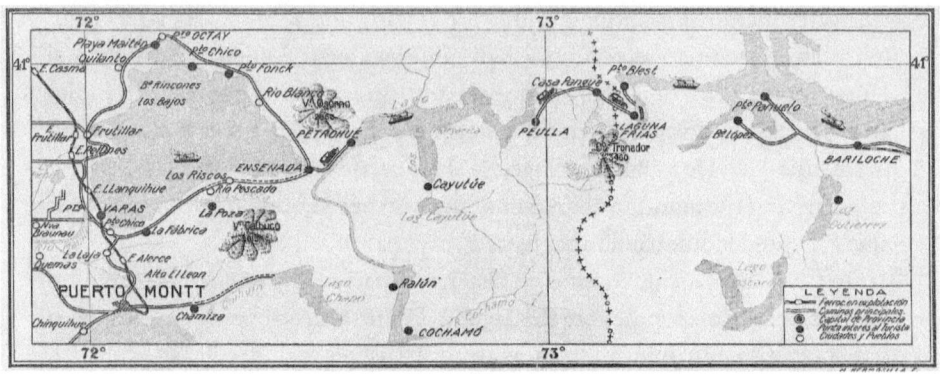

Figure 6.11. Map of Pérez Rosales Pass. *Guía del Veraneante*, 1937.

about crossing the Andes through the Pérez Rosales Pass, with references to modes of transportation, geographical landmarks in both countries, and names of towns (fig. 6.11). In contrast, all other maps did not include information on trans-Andean routes, not even in Filo Hua-Hum (Valdivia–San Martín de los Andes) or Mamuil Malal (Pucón–Junín de los Andes). *En Viaje* often published travel descriptions by contributors, some of whom included Nahuel Huapi as part of their itinerary to Lake Llanquihue. In doing so, these accounts integrated Bariloche and its surroundings as part of a continuous touristscape.[112] This did not mean that travelers saw Chilean and Argentine landscapes in the same way. Yet travelers' experiences did bring the views of the two countries together as part of a trans-Andean understanding of space rather than pull them apart, as much of the official literature did.

Perhaps the presence of a robust transportation company across the Pérez Rosales Pass explains why Nahuel Huapi was appended to Llanquihue travel itineraries more than any other Argentine region to any other Chilean touristscape. In 1914, the Chile-Argentina Trading and Cattle-Breeding Company, which monopolized traffic across this pass, sold its transportation business to a former employee, Ricardo Roth. He collaborated with Primo Capraro and Emilio Frey to expand travel opportunities between the two countries.[113] World War I decelerated these collaborations, yet they survived. Bariloche hoteliers advertised in Valdivian magazine *Turismo Austral*, suggesting both that they were aware of such publications and that they had the means and interest to market their business there.[114] Roth's services existed because trans-Andean trade had thrived; and trans-Andean trade had thrived because of an integrated understanding of space that dated back to Mapuche trading networks.

The mountaineering clubs in Chile and Argentine provided social spaces to reinforce the local character of the Patagonian Andes. In 1934, months before the foundation of the CAO, *Turismo Austral* praised the andinists that had hiked up Osorno volcano in search of ski slopes. The mountaineering club grew out of this group, and the spot they climbed on that day became the club's main location for winter sports for a decade.[115] During his visit to Mount Tronador in 1938, Adrián Patroni asked Emilio Frey, one of the founders of the CAB, whether "anyone practiced alpinism in the area." "Absolutely," responded Frey, "the only difference is that [we] call it andinism."[116] Years later, the CAB published a brief history and annotated bibliography on mountaineering in Argentina. In it, Fréderic Finó argued that the use of "andism" as opposed to "alpinism" was fitting because it clearly referred to mountain sports practiced in the Andes.[117] The difference did not lie in the technique but in the location. Geographical space provided an identity and sense of belonging.

CONCLUSION

The male tourist gaze offered the nationalist governments of Chile and Argentina a lens through which to recycle old tropes about the Patagonian Andes and to innovate in their application. It provided a means to generate "new" ways of seeing space, serving the same nationalizing interests. Undoubtedly, the work of the DPN and EFE, discussed in this and the previous chapter, left a significant mark on the built and symbolic landscapes. Particularly in Chile, travel publications used the peak-water-forest triad to represent the south of the country. Like the Bariloche aesthetic across the Andes, that composite served as a visual shorthand for tourists to recognize the nation in places they had never visited before.

EFE mobilized feminized tropes of nature to attract urban upper and middle classes to the South. Travel publications juxtaposed images of virgin landscapes and women's bodies to portray the southern provinces of, mainly, Valdivia and Llanquihue as desirable destinations for modern travelers. They also frequently equated the experience of traveling to the South to the promenades their readers took in Santiago or its neighboring cities like Viña del Mar, skillfully concealing that the train ride could be long and tiresome. In doing so, publications reaffirmed the notion that traveling for pleasure to escape busy city life constituted a modern endeavor. Tourists and local elites embraced the enjoyment of the outdoors as part of the modern experience of travel. In part, excursions marked class status, reserved for those who could afford them. In part, they also upheld a notion of nature separate from society, a harmonious space distant from the social conflict brought about by local feuds.

The re-signification of southern Chile, examined in this chapter, and Argentine northern Patagonia, analyzed mostly in the previous chapter, from agrarian to attractive remained incomplete. Despite national agencies' insistence on portraying the Andean valleys as pristine, farming, ranching, and sheepherding continued to be relevant activities for the majority of the population in the environs of Lakes Llanquihue and Nahuel Huapi. The fertility of the land was hard to replace with the virginity of nature. The tourist gaze did not uproot earlier understandings of space, at least not in the 1930s and 1940s. The incompleteness of these re-significations adds to the successive attempts by national authorities and local elites to landscape Patagonia.

Conclusion

THE OPPORTUNITY OF SPATIAL HISTORY

On the evening of November 30, 2018, the leaders of the G20 states attended an exclusive performance of *Argentum* at the world-renowned Colón Opera House in Buenos Aires.[1] Through a combination of dance, music, and projection mapping, the piece celebrated Argentina's five regions: North, Littoral, Cuyo, Center, and Patagonia.[2] Landscapes and cultural cues were projected on walls, ceilings, and onstage screens, immersing the audience in the region they were "visiting." The performance portrayed Argentina as a country of many peoples, many activities, and many places. For the segment on Patagonia, flute sounds accompanied the whistling wind in the background as dancers mimicked the movements of water and waves. The projected images showed penguins, whales, mountain peaks, and lakes and perhaps a very occasional fishing boat or oil pumping jack. The absence of humanity in these images contrasted heavily with the rich Indigenous cultures, urban customs, and economic activities of the other regions that filled the stage. Overwhelmingly, images and choreography portrayed Patagonia as empty of humans. The segment transitioned out of Patagonia with a man hiking over the Perito Moreno glacier, a popular destination for national and international tourists. Accompanied by triumphant melodies, this image evoked the victory

Figure 7.1. President Theodore Roosevelt poses in front of Mount Tronador in 1913 in Casa Pangue, Chile. Tulane University Digital Library, New Orleans, Louisiana.

of humans over nature.³ *Argentum* portrayed Patagonia as a beautiful, empty landscape. The trope of the Desert persists in new ways.⁴

The spatial discourses examined in this book, partly evident in *Argentum*, continue to set the beat for the way locals and outsiders experience the northern Patagonian Andes. The built environment and the enjoyment of the outdoors probably represent what most people, locals and visitors, would recognize today. Yet these experiences carry meanings often rooted in the chronology examined in this book. For instance, a popular excursion in Bariloche today is a catamaran ride on Lake Nahuel Huapi to Puerto Blest, the same route that sailboats full of cargo had once followed to Chile. When entering the lake's Blest Branch, the catamaran passes a small island, Sentinel Island, Francisco Moreno's final burial site. As a sign of respect, the captain blows the horn three times, which alerts tourists to this somber moment. Perhaps it also echoes the solemnity that enveloped the transportation of Moreno's remains from the Recoleta Cemetery to Nahuel Huapi in 1944. The armed forces had ousted president Ramón Castillo only six months earlier, on June 4, 1943. Military leaders, including vice president Edelmiro Farrell, gathered at the feet of Julio Roca's statue in Bariloche's Civic Center and attended mass, reaffirming Moreno's work in the interests of the nation, as outlined by the military.⁵ It must have been quite an image. To this day, people in Bariloche talk about Moreno as "the expert" (*el perito*), often forgetting that "perito" is

Figure 7.2. Unaware of Roosevelt's photograph, the author had her picture taken in the same spot a little over a century later, on a slightly cloudier day, 2017.

not part of his name (I myself had this confusion when learning about him as a child). People might certainly disagree on whether to remove Roca's statue or whether they undoubtedly disavow military regimes. Yet the spatial discourses around Moreno, both about what he did and how he is remembered, make it harder for us to see some of his work through a critical lens. He was the chief expert in the border negotiations with Chile, an accomplishment that every Argentine witnesses when they travel to any part of the Patagonian border. He is also remembered as the father of national parks, also an achievement people "see" when they visit any protected area. His name runs through streets, schools, stores, a glacier, and a national park. When travelers sail past Sentinel Island, a guide reminds them that everything they see around them was built on the work of Francisco Moreno. By juxtaposing Moreno's ideas and actions with those of his contemporaries, this book has cast him in a different light, offering texture to the flattened image visitors consume on their excursion through Lake Nahuel Huapi.

The spatial history of the border region that is the northern Patagonian Andes illuminates the proximate experiences of Chileans and Argentines. Argentines and Chileans need each other to get to the corners of their territories. Drivers going south along the Southern Highway in Chile can only make it to Villa O'Higgins, at a latitude between the Northern and Southern Ice Fields. The road ends there, and drivers are forced to cross into Argentina if they want to continue traveling south to Puerto Natales or Punta Arenas. Similarly, Argentine tourists driving to Ushuaia (Tierra del Fuego) need a Chilean ferryboat to cross a two-mile stretch of the Strait of Magellan (if they crossed it in Argentina, it would be a sixty-mile-long ride). Geography and national boundaries have shaped how Chileans and Argentines move around Patagonia. This book has shown that they have shaped more than modern-day travel. Doreen Massey has argued that space is not a "static slice through time" or just "representation" but rather a multiplicity of subjective trajectories.[6] In the preceding pages, explorers, authorities, farmers of different backgrounds, local police officers, judges, land inspectors, tourists, journalists, and hikers evidenced the multiplicity of trajectories in imagining and constructing the nation. The history of nation-making in the northern Patagonian Andes is not a "static slice" in the history of state formation. The six prisms examined in this book brought together the way different people, with varying degrees of power, created spatial discourses as a means to understand the space around them and construe it as part of the nation. For some, the highest peaks of the Andes represented an irrefutable borderline between two nation-states. For others, the Andean passes articulated self-evident resources for herding cattle to greener pastures in the east and using available land in the west for agriculture. For yet another group, both notions exemplified similar ways of constructing the nation. These trajectories show that the history of Patagonia is intrinsically transnational.

The persistence of old tropes in the northern Patagonian Andes points to the incompleteness of nation-making. Over the fifty years examined in the preceding pages, national authorities have imagined Patagonia as a Desert and acted accordingly. By assuming it was a site with no history in relation to the emerging nation-states, Chilean and Argentine political elites envisioned nation-making as a linear, positivist endeavor, one that scholars in the English-speaking world might recognize as a settler colonial project. Authorities sent armies and explorers, vanquished Indigenous peoples, collected data, and sanctioned legislation. They distributed land and brought foreign immigrants. None of these policies was completed. The military campaigns did not "conquer" all of Patagonia; not all Indigenous people disappeared (which does not take away from the fact that the military campaigns were genocidal) and their presence remains clear today; surveyors did not observe every square inch of the Andes; and the cluttered land distribution policies in both countries reinforced the trope of emptiness and backwardness in Patagonia. The fecund garden was slow to erase the Desert. In the 1930s, proximate experiences in both countries took similar turns when both governments recast the northern Patagonian Andes as a site to enjoy nature. This notion did not displace aspirations of productivity, as tourism served the economic model in both countries centered around exports. The re-signification of the northern Patagonian Andes into a postcard-ready landscape sought to elicit patriotic sentiments among visitors, where "seeing" the nation meant knowing it. All these instances of nation-making came like waves and crushed in northern Patagonia, leaving spatial resignifications incomplete. Perhaps at the core of nation-making lies a fundamental incompleteness, which, to me, opens up the possibility of understanding the past and envisioning a more inclusive future in terms of multiple trajectories.

Notes

Abbreviations

ADMP Archivo Documental del Museo de la Patagonia, Bariloche, Argentina
AGNE Departamento de Documentos Escritos, Archivo General de la Nación, Buenos Aires
AGNI Archivo Intermedio, Archivo General de la Nación, Buenos Aires
AHMRE Archivo Histórico del Ministerio de Relaciones Exteriores de la República de Chile, Santiago
AHN Archivo Histórico Nacional, Santiago

Introduction

1. Gutiérrez and Cock, *Americae sive qvartae orbis*; L'Isle and Guérard, *L'Amerique meridionale*. See also Drake et al., *World Encompassed by Sir Francis Drake*, 58, 72; and Livon-Grosman, *Geografías imaginarias*. Multiple depictions of Patagonia were also present in early modern cartography. While some maps defined Patagonia as everything south of the Captaincy of Chile and the Viceroyalty of Río de la Plata, others marked it as belonging to one jurisdiction or the other. See, e.g., Hondius, Jansson, and Mercator, *Americae pars meridionalis*; Jansson, *Americae pars meridionalis*; and Moll, *Map of Chili*.

2. Most scholars agree that the term "Patagonia," coined by Magellan, evoked the meaning of "big feet." However, a handful of linguists have hypothesized other origins. See Malkiel, "Para la toponimia argentina"; and Doura, "Acerca del topónimo 'Patagonia.'"

3. Torrano, "Ontologías de la monstruosidad," 2. For more on how monsters evoke a challenge to natural law and morality see Foucault, *Abnormal*.

4. For a critical examination of Latin American "Deserts" in colonial and national periods, see Trejo Barajas, *Los desiertos*. I've capitalized "Deserts" to signal the figure, not actual arid territories.

5. For references of Deserts as areas beyond state control, see Alejandro Malaspina's

expedition, which sailed from Spain around Cape Horn and to Alaska, making several stops not to make discoveries, as he put it, but to gain knowledge. See David Weber, *Barbaros*, 20–30; Glyn Williams, *Naturalists at Sea*, chap. 8; and Herda, "Ethnology in the Enlightenment."

6. Gattás Vargas, Núñez, and Lema, "La monstruosa cartografía patagónica," 124; Augé, *Non-places*, 77–78.

7. Hunt, *Genius Loci*, 186–87.

8. For an excellent trajectory of the legend, see Urbina Carrasco, *La frontera de arriba en Chile colonial*, 153–90. For more analyses of mythical cities in Latin America, see Ainsa, *Historia, utopía y ficción*; and Ainsa, "Myth, Marvel, and Adventure."

9. Hajduk et al., "De Chiloé," 246–47.

10. Rohde, "El Paso de Bariloche," 161–63.

11. For other mythical placemaking, see Jackson, "Subjection and Resistance"; Chapman, "Throwing the Explorer"; and Safier, "Fugitive El Dorado."

12. Bilbao, *La América en peligro*, 2; Zeballos, *Descripción amena*, 249. For another example, see Fonck, *Libro de los diarios*, 74–75.

13. Sarmiento, *Life in the Argentine Republic*, 15.

14. Gersdorf, *Poetics and Politics of the Desert*, 13–22.

15. Navarro Floria, "El desierto y la cuestión del territorio," 140–41. Other examples include Radding, *Landscapes of Power and Identity*, chap. 8; and Langer, *Expecting Pears*, introduction.

16. These included the robust Tehuelche, or Aónikenk (the "giants" spotted by Magellan's crew), who chased herds of guanacos in the continent; the Selk'am, who made large bonfires for warmth in Tierra del Fuego (which earned the island its name); and the canoeing Kawésqar, who inhabited the Chilean channels and lived almost exclusively on the water.

17. Bengoa, *Historia del pueblo mapuche*, 54–58.

18. Nicoletti and Navarro Floria, *Confluencias*, 46–50; Bengoa, *Historia del pueblo mapuche*, 52–54.

19. Lazzari and Lenton, "Araucanization and Nation," 45. See also Ortelli, "La 'araucanización' de las Pampas."

20. Olascoaga, "Memoria del gobernador de Neuquén," 571.

21. Navarro Floria, "El desierto y la cuestión del territorio," 140–41.

22. Lesser, *Immigration, Ethnicity, and National Identity*, 1.

23. Bade, "From Emigration to Immigration," 511–12. Bade also argues that the German government was anxious about losing German subjects. Hence, encouraging Germans to migrate to Central and South America, where they saw culture as "inferior," would have prompted migrants to maintain their language and traditions. Other examples include Seyferth, "German Immigration"; and Nobbs-Thiessen, *Landscape of Migration*, chap. 2.

24. Bengoa, *Historia del pueblo mapuche*, 321–34. Though military operations ended in the 1880s, state violence against the Mapuche did not. See Crow, *Mapuche in Modern Chile*.

25. David Weber, *Barbaros*, 273.

26. Delrio et al., "Discussing Indigenous Genocide," 141–46.

27. Eduardo Moreno, *Reminiscencias*, 19; Larson, *Conquest of the Desert*.

28. Harambour, *Soberanías fronterizas*, 97.

29. Olascoaga, *Estudio topográfico*, 49. See also Steffen, "On Recent Explorations in the Patagonian Andes," 57.

30. Jacob Bendle, "Patagonian Ice Sheet at the LGM," *Antarctic Glaciers* (blog), June 22, 2020, www.antarcticglaciers.org/glacial-geology/patagonian-ice-sheet/introduction-patagonian-ice-sheet.

31. In 1928, President Carlos Ibáñez del Campo introduced a number of administrative changes that gave the department of Osorno (the northern part of the province of Llanquihue) to the province of Valdivia. The two other departments north of Chaitén were merged with Chiloé until 1937, when Llanquihue was restored. This change also gave the area south of Chaitén to a new territory (then province), Aysén. See Ministerio del Interior [Chile], Decreto 2335; and Ministerio del Interior [Chile], "Divide la actual provincia de Chiloé."

32. Scholars have developed spatial history in two distinctive but overlapping ways: as an analytical framework, as I use it here, and as a method, which typically involves using digital tools to answer spatial questions. For examples of the latter, see Gregory, *Troubled Geographies*; and Yannakakis, "Digital Resources."

33. Carter, *Road to Botany Bay*, chap. 1.

34. Massey, *For Space*, chap. 5.

35. The studies in new spatial history that emerged from Russianists have been particularly influential here. See Baron, "New Spatial Histories"; and Bassin, Ely, and Stockdale, *Space, Place, and Power*.

36. The two most influential works in my research are Craib, *Cartographic Mexico*; and Appelbaum, *Mapping the Country of Regions*.

37. McCook, *States of Nature*; Wakild, *Revolutionary Parks*; Hecht, *Scramble for the Amazon*; Erbig, *Where Caciques and Mapmakers Met*.

38. See esp. William Zartman's argument that "the nature and conditions of the borderland is affected by the nature of the border itself." Zartman, "Identity, Movement, and Response," 5.

39. Schulten, *Geographical Imagination*, 3. Scholars have extensively examined the trope of differentiation and encounter through multiple angles. E.g., John Borneman and A. P. Cheater concluded that the tension between state control and its evasion resulted in a distinct border identity, particularly sensitive to gendered identities. Gabriel Popescu examined the tension between the "natural condition of earth [as] borderless" and the everyday, bounded world. See Borneman, "Grenzregime"; Cheater, "Transcending the State?"; and Popescu, *Bordering and Ordering*.

40. The concept of region has been central to the study of geography. As a result, geographers have examined what makes regions uniquely different from one another. For a more comprehensive discussion of the region as a unit of analysis in geography, see Hartshorne, *Perspective on the Nature of Geography*; Cresswell, *Place*, 32–33; and Turner, "Contested Identities."

41. Examples include Blanc, *Before the Flood*; and Nobbs-Thiessen, *Landscape of Migration*. See also Harambour, *Historia crítica*.

42. Greider and Garkovich, "Landscapes," 1–2.

43. Examples include Rogers, *Deepest Wounds*; Wakild, *Revolutionary Parks*; and De la Torre, *People of the River*.

44. Schulten, *Geographical Imagination*, 3.

45. Arneil, *Domestic Colonies*, 13. Laura Ogden's ethnographic work on Tierra del Fuego helps illustrate the entanglement of colonialism and environmental change. Ogden, *Loss and Wonder*.

46. Suriano, "El anarquismo"; Collier and Sater, *History of Chile*, 194–97; Craib, *Cry of the Renegade*; Edwards, *Carceral Ecology*, 86–89.

47. According to the 1920 census, 9,590 people lived in Puerto Montt, 6,038 people lived in Osorno, and 2,856 did so in Puerto Varas. Dirección General de Estadística, *Censo jeneral*, 96–98; Rey, "La economía del Nahuel Huapi," 37.

48. Romero, *History of Argentina*, 28.

49. Ortega Martínez, "La crisis de 1914-1924."

50. Korol, "La economía"; Collier and Sater, *History of Chile*; Belini, *Historia económica de la Argentina*.

51. Macor, "Partidos, coaliciones y sistemas de poder," 70–74.

52. Quiroga, "Notas sobre la historia de la democracia," 23.

53. Healey, *Ruins of the New Argentina*; Romero, *History of Argentina*, chap. 3.

54. Collier and Sater, *History of Chile*, 216–17.

55. Collier and Sater, *History of Chile*, chap. 9.

56. See McCook, *States of Nature*, chap. 2; Craib, *Cartographic Mexico*; Hecht, *Scramble for the Amazon*; and Appelbaum, *Mapping the Country of Regions*.

57. Craib, *Cartographic Mexico*, 8.

58. Urry, *Tourist Gaze*.

59. Paula Gabriela Núñez, "'She-Land.'"

60. Confino, *Nation as a Local Metaphor*.

Chapter One

1. Moreno to Dr. Amancio Alcorta (minister of foreign affairs), December 4, 1897, Fondo Fransico P. Moreno, box 3098, AGNE.

2. Vicuña Mackenna, *La Patagonia*, xxi.

3. The most comprehensive study of Chilean-Argentine border negotiations is Lacoste, *La imagen del otro*.

4. See esp. Raj, *Relocating Modern Science*, 8–9; and Carreras, "¿Un mismo origen con diferente destino?," 128. For a global discussion of the transnational origin of scientific inquiry as it relates to nation-making, see Sivasundaram, "Sciences and the Global." For a Latin American perspective, see Cañizares-Esguerra, *Nature, Empire, and Nation*, chap. 6.

5. Lafuente and López-Ocón, "Bosquejos de la ciencia nacional," 6; McCook, "Global Currents."

6. For two examples in the Caribbean and Mexico, see McCook, *States of Nature*; and Bueno, *Pursuit of Ruins*.

7. This is not to say that others did not use induction to reach generalizable conclusions. Particularly, scholars like Francisco José Calas influenced Humboldt's appreciation of the Andes as a space to study ecological and botanical diversity in a mountainous terrain. See Cañizares-Esguerra, *Nature, Empire, and Nation*, chap. 6.

8. Bello, "Exploración, conociemiento geográfico y nación," 68.

9. E.g., Rodolfo Hauthal, head of the geological and mineralogical section of the museum, led the Fifth Subcommission, and Gunardo Lange, director of the topographic section of the museum, participated in the Eighth Subcommission.

10. Bengoa, *Mapuche, colonos y el estado nacional*, chap. 3. State violence against the Mapuche persisted beyond military operations. See Crow, *Mapuche in Modern Chile*.

11. For a recent examination of the long-standing impact of this military campaign on Indigenous people's lives through the twenty-first century, see Larson, *Conquest of the Desert*.

12. Latour, "Drawing Things Together," 27–29.

13. The first Argentine representative was Octavio Pico, who passed away in 1892 while conducting negotiations in Chile. Juárez Celman, "Decreto nombrando."

14. Examples of studies of scientific networks in Latin America include Julia Rodríguez, "Beyond Prejudice and Pride"; and Sevilla and Sevilla, "Inserción y participación."

15. Chasteen and Castro-Klarén, *Beyond Imagined Communities*, xxviii.

16. For more on the initial "fantastical" Chile and Argentina, see Lacoste, *La imagen del otro*, chap. 9. For two voices in this debate, see de Angelis, *Memoria historica*; and Amunátegui, *Títulos de la República de Chile*.

17. "Tratado de 1881."

18. Montes de Oca, *Límites argentino-chilenos*, 4.

19. José Evaristo Uriburu, "Decreto nombrando perito"; Fonck, *Exámen crítico*, 11–12; Irarrázaval Larraín, *La Patagonia*, 127. Moreno was the fourth appointee for such a role, after Octavio Pico, Valentín Virasoro, and Norberto Quirno Costa. Barros Arana was succeeded by Arístides Martínez and Alejandro Bertrand. See Lagos Carmona, *Historia de las fronteras*, 88.

20. For the talks between Chileans and Argentines in the early 1890s, including the instructions they gave to crews, see Zeballos, *Demarcación de límites*; and Barros Arana, *La cuestion de límites*. Probably the best compendium of documents pertaining to the negotiations is Ministerio de Relaciones Exteriores y Culto and Sánchez, *La frontera argentino-chilena*.

21. Bertrand, *Memoria sobre la rejión central*, 132.

22. Lagos Carmona, *Historia de las fronteras*, 95–96.

23. Lacoste, *La imagen del otro*, 312–24.

24. Great Britain led the largest navy in the world, 1,065,000 tons, or 25.90 kilograms per inhabitant. For full details, see Lacoste, *La imagen del otro*, 324.

25. Numbers and reasons vary among sources. See "American Sailors Alleged in Chile: Four or the Baltimore's Men Slain in a Fight with Junta Blue Jackets," *Chicago Daily Tribune*, October 17, 1891; "Fiery Chileans: Additional Details of the Affray at Valparaiso. An Armed Force to Be Landed Today to Bury the Sailor Who Was Killed—Fears of More Trouble," *Los Angeles Times*, October 18, 1891; "Chile Weakens: The Right of Asylum Is Recognized Investigating the Street Riots Burial of the Victims of the Valparaiso Mob—Public Sentiment EXPECT TO MEET HIM Two Chileans Who Assert That . . . Is Not . . . ," *San Francisco Chronicle*, October 20, 1891.

26. Lacoste, *La imagen del otro*, 298–99.

27. The final report on the incident can be found in United States House of Representatives, *Message of the President of the United States*.

28. Holdich, *Countries of the King's Award*, 30.

29. Barros Arana, *Esposicion de los derechos*, 2.

30. For examples in the Americas, see López-Ocón, "La Sociedad Geográfica de Lima"; Bernal and Fernanda, "La Sociedad Mexicana de Geografía y Estadística"; Craib, *Cartographic Mexico*; and Appelbaum, *Mapping the Country of Regions*.

31. Schell, *Sociable Sciences*, 43.

32. Schell, *Sociable Sciences*, 43.

33. Museo Nacional de Historia Natural, *Guía del museo nacional de Chile*, 4.

34. Museo Nacional de Historia Natural, *Guía del museo nacional de Chile*, 4; Castro et al., "Rodulfo Amando Philippi"; "Rodulfo Amando Philippi Krumwilda (1808-1904)," Memoria Chilena, Biblioteca Nacional de Chile (website), accessed April 11, 2018, www.memoriachilena.cl/602/w3-article-795.html.

35. Bello, "Discurso pronunciado en la instalación de la Universidad de Chile."

36. Scholars have long debated whether a national science was truly independent, and not simply mimetic, from that of European academic centers. See Raj, *Relocating Modern Science*; and Sanhueza Cerda, *La movilidad del saber científico*.

37. Barros Arana, *La cuestion de límites*, 5.

38. Barros Arana, *La cuestion de límites*, 6–7; Oficina de Mensura de Tierras, *Primera memoria*, 28–29. However, Pissis focused less on the regions south of Araucanía, often relying only on astronomical measurements, second-grade triangulations, and observations by others, including Claude Gay. For some reminders of this from the time, see Fonck, *Introducción a la orografía y a la jeolojía*, vii; and Steffen, "On Recent Explorations in the Patagonian Andes," 66. See also González Leiva and Andrade Johnson, "Geografía física de la República de Chile," xxxii.

39. "Rectoras y rectores de la U. de Chile," Universidad de Chile (website), accessed April 8, 2018, www.uchile.cl/portal/presentacion/historia/rectores-de-la-u-de-chile.

40. See, e.g., Philippi, "Espedición al volcán de Osorno"; Fonck and Hess, "Informe"; and Döll, "Exploracion del territorio de Osorno."

41. "Oficina Hidrográfica," Memoria Chilena, Biblioteca Nacional de Chile (website), accessed April 11, 2018, www.memoriachilena.cl/602/w3-article-96473.html.

42. See, e.g., García, "Diario del viaje y navegación"; "Viajes del padre Francisco Menéndez"; and "Viaje de Enrique Brouwer."

43. See Saldivia Maldonado and De la Jara Nova, "La Sociedad Nacional de Agricultura"; and Correa, "¿Quiénes son los profesionales?"

44. Olascoaga, *La conquista del desierto*.

45. For more than three decades, scholars have revisited the ill-termed "Conquest of the Desert," disarticulating its meaning into the uneven, genocidal war against Indigenous communities. See Navarro Floria, "El desierto y la cuestión del territorio"; Briones and Delrio, "La 'Conquista del Desierto'"; Pilar Pérez, "Historia y silencio"; and Larson, *Conquest of the Desert*.

46. Olascoaga, *Plano del territorio de La Pampa y Río Negro*.

47. Rohde, *Descripción de las gobernaciones nacionales*; Francisco P. Moreno, *Apuntes preliminares*.

48. Babini, *Historia de la ciencia en la Argentina*, 140–41.

49. Sociedad Científica Argentina, "Estatutos fundamentales," 7.

50. Sociedad Científica Argentina, *Anales*, 3:283–84; Arata, "Contribuciones al conocimiento higiénico"; Iturbe and de Candioti, "Fábrica nacional de sombreros."

51. Sociedad Científica Argentina, "Estatutos fundamentales," 7.

52. Sociedad Científica Argentina, *Anales*, 2:20, 3:61. See also Babini, *Historia de la ciencia en la Argentina*, 142. Lista's 1877 expedition did not come to fruition because of a military mutiny in Punta Arenas. See Lista, *Mis esploraciones y descubrimientos*, 12–13.

53. Dodds, "Geography, Identity and the Creation of the Argentine State," 311.

54. Instituto Geográfico Argentino, *Boletín del Instituto Geográfico Argentino* 3 (1882): 159; "Expedición al Neuquén." This rhetoric was also present in the explorations of the Chaco region in northern Argentina, see Seguí, "Expedición del Bermejo"; and Seguí, "Expedición al Chaco."

55. Edney, *Mapping an Empire*, 302.

56. Zeballos, *La conquista de quince mil leguas*, 327.

57. Zusman and Minvielle, "Sociedades geográficas y delimitación del territorio," 6.

58. E.g., Fontana, "Expedición al Río Pilcomayo"; Villegas, "Diario general"; and Furque, "Descripción del Pueblo General Roca." For Chaco as a desert, see Lois, "La invención del desierto chaqueño."

59. See, e.g., Gabriel Carrasco, "La provincia de Santa Fe y el Chaco"; and Ganeval, "La colonización moderna," cited in Fernández, "Crónica geográfica," 213–17.

60. Instituto Geográfico Argentino, "Procedimientos del Instituto Geográfico Argentino: El mapa de la república," *Boletín del Instituto Geográfico Argentino* 3 (1882): 63–64.

61. Instituto Geográfico Argentino, *Atlas de la República Argentina*, 3.

62. Serrano Montaner, "Discusión sobre los Andes australes," 196. For a summary of Moreno's early expeditions, see Frederico Freitas, "The Journeys of Francisco Moreno," *Frederico Freitas* (blog), August 18, 2009, https://fredericofreitas.org/2009/08/18/the-journeys-of-francisco-moreno.

63. Barros Arana, *Esposicion de los derechos de Chile*, 42n36.

64. Andermann, "Reshaping the Creole Past," 147.

65. Roca, "Proyecto de creación de un museo nacional," 484.

66. Escobedo, "El autóctono sud-americano," 149.

67. Pesoa and Sabate, "La Plata y la construcción de un país."

68. Francisco P. Moreno, "El Museo de La Plata," 39.

69. Podgorny, "De razón a facultad," 91–92.

70. Francisco P. Moreno, "El Museo de La Plata," 42.

71. Ten Kate, "Matériaux pour servir à l'anthropologie des Indiens de la République Argentine," 36. Here, ten Kate, a Dutch anthropologist who curated part of the exhibit in the La Plata Museum, reproduces his conversations with Emilio Beaufils, a botanist and colleague of his who had participated in expeditions in Patagonia.

72. Oldani, Añon Suarez, and Pepe, "Las muertes invisibilizadas del Museo de La Plata." Beginning with Inacayal's remains in 1994, the La Plata Museum has made sixteen restitutions of human remains to Native communities. Thirteen were in the last ten years (2014–23). See "Restituciones realizadas por el Museo de La Plata," Museo de La Plata (website), accessed October 8, 2023, www.museo.fcnym.unlp.edu.ar/home/restituciones-realizadas-por-el-museo-de-la-plata-372.

73. Newkirk, *Spectacle*, xiii.

74. Bueno, *Pursuit of Ruins*.

75. Azar, Nacach, and Navarro Floria, "Antropología, genocidio, y olvido"; Escolar and Saldi, "Castas invisibles de la nueva nación"; "Restos humanos en el Museo de Ciencias Naturales de La Plata," Cayu, May 12, 2006, https://cayu.com.ar/index.php/2006/12/05/restos-humanos-en-el-museo-de-ciencias-naturales-de-la-plata. For further analyses of nineteenth-century studies of Indigenous bodies, see Kerr, *Sex, Skulls, and Citizens*.

76. See, for example, Julia Rodríguez, "Beyond Prejudice and Pride"; and Sevilla and Sevilla, "Inserción y participación."

77. Steffen, *Patagonia occidental*, 1:12.

78. Francisco P. Moreno, "Explorations in Patagonia," 244.

79. República Argentina, *Argentine-Chilian Boundary*, vi–xiii.

80. Hilgartner, *Science on Stage*; Vandendriessche, Peeters, and Wils, *Scientists' Expertise as Performance*.

81. The other two members of the arbitral tribunal were Baron Edward Macnaghten, member of the Privy Council, and Maj. E. H. Hills, head of the topographical section of the Intelligence Division of the British War Office. See United Nations, *Cordillera of the Andes Boundary Case*, 43; and Goodenough and Dalton, *Army Book for the British Empire*, 569.

82. See, e.g., Church to Steffen, February 14, 1901, Nachlass [Korrespondenz Von Personen Und Körperschaften A-Z], Ibero-Amerikanisches Institut - Preußischer Kulturbesitz, Germany. 1892, https://digital.iai.spk-berlin.de/viewer/image/755399633/2/.

83. For the purpose of clarity, I am using the same name that the sources use. Today, the Palena River is known as Carrenleufu in Chile and Corcovado in Argentina. See Roig et al., "Informe de las subcuencas de los ríos Carrenleufú y Pico," 14.

84. A hydroelectric dam on the edge of Lake Situación has created another, larger lake, Amutui Quimey, that has swallowed the other four. Today, Futaleufú flows from this dam.

85. Taylor, "Welsh Way of Colonisation," 1077.

86. Steffen, *Patagonia occidental*, 1:51.

87. For gold mining in Neuquén, see Lavandaio and Catalano, *Historia de la minería argentina*, 238. For gold mining in Tierra del Fuego, see Bandieri, *Historia de la Patagonia*, 187–89.

88. Francisco P. Moreno, *Apuntes preliminares*, 91.

89. Departamento Nacional de Minas y Geología, "Autorizando a Paul Ahehelm a verificar la existencia de minerales auríferos"; Guenem, "Manifestación de Alberto Wecker"; Steffen, *Problemas limítrofes*, 55.

90. Luis Jones, *Hanes y wladva Gymreig Tiriogaeth Chubut*, 175.

91. Steffen, *Problemas limítrofes*, 57. For references on the agricultural potential of the Palena valley, see Steffen, "On Recent Explorations in the Patagonian Andes," 63; and "Record of Geographical Progress," 72. For Argentine reports on gold deposits in Chubut, see Cobos, "Expedición minera al Territorio Nacional del Chubut." For Steffen's references to Argentines' knowledge of these deposits, including Cobos's report, see Steffen, "Memoria jeneral," 165–66; and Steffen, *Problemas limítrofes*, 55–56.

92. Steffen, *Problemas limítrofes*, 55.

93. "Geographischer Monatsbericht," 95. See also Steffen, "On Recent Explorations in the Patagonian Andes," 61.

94. Steffen, "Memoria jeneral," 151–52; Steffen, *Problemas limítrofes*, 58.

95. Steffen, "Memoria jeneral," 151–52; Steffen, *Problemas limítrofes*, 58. See also "Geographischer Monatsbericht," 95.

96. Francisco P. Moreno, *Apuntes preliminares*.

97. Francisco P. Moreno, *Apuntes preliminares*, 92.

98. Barros Arana, *Historia jeneral de Chile*, 30. See also Ministerio de Relaciones Exteriores y Culto and Sánchez, *La frontera argentino-chilena*, 709.

99. Moreno to Alcorta, December 4, 1897, AGNE.

100. Franciso P. Moreno, "Explorations in Patagonia," 244.

101. Ministerio de Relaciones Exteriores y Culto and Sánchez, *La frontera argentino-chilena*, 672–75.

102. Craib, "Cartography and Decolonization," 19.

103. Steffen, *Patagonia occidental*, 2:301.

104. Del Castillo, "Cartography in the Production (and Silencing) of Colombian Independence History," 111.

105. García, "Diario del viaje y navegación." See in particular the map at the end, where the priest gave the label "Calén Nation" to the area Moreno visited in late 1897.

106. Serrano Montaner, *Derrotero del Estrecho de Magallanes*, 309. For Serrano's discussion of García's naming of Calén, see Serrano Montaner, "Reconocimiento del Río Buta-Palena," 148.

107. Francisco P. Moreno, "Dr. Steffen's Exploration," 219.

108. Steffen, *Patagonia occidental*, 2:169.

109. Steffen, *Patagonia occidental*, 2:170.

110. Martin, "Dr. Hans Steffens," 124–25.

111. Francisco P. Moreno, "Dr. Steffen's Exploration," 219. *Toro* was the Chilean steamer commandeered by Ramón Serrano Montaner among the Chilean fjords in 1885.

112. Royal Geographical Society, "Meetings of the Royal Geographical Society," 102.

113. Francisco P. Moreno, "Explorations in Patagonia."

114. Steffen, *Patagonia occidental*, 2:164.

115. Francisco P. Moreno, "Dr. Steffen's Exploration"; Moreno to Alcorta, December 4, 1897, AGNE.

116. República Argentina, *Argentine-Chilian Boundary*, 351.

117. For the name change of Lake Buenos Aires, see Ministerio del Interior [Chile], "Crea y fija los límites de los departamentos."

118. Barros Arana, *Esposicion de los derechos*, 89–90.

119. Steffen, *Problemas limítrofes*, 98–99.

120. Steffen, *Patagonia Occidental*, 1:353.

121. Steffen, *Problemas limítrofes*, 99.

122. Steffen, *Viajes de exploración y estudio*, 396. Holdich did not provide any account of such gathering.

123. Steffen, *Patagonia occidental*, 1:351.

124. Holdich, *Countries of the King's Award*, 354.

125. HLC, "Col. Sir Thomas Holdich, K.C.M.G., K.C.I.E."

126. Holdich, *Countries of the King's Award*, 30.

127. For examples of studies that move away from the center/periphery paradigm to write a history of science in Latin America, see Blanco and Page, *Geopolitics, Culture, and the Scientific Imaginary*.

Chapter Two

1. Viotti da Costa, *Brazilian Empire*, 78–79; Tutino, "From Involution to Revolution," 811.

2. Delrio and Pérez, "Territorializaciones y prácticas estatales."

3. Diacon, *Millenarian Vision, Capitalist Reality*, chap. 4; Gordillo, *Landscapes of Devils*, chap. 3; Reeves, *Ladinos with Ladinos*, chap. 3.

4. Castro and Lavinas Picq, "Stateness as Landgrab." For another example in Patagonia, see Soluri, *Creatures of Fashion*, chap. 2.

5. I follow Michael Goebel's nuanced use of "settler colonialism" for postcolonial Chile and Argentina; see Goebel, "Settler Colonialism in Postcolonial Latin America." Nancy Shoemaker has analyzed the tension between settler colonialism as a grand theory and as English heritage; see Shoemaker, "Settler Colonialism." For other discussions on settler colonialism in Chile and Argentina, see Castellanos, "Introduction"; Ugarte, Fontana, and Caulkins, "Urbanisation and Indigenous Dispossession"; and Greenwald, "Now I Walk on Foreign Soil."

6. Congreso Nacional de la República de Chile, "Provincia de Arauco."

7. For more on the territorial shifts in Chile, see González Leiva and Bernedo Pinto, "Cartografía de la transformación de un territorio"; and Estefane, "Estado y ordenamiento territorial."

8. For more on the Chilean incorporation of Araucanía, see González Leiva and Bernedo Pinto, "Cartografía de la transformación de un territorio"; and Crow, *Mapuche in Modern Chile*, chap. 1.

9. For a discussion on these estimates in Araucanía, see Bengoa, *Historia del pueblo mapuche*, 251–53. For a similar discussion about the estimates in Valdivia (which included Osorno and La Unión), see Vergara del Solar, *La herencia colonial del Leviatán*, 140–42.

10. Almonacid Zapata, "El desarrollo de la propiedad rural," 28–29.

11. This denoted a change in policy from protection to subjugation, examined by numerous historians. For Arauco, see Pinto Rodríguez, "Al final de un camino." For Llanquihue and Valdivia, see Vergara del Solar, *La herencia colonial del Leviatán*. For Magallanes, see Harambour, *Soberanías fronterizas*. For a comparative approach to legislation in southern and northern Chile, see Gundermann Kröll, "Los pueblos originarios del norte de Chile."

12. "Mapuche" is a broad term to describe a shared social and economic structure. Mapuche groups are differentiated from one another through an array of markers, especially the spaces they lived in. For an excellent introductory summary, see Bengoa, *Historia del pueblo mapuche*, chap. 2; and Herr, *Contested Nation*, 4–7.

13. Pinto Rodríguez, *De la inclusión a la exclusión*, 131–49.

14. A good in-depth account of the occupation of Araucanía can be found in the central chapters of Bengoa, *Historia del pueblo mapuche*.

15. For more on the immigrant occupation of Valdivia and Llanquihue through landed property in the first half of the nineteenth century, see Almonacid Zapata, "El desarrollo de la propiedad rural"; Almonacid Zapata, "El mercado de tierras"; Lespai Silva, "Consolidación del capitalismo agrario"; Almonacid Zapata, "El problema de la propiedad de la tierra"; and Muñoz Sougarret, "Apropiación pública y privada."

16. Pinto Rodríguez, *De la inclusión a la exclusión*, 152–60.

17. Campos, "Territorial Conflicts, Bureaucracy, and State Formation," 70–71.

18. E.g., Luis Sáenz Peña, "Decreto del 22 de mayo de 1894," Fondo Ministerio del Interior, Oficina de Tierras y Colonias (Actos Dispositivos), box 1, AGNI; letter to the director of the Oficina de Tierras y Colonias (Office of Lands and Colonies), December 12, 1894, folio 31r–v, Fondo Ministerio del Interior, Oficina de Tierras y Colonias, copybook 2, 1894–96, AGNI; and Bernal, "Memoria de la gobernación de Río Negro," 497–98.

19. For varying examples of the occupation of northern Patagonia, see Olascoaga, *La conquista del desierto*; and Zeballos, *Descripción amena*. For dissenting voices, see Lenton, "Relaciones interétnicas."

20. For narratives of the military campaign, see Delrio, *Memorias de expropriación*, chap. 2; and Escolar, Salomon Tarquini, and Vezub, "La 'Campaña del Desierto.'" For ethnohistories on the experience of displacement, removal, and genocide in the 1880s and 1890s, see Delrio, *Memorias de expropriación*; and Ramos, *Los pliegues del linaje*. For a discussion on the shorthand term "Conquest of the Desert," see note 45 in chapter 1 of this book.

21. Quijada, "La ciudadanización del 'indio bárbaro,'" 698–701; Podgorny and Lopes, *El desierto en una vitrina*, 176–77; Bayer, "Proyecto de ley," 24; Herner, "La invisibilización del otro indígena," 123; Escolar, Salomon Tarquini, and Vezub, "La 'Campaña del Desierto,'" 236–37.

22. For the distribution of Indigenous peoples to various locations, see Lenton, "Relaciones interétnicas," 29; Azar, Nacach, and Navarro Floria, "Antropología, genocidio, y olvido"; Lenton, "La 'cuestión de los indios'"; and Escolar and Saldi, "Apropiación y destino de los niños indígenas." For concentration camps, see Papazian and Nagy, "La Isla Martín García como campo de concentración"; and Pilar Pérez, "Futuros y fuentes."

23. Congreso Nacional de la República de Chile, "Colonias de naturales i estranjeros."

24. Varas, *Colonización de Llanquihue, Valdivia i Arauco*, 24–25.

25. Zenteno Barros, *Recopilacion de leyes i decretos supremos*, 2:427.

26. Baeza and Silva, "Imaginarios sociales del Otro," 33.

27. For the project of colonies of single or widowed men, see Vega, "Memoria del ajente jeneral de colonizacion," 117–19. For the results, see Rehren, "Anexo no 5," 104.

28. Almonacid Zapata, "El desarrollo de la propiedad rural," 29–30.

29. For a brief analysis of this legislation, see Fernández, "La ley argentina de inmigración de 1876."

30. Almonacid Zapata, "El desarrollo de la propiedad rural," 27. For an in-depth study of mestizo Chileans in southern Chile, see Órdenes Delgado, "La experiencia de los sin voz."

31. For a summary of this law, see Aylwin, *Estudio sobre tierras indígenas*. An 1873 legal case brought this interpretation to the Supreme Court, which ruled that Valdivia was "under ordinary rules" and not Indigenous territory, because Indigenous people living there were "now civilized, they [did not] belong to untamed tribes." See Donoso and Velasco, *Historia de la constitución de la propiedad austral*, 317.

32. Riso Patrón, "Tierras de colonización," 59.

33. Congreso Nacional de la República de Chile, "Fundación de poblaciones."

34. Bengoa, *Historia del pueblo mapuche*, 159–61.

35. Almonacid Zapata, "El problema de la propiedad de la tierra," 8.

36. For more on the exclusion of the Mapuche, see Pinto Rodríguez, *De la inclusión a la exclusión*.

37. Almonacid Zapata, "El desarrollo de la propiedad rural," 31.

38. Bengoa, *Historia del pueblo mapuche*, 350. For a historical analysis on how authorities demonized the Mapuche, see Pinto Rodríguez, "Bárbaros, demonios y bárbaros de nuevo."

39. For more on political and economic relations between the Mapuche in Araucanía and the Chilean state, see Herr, *Contested Nation*, chap. 5.

40. Almonacid Zapata, "El desarrollo de la propiedad rural," 28.

41. Almonacid Zapata, "El desarrollo de la propiedad rural," 31.

42. Campos, "Territorial Conflicts, Bureaucracy, and State Formation," 129.

43. Almonacid Zapata, "El desarrollo de la propiedad rural," 30.

44. Gana, "Conferencia hecha en la Seccion Topográfica."

45. For rural property development in Valdivia and Llanquihue, including conflict among settlers, see Almonacid Zapata, "El desarrollo de la propiedad rural"; and Almonacid Zapata, "El problema de la propiedad de la tierra."

46. For more on polygamy in Mapuche society for alliances, see Bengoa, *Historia del pueblo mapuche*, 71–72; and Olea Rosenbluth, *La mujer en la sociedad mapuche*.

47. Vicuña Mackenna, *La conquista de Arauco*, 7.

48. Baeza Espiñeira, "Memoria de la inspeccion jeneral," 16.

49. Bórquez, "Memoria de la Prefectura Apostolica," 293–97.

50. Domingo Carrasco, "Memoria de la Prefectura Apostólica," 300.

51. Maldonado Coloma, *Estudios geográficos é hidrográficos*, lxxxii.

52. Baeza Espiñeira, "Memoria de la inspeccion jeneral," 17.

53. González-Caniulef, "Mujeres mapuche en manos de primitivos dueños."

54. González-Caniulef, "Mujeres mapuche en manos de primitivos dueños."

55. Briones Luco, "Radicación de indíjenas," 668.

56. Baeza Espiñeira, *Memoria de la inspección jeneral*, 11.

57. Álvarez Correa, "Cartografía y geodesia."

58. For an analysis on this work in Araucanía, see Campos, "Territorial Conflicts, Bureaucracy, and State Formation," chap. 2.

59. Rehren, "Anexo no 5," 110–11.

60. Alfredo Weber, "Memoria de la inspección," 44–45.

61. Alfredo Weber, "Memoria de la inspección," 45.

62. Rehren, "Anexo no 5," 95.

63. Vega, "Memoria del ajente jeneral de colonizacion," 153–54.

64. Rehren, "Anexo no 5," 99.

65. Raimundo Silva Cruz, "Nota no 43 al Ministro de Instrucción Pública," February 22, 1898, vol. 258, folder 49, Copiador de correspondencia enviada a autoridades y ministerios de Chile sobre asuntos de Colonización, AHMRE.

66. Vega, "Memoria del ajente jeneral," 128–31.

67. Rehren, "Anexo no 5," 94–95; Baeza Espiñeira, *Memoria de la inspección*, 142.

68. Whiteside Toro, "Chiloé i sus colonias," 18.

69. Rehren, "Anexo no 5," 104.

70. Rehren, "Anexo no 5," 95.

71. Alfredo Weber, "Inspeccion de colonizacion de Llanquihue i Chiloé," 222–23.

72. Alfredo Weber, "Inspeccion de colonizacion de Llanquihue i Chiloé," 223.

73. Alfredo Weber, "Inspeccion de colonizacion de Llanquihue i Chiloé," 223.

74. Oficina Central de Estadística, *Sétimo censo jeneral de la poblacion de Chile*, 243; República Argentina, *Segundo censo de la República Argentina*, 2:iv.

75. For some examples, see Baeza Espiñeira, *Memoria de la inspección jeneral*, 113–39. For examples of concerns about hardworking populations beyond what I have cited thus far, see Briones Luco, "Indíjenas"; Briones Luco, "Fueguinos."

76. Muñoz Sougarret, "Apropiación pública y privada"; Almonacid Zapata, "El desarrollo de la propiedad rural," 35.

77. República de Chile, "Contardi Juan B.," 309.
78. Briones Luco, "Ocupación de terrenos fiscales," 517–18.
79. Almonacid Zapata, "El desarrollo de la propiedad rural," 33.
80. Harambour, *Soberanías fronterizas*, loc. 392 of 6579, Kindle.
81. I have maintained original spelling for names. Oficina de Límites and Patrón, *La linea de frontera*, 76.
82. Oficina de Mensura de Tierras, *Primera memoria*, 133.
83. *Documentación de los contratos de colonización*.
84. Almonacid Zapata, "El desarrollo de la propiedad rural," 33.
85. Oficina de Mensura de Tierras, "Concesion Woodhouse," 106.
86. *Documentación de los contratos de colonización*, 29.
87. República de Chile, "Contrato entre la República de Chile y Charles Colson," 1438.
88. Vega, "Memoria del ajente jeneral de colonizacion," 103–4.
89. Fonck, Francisco. "Carta de Francisco Fonck a Adolfo Guerrero," Libro copiador de Francisco Fonck, 1895-1899, Fondo Fonck, Caja 2, Estuche 8, Biblioteca "Emilio Held," Deutsch-Chilenischer Bund, Santiago (Chile), March 20, 1896.
90. Essinger, "La emigración danesa," 90. For descriptions of Lindholm as *asimilado*, see Congreso Nacional de la República de Chile, "Ley 2757."
91. Baeza Espiñeira, *Memoria de la inspección jeneral*, 26. The report praises García for bringing ninety-six families, but other documents listed above confirm it was eighty-six. For more on the *asimilados* in the Chilean military, see Nunn, *Yesterday's Soldiers*, 100–112.
92. Agustín Baeza Espiñeira, "Nota no. 1283 del inspector jeneral de Tierras i Colonización," November 27, 1902, Fondo Histórico, Oficios Dirigidos por la Inspección Jeneral de Tierras i Colonización a Este Ministerio, 1896–1902, AHMRE.
93. For Tuza's national origin, see Ministerio del Interior [Chile], *Rol de cartas de naturalización*, 138. For his store in Arica, see Silva Narro, *Guía administrativa, industrial y comercial*, 78.
94. Ricardo del Río Pinochet, "Nota no. 934 del inspector jeneral de Tierras i Colonización," August 21, 1902, Fondo Histórico, Oficios Dirigidos por la Inspección Jeneral de Tierras i Colonización a Este Ministerio, 1896–1902, AHMRE.
95. Del Río Pinochet, "Nota no. 934," AHMRE.
96. We know of Deffarges's trajectory from piecing together reports about Chiloé Island and decrees from the Ministry of Foreign Affairs. See Baeza Espiñeira, *Memoria de la inspección jeneral*, 98, 102; Alfredo Weber, *Chiloé*, 112; Rehren, "Anexo no 5," 102; and *Indice de decretos i leyes del Ministerio de Relaciones Esteriores Culto i Colonización*, 602, 667, 672.
97. Baeza Espiñeira, *Memoria de la inspección jeneral*, 27.
98. Ministerio de Relaciones Exteriores, Culto y Colonización, "Memoria del ajente jeneral," 175.
99. Ministerio de Relaciones Exteriores, Culto y Colonización, "Memoria del ajente jeneral," 176.
100. Ministerio de Relaciones Exteriores, Culto y Colonización, "Memoria del ajente jeneral," 176–77.
101. Baeza Espiñeira, *Memoria de la inspección jeneral*, 30.
102. Alfredo Weber, *Chiloé*, 182.
103. N. Vega, "Oficio no 259. Respuesta al pedido de John Öhlander para ser agente de

colonización," July 10, 1899, 3, Fondo Histórico, Inspeccion Jenederal de Tierras i Colonizacion—Oficios Dirigidos, 1896–1902, AHMRE.

104. John Öhlander, "Pedido de John Öhlander para ser agente de colonización," April 19, 1899, Fondo Histórico, Inspeccion Jenederal de Tierras i Colonizacion—Oficios Dirigidos, 1896–1902, AHMRE.

105. N. Vega, "Oficio no 259," 2–3, AHMRE.

106. John Öhlander, "Pedido de John Öhlander para ser agente de colonización," April 19, 1899, Fondo Histórico, Inspeccion Jenederal de Tierras i Colonizacion—Oficios Dirigidos, 1896–1902, AHMRE.

107. Montt, "Introduccion e instalacion de cinco mil familias de colonos," 1433.

108. Congreso Nacional de la República de Chile, *Comisión Parlamentaria de Colonización*, 320.

109. They nicknamed him Pichi Juan, which means "Little John," partly as a reference to his youth but also probably infantilizing his Indigenous background. Indeed, Juan Currieco was immortalized, like many other men, by having a geographical feature named after him. For other honorees, the government used their last name, like in naming Glacier Steffen, Mount Fonck, and Pérez Rosales Pass. For Currieco, they chose a low hill and named it Pichi Juan, lessening his legacy.

110. Coihueco Island should not be confused with a homonymous locality in Chillán (Ñuble), a Chilean province farther north.

111. An 1891 document recognized this land on Coihueco Island as the Currieco estate "by paternal inheritance." "Adjudicación," *El Reloncaví*, March 2, 1891, Avisos.

112. "Presentacion de algunos habitantes del departamento de Osorno," 1292.

113. Munizaga, "Informe el injeniero don Enrique Munizaga," 1294.

114. Pizarro to Llanquihue intendant, May 14, 1895, in Zenteno Barros, *Recopilacion de leyes i decretos supremos*, vol. 3, 1295.

115. Barros Borgoño to Llanquihue intendant, June 10, 1895, in Zenteno Barros, *Recopilacion de leyes i decretos supremos*, vol. 31297.

116. "Homicidios en Rupanco," *Corre Vuela*, January 1, 1908, 43–44.

117. Bernardino García, "Compraventa de Federico Hechenleitner a Reinaldo Olivares," in *Registro de Conservatorio de Propiedad* (Puerto Montt, Chile: Consservador de Bienes de Puerto Montt, 1899), Archivo Nacional de la Administración, Santiago. For the auctions held on Chanchan estate, where these farmers had their lots, see "Adjudicación," *El Reloncaví*, March 16, 1898, Avisos. For a longer history of Indigenous expropriation in Chanchan, see Vergara del Solar, *La herencia colonial del Leviatán*, 189–92.

118. Hederra, "Francisco Huaiquipan i otro."

119. We do know Hechenleitner sold his estate a year after the incident to his brother and their brother-in-law Carlos Ebensperger. See Abraham Gajardo, "Compraventa de Francisco Hechenleitner y Carlos Ebensperger a Federico Hechenleitner," in *Registro de Conservatorio de Propiedad* (Puerto Montt, Chile: Consservador de Bienes de Puerto Montt, 1909), Archivo Nacional de la Administración, Santiago.

120. "Isla de Coihueco: Nuevo Gravísimo Despojo. El Colmo de la barbarie," *El Llanquihue*, April 11, 1911. The land title belonged to their son-in-law Adolfo Mardof. See Bernardino García, "Hipoteca de Adolfo Mardorf a Jenaro Saldivia y Candelaria Mancilla," in *Registro*

Conservatorio de Hipotecas y Gravámenes (Puerto Montt, Chile: Conservador de Bienes Raíces de Puerto Montt, 1903), Archivo Nacional de la Administración, Santiago.

121. The cemetery was declared a national historic monument in 2014. Cánovas, "A cien años de la matanza de Forrahue," 253–54.

122. José Esteban Canuipan, "Solicitud de amparo y puesta en posesión de un terreno," Fondo Intendencia de Llanquihue, Solicitudes Varias 1908–9, vol. 211, AHN.

123. Canuipan, "Solicitud de amparo y puesta," AHN.

124. Canuipan, "Solicitud de amparo y puesta," AHN.

125. For a discussion on *indios chilenos*, see Delrio, "De 'salvajes' a 'indios nacionales.'"

126. Ceferino Catrilef, "Solicitud de Caferino Catrilef para que la fuerza pública intervenga," Fondo Intendencia de Llanquihue, Solicitudes Varias 1912–13, vol. 227, AHN.

127. Catrilef, "Solicitud de Caferino Catrilef," AHN.

128. "Solicitud para remover colonos nacionales de tierras indígenas," March 31, 1913, Fondo Intendencia de Llanquihue, Solicitudes Varias 1912–13, vol. 227, AHN.

129. "Solicitud para remover colonos nacionales de tierras indígenas," AHN.

130. Enrique Rodríguez, "Constitución de la Sociedad Agrícola y Ganadera Ñuble y Rupanco," in *Registro conservatorio de comercio*, 1906, 214:1649v–1671r, Conservador de Bienes Raíces de Santiago, Archivo Nacional de la Administración, Santiago.

131. Agustín Baeza Espiñeira to Minister of Foreign Affairs, October 9, 1902, expediente 1095, Fondo Histórico—Sección Colonización, Oficios Dirijidos por la Inspección de Tierras i Colonización a Este Ministerio, 1896–1902, AHMRE.

132. "Isla de Coihueco—especulacion de sociedades en perjuicio de sus habitantes," *El Llanquihue*, March 10, 1905.

133. Fritz Gädicke, "Angelegenheit der Ñuble y Rupanco Kolonisation," *Deutsche Zeitung*, September 5, 1905.

134. *La Prensa* article quoted in "Isla de Coihueco."

135. "Homicidios en Rupanco," *Corre Vuela*, January 1, 1908.

136. Congreso Nacional de la República de Chile, *Comisión Parlamentaria de Colonización*, v. See also Almonacid Zapata, "El problema de la propiedad de la tierra," 19–26.

137. Congreso Nacional de la República de Chile, *Comisión Parlamentaria de Colonización*, 456.

138. Congreso Nacional de la República de Chile, *Comisión Parlamentaria de Colonización*, 466–68.

139. Almonacid Zapata, "El problema de la propiedad de la tierra," 24–25.

140. Applicants needed to be at least twenty-two years old, make a commitment to live in the lot for at least five years, have no other property, and work the land by planting trees, growing crops, and especially raising cattle within the first year. After that, they could purchase the title at the low price of M$N500. See República Argentina, "Ley no 1501"; and José Evaristo Uriburu, "Decreto no 300." For priority given to Indigenous settlers in Catriel, Valcheta, Cushamen, and San Martín, see Ministerio de Agricultura, "Decreto fundando colonias pastoriles"; and Ministerio de Agricultura, "Decreto fundando una colonia pastoril y reservando tierras."

141. Bandieri, "Del discurso poblador a la praxis latifundista," 5.

142. Delrio, "De 'salvajes' a 'indios nacionales.'" See also Greenwald, "'Improve Their Condition.'"

143. Ministerio de Guerra, "Ley 1628," 250–52.

144. Ministerio de Guerra, "Ley 1628," 251.

145. Bandieri et al., "Los propietarios de la nueva frontera."

146. Délia Posse de Possolo, e.g., applied for a land title with the certificate from her late husband. See Antonio Bermejo and José Evaristo Uriburu, "Decreto asignando lote a Delia Posse de Possolo," June 5, 1895. Similarly, Emilio and Julia López, beneficiaries of their father's service, applied for 4,000 hectares in La Pampa. See Luis Sáenz Peña, "Decreto asignando lote a Emilio y Julia López," October 6, 1894. Both documents in Fondo Ministerio del Interior, Oficina de Tierras y Colonias (Actos Dispositivos), box 1, AGNI.

147. Ministerio de Agricultura, *Memoria presentada al Honorable Congreso*, xxvi.

148. Examples include Departamento de Interior, "Nombramiento de una comision"; Sáenz Peña, "Decreto del 22 de mayo de 1894," AGNI; and Antonio Bermejo, "Decreto del 4 de agosto de 1896," Fondo Ministerio del Interior, Oficina de Tierras y Colonias (Actos Dispositivos), box 1, AGNI.

149. Departamento de Tierras y Colonias, "Decreto aprobando un contrato de arrendamiento de tierras."

150. See, e.g., Departamento de Interior, "Decreto no 15580"; and Departamento de Interior, "Decreto no 15581."

151. Bernal, "Memoria de la gobernación," 497–98.

152. Quirno Costa, "Decreto comisionando al agrimensor González." González surveyed the lot of Warren Lowe, who had not provided his own surveyor. See Ministerio de Agricultura, "Decreto declarando en vigencia el de 29 de marzo de 1894."

153. Departamento de Interior, "Decreto no 17597"; Departamento de Interior, "Resolución concediendo á D. José Maria Saavedra"; Departamento de Interior, "Decreto no 17082."

154. Bernal, "Memoria de la gobernación," 497–98.

155. Congreso Nacional de la República Argentina, *Diario de sesiones de la Cámara de Diputados*, 18.

156. García and Valverde, "Políticas estatales y procesos de etnogénesis"; Collinao et al., *Lof Paichil Antreao*, 35–37.

157. Geodesy in Argentina had its origins in the barracks. The first topographic materials were tied to the military interests of expanding the nation into the interior in the 1870s and 1880s. Interest in geography and mapmaking existed among college students (a young Estanislao Zeballos founded the Argentine Geographical Institute in 1879). Some of those who took professional exams in the 1880s, like Federico Bazzano, were later appointed to the Geodesy Division that put together the *Plano demostrativo*. See Departamento de Gobierno, *Memoria presentada por el ministro secretario*, 97–98; Antonio Bermejo, "Nombramientos en el Departamento de Tierras, Colonias y Agricultura," February 1, 1895, Fondo Ministerio del Interior, Oficina de Tierras y Colonias (Actos Dispositivos), box 1, AGNI; and Pedro Ezcurra, "Nombramientos en el Ministerio de Agricultura," March 5, 1909, Fondo Ministerio del Interior, Oficina de Tierras y Colonias (Actos Dispositivos), box 7, AGNI. For technical education and cartography in Argentina, see Lois, "La Patagonia en el mapa de la Argentina moderna"; and Mazzitelli Mastricchio, *Imaginar, medir, representar y reproducir el territorio*.

158. Lois, *Mapas para la nación*, chap. 8.

159. Dirección General de Correos y Telegráfos, "República Argentina."

160. "New Maps."

161. Bermejo, *Memoria*, lxiii.
162. Bermejo, *Memoria*, c.
163. Ministerio del Interior [Argentina], *Memoria del ministro [. . .] 1895*, 2:497–98.
164. Benjamín Zorrilla, "Nota no. 125," May 16, 1894, Fondo Ministerio del Interior, Oficina de Tierras y Colonias (Actos Dispositivos), AGNI.
165. Bermejo, *Memoria*, c–ci.
166. Zorrilla, "Nota no. 125," AGNI.
167. Harambour, *Soberanías fronterizas*, 392.

Chapter Three

1. Sociedad Agrícola i Frigorífica de Cochamó, *Prospecto de la Sociedad Agrícola*, 61.
2. Intendencia de Llanquihue, "Oficio no 520 al Ministerio de Colonización," December 18, 1906, Oficios Despachados a Ministerios: 1906, Archivo Histórico de Puerto Montt, Chile.
3. Intendencia de Llanquihue, "Oficio no 520."
4. Manara, "La disputa por un territorio indígena," 21–22.
5. For more on the legibility of nature, see Scott, *Seeing Like a State*, 54–63. For spatial sites as experience, see Bassin, Ely, and Stockdale, *Space, Place, and Power*, 7.
6. Aguiar, *Tracking Modernity*, 6–7.
7. Graciela Blanco, "Neuquén en el espacio patagónico."
8. Bandieri et al., "Los propietarios de la nueva frontera"; Graciela Blanco, "Las sociedades anónimas cruzan los Andes"; Méndez and Muñoz Sougarret, "Alianzas sectoriales en clave regional"; Harambour, "Soberanía y corrupción"; Barbería, *Los dueños de la tierra*, chap. 8.
9. Yofré, *Memoria del Ministro del Interior*, 69.
10. Bess, *Routes of Compromise*, 3.
11. Briones Luco, *Glosario de colonización*, 70.
12. Ministerio de Relaciones Esteriores, Culto i Colonización, *Anexos a la memoria del ministro de colonización i culto*, 148.
13. In Chile, the Chilean state developed transportation infrastructure from the Central Valley to the South while private capital invested in roads and railroads to get exports from the mines to the Pacific ports. See "Guajardo Soto, "Infraestructura y movilidad," 157–58.
14. Andrés Núñez, "El país de las cuencas."
15. For the tension between the ideas and realities of roads, see Bess, *Routes of Compromise*, 3.
16. Ministerio de Obras Públicas [Chile], "Título XI," chap. 3.
17. Rehren, "Memoria correspondiente al año 1905," 229.
18. Intendencia de Llanquihue, "Libro copiador—memoria de la Intendencia de Llanquihue," 1907, folios 1r–2r, Fondo Intendencia de Llanquihue, AHN.
19. Cavada, "Memoria del administrador de la colonia," 302.
20. Ilustre Corporación de Osorno, "Oficio no 23," August 7, 1899, Libro copiador de Sesiones Municipales de Osorno, 1899–1902, Museo Histórico de Osorno, Chile.
21. Intendencia de Llanquihue, "Oficio al Ministerio de Obras Públicas," October 26, 1906, folio 64r, Oficios Despachados a Ministerios, Archivo Histórico de Puerto Montt, Chile.
22. Cavada, "Memoria del administrador de la colonia," 303.

23. Ministerio de Relaciones Exteriores y Colonización, *Inspección jeneral de tierras i colonizacion, 1896–1902*, file 1153, AHMRE.

24. Agustín Baeza Espiñeira, "Nota no 1153 del inspector jeneral de tierras i colonización," November 14, 1901, Oficios Dirigidos por la Inspección Jeneral de Tierras i Colonización a Este Ministerio, 1896–1902, AHMRE. Like other individuals who received vast land leases, Christie sold it to a landholding company, the Taitao Lumber and Industrial Company, in 1904 or 1905. See Oficina de Mensura de Tierras, *Primera memoria del director*, 132.

25. "Abelardo Almonacid contra Santiago Olavarria," November 1908, folio 2r, Fondo Intendencia de Llanquihue, Solicitudes Varias 1908–9, vol. 211, folio 2r, AHN.

26. "Reglamento para colonos nacionales o estrangeros radicados en territorio dependiente de la Inspección Jeneral de Tierras y Colonización del Ministerio de Relaciones Exteriores," July 30, 1902, folio 1r, Inspeccion Jeneral de Tierras i Colonizacion—Oficios Dirigidos, 1896–1902, vol. 237, AHMRE.

27. Intendencia de Llanquihue, "Solicitud de Juan Antonio Cárdenas y Manuel Oyarzún sobre el camino bloqueado por Juan Linde," Fondo Intendencia de Llanquihue, Solicitudes Varias 1912–13, vol. 227, AHN.

28. Intendencia de Llanquihue, "Solicitud de Juan Antonio Cárdenas," AHN.

29. Intendencia de Llanquihue, "Solicitud de Juan Antonio Cárdenas," AHN.

30. Ilustre Corporación de Osorno, "Oficio no 24," October 26, 1900, Libro copiador de Sesiones Municipales de Osorno, 1899–1902, Museo Histórico de Osorno, Chile.

31. Ilustre Corporación de Osorno, "Oficio no 5," March 30, 1901, Libro copiador de Sesiones Municipales de Osorno, 1899–1902, Museo Histórico de Osorno, Chile.

32. Antonio Staforelli, "Solucitud de intervención de Angelino Alvarado contra Bernardo Barría—informe del ingeniero," October 14, 1912, folios 2v–3r, Fondo Intendencia de Llanquihue, Solicitudes Varias 1912–13, vol. 227, AHN.

33. In his petition, Huenchuman claimed he had been on this lot for thirty years. There is evidence he was in the area before 1892, probably working for Antonio Emhardt, but the official land lease was issued in 1892. Zenteno Barros, *Recopilacion de leyes i decretos supremos*, 2:724.

34. Francisco Huenchuman, "Reclamo Francisco Huenchuman contra Lorenzo y Guillermo Möding," 1908, Fondo Intendencia de Llanquihue, Solicitudes Varias 1908–9, vol. 211, AHN.

35. Even if Gallardo had no legal standing to sell the lot he inhabited, an 1892 decree did include it as part of Huenchuman's lease. I could not find any subsequent lease renewals to see whether the state continued to recognize the two lots as part of one contract. Despite this, Huenchuman appears as a landowner in a 1908 census of landowners. See Oficina de Estadística e Informaciones Agrícolas, *Indice de propietarios rurales*, 822.

36. Huenchuman, "Reclamo Francisco Huenchuman," AHN.

37. Francisco Steeger, "Informe sobre Augusto Minte, Albino Bayer, Mödinger, Gebauer Tampe y otros y el camino cerrado por Francisco Huenchuman," April 1908, Fondo Intendencia de Llanquihue, Solicitudes Varias 1908–9, vol. 211, AHN.

38. Steeger, "Informe sobre Augusto Minte," AHN; Fancisco Huenchuman, "Pide amparo como garantía de mi persona, bienes i propriedad," April 1908, Fondo Intendencia de Llanquihue, Solicitudes Varias 1908–9, vol. 211, AHN.

39. Huenchuman, "Pide amparo como garantía de mi persona," AHN.

40. Steeger, "Informe sobre Augusto Minte," AHN.
41. Lefebvre, *Production of Space*, chap. 2.
42. Dalakoglou and Harvey, "Roads and Anthropology."
43. Examples are numerous; see, e.g., Aguiar, *Tracking Modernity*, chaps. 1–2.
44. Studies on the developmental impact of railroads on national economies include Coatsworth, *Growth against Development*; and Schvarzer, Regalsky, and Gómez, *Estudios sobre la historia*. Studies on railroad history in Latin America have certainly expanded beyond economic impacts into the realm of cultural and social history. See Clark, *Redemptive Work*; Van Hoy, *Social History of Mexico's Railroads*; and Matthews, *Civilizing Machine*.
45. For a comprehensive history of railroads in Latin America, see Kuntz Ficker, *Historia mínima de la expansión ferroviaria*.
46. Guajardo Soto, *Tecnología, estado y ferrocarriles*, 42. Cuba, Brazil, and Uruguay are, to varying degrees, other examples of places with first railroads from an "extraction" site to a port. See Zanetti Lecuona, *Sugar and Railroads*, chap. 2; Gastón Díaz, "Uruguay"; and Marchant, *Viscount Maua and the Empire of Brazil*, chap. 6.
47. Guajardo notes that the state did maintain some participation in the North, either by being a stakeholder (like in the Santiago-Valparaíso line) or by later purchasing branches (like the Copiapó-Caldera line). See Guajardo Soto, *Tecnología, estado y ferrocarriles*, 42.
48. Guajardo Soto, *Tecnología, estado y ferrocarriles*, chap. 3.
49. An excellent summary of railroad policy, land distribution, expropriation, and debt in Argentina can be found in Schvarzer, Regalsky, and Gómez, *Estudios sobre la historia de los ferrocarriles argentinos*.
50. Rocchi, "El péndulo de la riqueza," 20.
51. Martinic B., "Ferrocarriles en la zona austral de Chile"; Taranda, "Papel del estado y del capital británico."
52. Andrés Núñez, "El país de las cuencas."
53. Pinto Rodríguez, "Al final de un camino," 290.
54. To obtain this number, I compared the import fees from all ports with a customs office in Chile for the period between 1912 and 1915. See Oficina de Estadística de la Superintendencia de Aduanas, *Estadística comercial de la república de Chile, 1913*; Oficina de Estadística de la Superintendencia de Aduanas, *Estadística comercial de la república de Chile, 1914*; and Oficina de Estadística de la Superintendencia de Aduanas, *Estadística comercial de la república de Chile, 1915*.
55. Carlos Wiederhold, "Pide concesión de playa en 'El Desagüe' del Lago Llanquihue," September 23, 1912, Fondo Intendencia de Llanquihue, Solicitudes Varias 1912–13, vol. 227, AHN. See also "Llanquihue—la industria del lino," March 14, 1919, *La Nación* (Santiago).
56. Propietarios y vecinos de Pelluco, "Solicitan la reposición de un camino," March 31, 1913, Fondo Intendencia de Llanquihue, Solicitudes Varias 1912–13, vol. 227, AHN.
57. "Solicitud de permiso para cargar armas por Daniel Maldonado," April 23, 1908, Fondo Intendencia de Llanquihue, Solicitudes Varias 1908–9, vol. 211, AHN.
58. "Solicitud de permiso para cargar armas por Reinaldo Yunge," May 23, 1908, Fondo Intendencia de Llanquihue, Solicitudes Varias 1908–9, vol. 211, AHN.
59. Saus, "La 'britanización' de Bahía Blanca." For tension with other capitalists, see Chalier, "El puerto comercial de Punta Alta."
60. Álvarez, "Industrias y proyectos de desarrollo."

61. In Patagonia, Law 5559 gave the president authority to build two railroads from the Atlantic to the Andes: one from the port of San Antonio Oeste to Bariloche, and another from Puerto Deseado (Santa Cruz) to the railroad to Bariloche, branching off to different colonies, including Lake Buenos Aires (Santa Cruz) and Colony 16 de Octubre (Chubut). See Ministerio de Obras Públicas [Argentina], *Ley no. 5559*, 3.

62. Compañia del Gran Ferrocarril del Sud de Buenos Aires, *Ferrocarril del Sud*, 34. See also Francisco P. Moreno, "Memorandum sobre una vía férrea a través de los Andes en el Territorio de Neuquén," August 1908, folio 9r, Fondo Francisco P. Moreno, box 3099, AGNE.

63. Julián Cacerez, "Maniobras chilenas en 1902," 1LT, Servicio Histórico del Ejército, Buenos Aires.

64. Congreso Nacional de la República Argentina, *Diario de sesiones de la Camara de Senadores*, 741.

65. Olascoaga, *Regiones australes*, 119.

66. This recalls the fraternity/enmity trope that Alex Bowen explores in his 2005 film, *Mi mejor enemigo*, set in a later, almost-armed conflict in Chilean-Argentine Patagonia. Freeman, "Identity and the Militarized Border."

67. Ramos Mexía, *Memoria presentada al Honorable Congreso*, 24.

68. Francisco P. Moreno, "Memorandum," AGNE.

69. Francisco P. Moreno, "Memorandum," folios 12r–13v, AGNE.

70. Francisco P. Moreno, "Memorandum," folios 14r–15v, AGNE.

71. Lacoste, "Las propuestas de integración económica sudamericana," 110; Almonacid Zapata, "Comercio entre Chile y Argentina en la zona sur," 184–85; Harvey, "Engineering Value."

72. Paula Gabriela Núñez, "Naturaleza ajena en un territorio a integrar," 130–31; Graciela Blanco, "Las sociedades anónimas cruzan los Andes."

73. The Chilean census of 1895 shows 156,065 living in Llanquihue and Chiloé that year while the 1914 Argentine Census had 106,625 in the national territories of Neuquén, Río Negro, Chubut, Santa Cruz, and Tierra del Fuego. See Comisión Central del Censo, *Memoria presentada al Supremo Gobierno*, 1050; and República Argentina, *Tercer censo nacional*, 65.

74. Ministerio del Interior [Argentina], *Memoria del ministro [. . .] 1904–1905*, 101. See also Dirección General de Territorios Nacionales [Argentina], "Proyecto para continuar en una forma económica la construcción del ferrocarril de San Antonio Oeste al Lago Nahuel Huapi," September 18, 1915, folio 1r, Fondo Ruiz Moreno, box 3095, AGNI.

75. Escalante, "Neuquén," 88.

76. Ministerio del Interior [Argentina], *Memoria del ministro [. . .] 1883*, lxxx–lxxxiii. See also Quintana, *Memoria del ministro del interior*, 88–89.

77. Gobernación de Neuquén, "Expediente 751N—camino internacional," October 1895, Fondo Ministerio del Interior, Expedientes Generales, 1895, box 3, AGNI.

78. Godio, "Tierra adentro," 393.

79. The territory of Neuquén had a total of 14,512 inhabitants. Of these, 62 percent were born outside Argentina, and 98 percent of these foreigners were Chileans. The foreign population totaled 9,012, with 8,861 Chileans, forty-one Spaniards, thirty French, twenty-nine Italians, eighteen Uruguayans, thirteen Germans, five English, four US Americans, three Swiss, one Brazilian, one Paraguayan, one Austrian, and two listed as "other." In the census of 1895, the total of foreigners in table 3 does not match the breakdown by nationality in tables

7a and 7b (population by district by nationality). See República Argentina, *Segundo censo,* 2:644, table 3, 658, table 7a, 661, table 7b.

80. República Argentina, *Segundo censo,* 659.

81. Gobernación de Neuquén, "Expediente 751N," AGNI.

82. Varela and Manara, "Tiempos de transición," 40–41.

83. Gobernación de Neuquén, "Expediente 751N," AGNI.

84. Gobernación de Neuquén, "Expediente 751N," AGNI.

85. Gobernación de Neuquén, "Expediente 751N," AGNI.

86. The currency abbreviation "M$N" refers to "pesos moneda nacional," Argentina's currency between 1881 and 1970. See Santacreu Soler, "Unidad monetaria, vertebración territorial y conformación nacional."

87. Gobernación de Neuquén, "Expediente 751N," AGNI.

88. Gobernación de Neuquén, "Expediente 751N," AGNI.

89. Gobernación de Neuquén, "Expediente 751N," AGNI.

90. For the decree authorizing the construction of the road, see Ministerio del Interior [Argentina], "Acuerdo autorizando á la gobernación del Neuquén," 370–71.

91. Rawson, "Neuquén," 486.

92. Francisco P. Moreno, *Apuntes preliminares,* 29.

93. Bengoa, *Historia del pueblo mapuche,* chap. 1.

94. Rawson, "Neuquén," 452–57.

95. For a thorough analysis of the principle of *cordillera libre* and the 1905–10 trade agreement, see Lacoste, "Vinos, carnes, ferrocarriles y el tratado de libre comercio."

96. Ministerio de Relaciones Exteriores, *Memoria del Ministerio de Relaciones Exteriores, Culto y Colonización,* 9.

97. See, e.g., Christie, *El camino de Vuriloche,* 30.

98. The tax on Argentine cattle caused legal imports to decline so severely that the Chilean Congress repealed this law in 1907. See Huergo, *Conversación,* 19–20; and Ministerio de Hacienda, "Que suspende el impuesto."

99. Superintendencia de Aduanas, *Memoria del superintendente,* 22–26.

100. Astorga Pereira, "Contestacion a la circular número 5."

101. Biedma, *Crónica histórica del lago Nahuel Huapi,* 155.

102. Muñoz Sougarret, "Empresariado y política," 139–40.

103. Cariola and Sunkel, *Un siglo de historia económica de Chile,* 175–85.

104. República Argentina, "Decreto reservando para fundación de pueblos," 134.

105. Speech delivered by Carlos Wiederhold on February 3, 1925, at a celebration of the thirtieth anniversary of his arrival in Nahuel Huapi to build a store, cited in Biedma, *Crónica histórica del lago Nahuel Huapi,* 160.

106. *Die familie Achelis.*

107. Wiederhold withdrew from the Argentine side of the business due to health issues, and his son-in-law expanded the company. See Vallmitjana, *San Carlos,* 5. In Puerto Montt, Wiederhold continued his business of importing toys, furniture, fabrics, and jewelry. See "La Casa Wiederhold y C.," advertisement, *El Llanquihue,* January 13, 1910.

108. Dirección General de Navegación y Puertos [Argentina], "Expediente 1579O—carta de apoyo a la propuesta de Federico Hube," June 14, 1899. See also Valentín Balbín, "Expediente 1579O—carta de apoyo a la propuesta de Federico Hube del Director General de Obras

Hidráulicas," June 17, 1899, both in Fondo Ministerio del Interior, Expedientes Generales, 1899, box 9, AGNI.

109. Ministerio de Relaciones Exteriores [Argentina], "Expediente 3127R—acompaña copia autorizada de una nota recibida de nuestro consulado en Puerto Montt, a propósito del establecimiento de varios servicios en la cordillera y Lago Nahuel Huapi," October 16, 1899, Fondo Ministerio del Interior, Expedientes Generales, 1899, box 9, AGNI; Ministerio de Relaciones Exteriores y Culto, *Memoria de relaciones exteriores y culto*, 457.

110. Dirección General de Correos y Telégrafos, "Expediente 15790—carta de objeción a la propuesta de Federico Hube," May 28, 1900, Fondo Ministerio del Interior, Expedientes Generales, 1899, box 9, AGNI.

111. Lacoste, *La imágen del otro*, 324.

112. Méndez, *Estado, frontera y turismo*, 118–19; Muñoz Sougarret, "Empresariado y política," 179–80.

113. Tagle y Jordan, *El tratado de comercio*.

114. República de Chile, *Anuario estadístico de la República de Chile, año 1910*, 296–97; República de Chile, *Anuario estadístico de la República de Chile, año 1916*, 2–3.

115. Palacio, "La antesala de lo peor," 103–6.

116. Bandieri, *Historia de la Patagonia*, 337–39.

117. David Burr, "Compraventa Augusto Minte y Ricardo Roth a la Compañía Agrícola Ganadera Chile-Argentina (Ensenada)," 60:263v–65, annotation 188; David Burr, "Compraventa Augusto Minte y Ricardo Roth a la Compañía Agrícola Ganadera Chile-Argentina (Casa Pangue)," 60:266–67, annotation 190; and David Burr, "Compraventa Augusto Minte y Ricardo Roth a la Compañía Agrícola Ganadera Chile-Argentina (Peulla)," 60:265–66, annotation 189, all in *Registro Conservatorio de Propiedad* (Puerto Montt, Chile: Conservador de Bienes Raíces de Puerto Montt, 1914), Archivo Nacional de la Administración, Santiago.

118. Méndez and Muñoz Sougarret, "Alianzas sectoriales en clave regional."

119. Andrés Núñez, "El país de las cuencas."

120. Ruffini, "La Patagonia en el pensamiento y la acción."

121. Méndez, *Estado, frontera y turismo*, 194–95.

122. For some examples, see Gobernación de Tierra del Fuego, "Expediente 1832T—eleva mapa e informe sobre ese territorio," April 1910, 6–7; and Gobernación del Neuquén, "Expediente 1643N—plano e informe del Territorio de Neuquén solicitado por el Ministro del Interior," 3–5, both in Fondo Ministerio del Interior, Expedientes Generales, 1910, box 7, AGNI.

123. Gobernación del Neuquén, "Expediente 1643N," AGNI.

124. Robert Runciman, "Expediente 1760R—sobre permiso para establecer una linea telefónica en el Territorio de Río Negro," April 25, 1910; and Robert Runciman, "Expediente 4989R—en representación de la Compañía de Tierras Sud Argentino, Limitada, presenta plano de la línea telefónica entre la Estancia Pilcañeu y csa de Don Manuel Acosta," October 27, 1910, both in Fondo Ministerio del Interior, Expedientes Generales, 1910, box 7, AGNI.

125. Pedro Serrano to Isidoro Ruiz Moreno, August 18, 1914, Fondo Ruiz Moreno, box 3093, AGNE.

126. Eduardo Elordi to Isidoro Ruiz Moreno, April 10, 18, 1914, Fondo Ruiz Moreno, box 3092, AGNE.

127. Ramos Mexía, *Memoria presentada al Honorable Congreso*, 10.

128. Francisco P. Moreno, "Memorandum," folios 6r–7r, AGNE.

129. Ramos Mexía, *Memoria presentada al Honorable Congreso*, 10, 20.

130. Ramos Mexía, *Memoria presentada al Honorable Congreso*, 147.

131. Ramos Mexía, *Memoria presentada al Honorable Congreso*, 2.

132. Blackwelder, "Bailey Willis," 334.

133. Comisión de Estudios Hidrológicos, *Northern Patagonia*, 419.

134. Willis to Ezequiel Ramos Mexía, telegram, January 6, 1913, Colección Frey, binder 4, ADPM.

135. Greenberg, "Reassessing the Power Patterns."

136. Comisión de Estudios Hidrológicos, *Northern Patagonia*, 433.

137. Willis and Comisión de Estudios Hidrológicos, *El norte de la Patagonia*, 8.

138. Gobernación de Río Negro, "Expediente 5364—remite actuaciones sobre denominación del Pueblo 'Ingeniero Jacobacci,'" January 1944, Fondo Ministerio del Interior, Expedientes Generales, 1940, box 39, AGNI. See also Varios vecinos del pueblo isla (Choele Choel), "Expediente 4591V: Piden oficina telegráfica," March 18, 1915, Fondo Ministerio del Interior, Expedientes Generales, 1915, box 16, AGNI.

139. Ortiz, *Memoria presentada al Honorable Congreso*, 688–89.

140. The other railroads were Comodoro Rivadavia (Chubut), Puerto Deseado (Santa Cruz), Del Este (Entre Ríos and Corrientes), and Formosa, in the homonymous territory. Ortiz, *Memoria presentada al Honorable Congreso*, 681.

141. Ortiz, *Memoria presentada al Honorable Congreso*, 711–16.

Chapter Four

1. Policía Fronteriza, "Copia de sumario instruido con motivos de denuncia formulada contra el juez de paz de El Bolsón Vicente Fernandez Palacios y comisario Laudalde: Testimonio de Valentín Bustamente," March 12, 1913, Ministerio del Interior, Expedientes Generales, 1913, box 46, AGNI.

2. Sánez Peña, "Decreto aceptando renuncias," 1061; José Felix Uriburu, "Decreto nombrando," 902.

3. In addition to the works cited elsewhere in this chapter, some examples include Ablard, Di Liscia, and Bohoslavsky, *Instituciones y formas de control social*; Salvatore, "Burocracias expertas y exitosas en Argentina"; and Allevi, "La creación clínica de normas sexuales."

4. Ruggiero, *Modernity in the Flesh*, 2.

5. Caimari, *Apenas un delincuente*, 64. Ryan Edwards best analyzes the development of a penal colony in Tierra del Fuego in relation to the frontier environment. See Edwards, *Carceral Ecology*.

6. Examples include Elsey, *Citizens and Sportsmen*, chaps. 1–2; Donoghue, "Roberto Durán"; Allen, *History of Boxing*; Bocketti, *Invention of the Beautiful Game*, chap. 1; and Blakeslee, "Foot-Ball!"

7. Stern, "'Professionals, Merchants, and Industrialists Unite!'" See also Mosse, *Image of Man*, 4–5.

8. Historians accept Mardoqueo Navarro's estimate of 13,614 deaths in a city of 177,787, per the 1869 census. For an analysis of the epidemic and its accounts, see Fiquepron, *Morir en las grandes pestes*, chap. 6. Note that the city of Buenos Aires at the time of the 1869

census did not include the districts of Belgrano and Flores. See República Argentina, *Primer censo*, 28–29.

9. Salessi, *Médicos, maleantes y maricas*; Rodriguez, Rivero, and Carbonetti, "Convicciones, saberes y prácticas higiénicas."

10. Armus, *Ailing City*, 227–28.

11. Stepan, *Hour of Eugenics*, 17.

12. For the influence of neo-Lamarckism in Latin America, see Stepan, *Hour of Eugenics*, chap. 3.

13. Martínez de Ferrari, *Contribución al estudio del problema pendiente*, 6.

14. For some examples of the intersection between inebriation and public order in Latin America, see Piccato, *City of Suspects*, chap. 4; and Bouret, "Lo sano y lo enfermo."

15. For some recent studies of anarchism in Buenos Aires and Santiago see Craib, *Cry of the Renegade*; and de Laforcade, "Memories and Temporalities."

16. Edwards, *Carceral Ecology*, 87.

17. Ministerio del Interior [Argentina], "Ley de Defensa Social," 149.

18. Zimmermann, *Los Liberales Reformistas*; Favaro and Morinelli, "Los reformistas de la clase dominante." Demands for political and social reforms also came from socialist and Catholic groups. See Ruffini, "La Patagonia en el pensamiento y la acción."

19. Ramos Mexía, "Mensaje y projecto," 20.

20. Argeri, *De guerreros a delincuentes*, 260.

21. "Sin alcoholizar," *La Nueva Era*, April 20, 1919. For other examples, see "Policía de Patagones," *La Nueva Era*, November 30, 1913; and "El atentado del miércoles," *La Nueva Era*, December 28, 1919.

22. Enrique Feinmann, "Enfermedades sociales: El peligro alcohólico," *La Nueva Era*, October 18, 1914.

23. "La corrupción en auge," *Ambas Márgenes*, January 19, 1918, clipping in Fondo Ministerio del Interior, Expedientes Generales, box 5, expediente 875V: "Varios vecinos de Viedma (Río Negro) formulan quejas contra la policía de esa gobernación," AGNI.

24. Policía Fronteriza de Chubut, "Testimonio de Manuel Sales," May 9, 1913, Fondo Ministerio del Interior, Expedientes Generales, box 46, expediente 9238C: "Copia de sumario instruido con motivos de denuncia formulada contra el juez de paz de El Bolsón Vicente Fernandez Palacios y comisario Laudalde," AGNI.

25. Feinmann, "Enfermedades sociales." For a thorough analysis of hygienists' writings as they pertained to Buenos Aires, see Julia Rodriguez, *Civilizing Argentina*.

26. M. B. Moore, "El alcoholismo," *La Nueva Era*, March 25, 1917. Based on the references, it looks like this article was published originally in England, but I could not find out where *La Nueva Era* picked it up.

27. Leon Bourgeois, "El hombre sociable (colaboración)," *La Nueva Era*, October 18, 1914.

28. "Gobernación de Río Negro: Resolución sobre el alcoholismo," *La Nueva Era*, August 5, 1917.

29. For other prohibitionist movements in Latin America, see Fernández Lobbé, "La virtud como militancia"; Meza Bazán, "El enfoque médico social"; and Autrique Escobar, "Los orígenes de los movimientos."

30. For a comparative analysis on temperance organizations in Chile and Argentina, see Lavrín, *Women, Feminism, and Social Change*, 137–40; and Sánchez Delgado, "Chile y Argentina," 112–14.

31. República Argentina, *Código rural de los territorios nacionales*, 45. The code suffered minor modifications of articles 197 and 198, which pertained to taxation of agricultural production. See República Argentina, *Leyes nacionales*, 89.

32. McGee Deutsch, *Counterrevolution in Argentina*, 67.

33. For the National Association of Labor, see Ospital, "Patrones e inmigrantes."

34. Rein, *Argentine Jews or Jewish Argentines?*, 21.

35. McGee Deutsch, *Counterrevolution in Argentina*, chap. 3.

36. Devoto, *Nacionalismo, fascismo y tradicionalismo*, 154; Gallucci, "Nación, república y constitución," 311–13. Personally, I think Gallucci takes a narrow interpretation of McGee Deutsch's idea of counterrevolution as a simple revolutionary rejection of a revolution.

37. McGee Deutsch, *Counterrevolution in Argentina*, 112–20.

38. Bohoslavsky, *El complot patagónico*, 99.

39. Ruffini, *La Patagonia mirada desde arriba*, 83–85.

40. Bohoslavsky, *El complot patagónico*, chap. 3.

41. Bohoslavsky, *El complot patagónico*, 89.

42. "Los juegos de azar," *La Nueva Era*, February 14, 1925.

43. "The Murder of Llwyd Ap Iwan: Details of a Welshman's Fate," *Cardiff Times*, February 5, 1910; "El bandorelismo en el sur: Los asesinos de Ap Iwan; Carencia de policía," *La Prensa*, January 7, 1910; "Bandidos misteriosos," *El Llanquihue*, February 22, 1910.

44. R. Bryn Williams, *Y Wladfa*; Pearson, "One Letter and 55 Footnotes," 63.

45. Pearson, "One Letter and 55 Footnotes," 67.

46. "Bandidos misteriosos."

47. "Murder of Llwyd Ap Iwan"; Pearson, "One Letter and 55 Footnotes," 69–70; Maggiori, *La cruzada patagónica*, 16; Jameson, *Butch Cassidy*, 95–103.

48. Buck and Meadows, "Neighbors on the Hot Seat," 6.

49. For more on the chase of the two robbers, see Maggiori, *La cruzada patagónica*, 29–31.

50. "Ap Iwan Avenged: Two Bandits Shot While Resisting Arrest," *Druid*, January 25, 1912.

51. "El bandolerismo en el territorio," *La Nueva Era*, June 4, 1911.

52. "El bandolerismo en el territorio."

53. República Argentina, *Código rural de los territorios nacionales*, 45.

54. "Pathetic Letter from Mr. Llwyd Ap Iwan," *Weekly Mail*, February 19, 1910.

55. "Bandidos misteriosos."

56. "Murder of Llwyd Ap Iwan."

57. "Llofruddiaeth Mr. Llwyd Ap Iwan," *Y Rhedegydd*, March 12, 1910.

58. "Murdered by a Cowboy," *Evening Express*, January 10, 1910.

59. "El bandolerismo en el territorio."

60. "Bandidos misteriosos."

61. "Archivo municipal de Patagones: Necesidad de su organización," *La Nueva Era*, January 1, 1911.

62. "Exageraciones informativas: Antropófagos," *La Nueva Era*, December 4, 1910.

63. Bandieri, "Condicionantes históricos," 140. See also Cárdenas Palma, "El conflicto por la tierra," 195–207.

64. "Argentina," *El Comercio*, January 6, 1910; "Murder of Llwyd Ap Iwan."

65. "Murdered by a Cowboy"; "Welshman Murdered," *Evening Express*, January 24, 1910; "Ap Iwan Avenged."

66. "Murder of Llwyd Ap Iwan."

67. "Welshman Murdered."

68. H. S. Orde, "The Bandits of the Argentine," *Wide World Magazine*, 1911, 65–66.

69. Orde, "Bandits of the Argentine," 64.

70. Ardüser, *Un Suizo en la Patagonia*, 46.

71. "La criminalidad en el sur," *La Prensa*, January 1910.

72. "La persecución de los criminales," *La Nación* (Buenos Aires), January 9, 1910; "La policía del territorio," *La Nueva Era*, January 23, 1910; "Falta de policía," *La Prensa*, January 29, 1910; "La criminalidad en el sur."

73. "La criminalidad en el sur."

74. Maggiori, *La cruzada patagónica*, 33–38.

75. Gobernación del Neuquén, "Expediente 1643N—plano e informe del Territorio de Neuquén solicitado por el ministro del interior," April 18, 1910, Fondo Ministerio del Interior, Expedientes Generales, 1910, box 7, AGNI.

76. Ministro del Interior [Argentina], *Memoria del ministro [. . .] 1912–1913*, 154.

77. Ministro del Interior [Argentina], *Memoria del ministro [. . .] 1912–1913*, 154.

78. Ministro del Interior [Argentina], *Memoria del ministro [. . .] 1912–1913*, 140. For more on the conference, see Ruffini, "Ecos del centenario."

79. Ministro del Interior [Argentina], *Memoria del ministro [. . .] 1910–1911*, 59.

80. Ministro del Interior [Argentina], *Memoria del ministro [. . .] 1912–1913*, 159.

81. Ministro del Interior [Argentina], *Memoria del ministro [. . .] 1910–1911*, 90.

82. Gobernación del Neuquén, "Expediente 1643N," AGNI; Alfredo Weber, "Mapa geográfico-comercial."

83. The prison had a total of thirty-nine officers, while the *jefatura* had thirty-six.

84. José Antonio Lafquen, "Denuncia al carabinero que indica y se le ordene entregue el caballo que expresa," 1913, Fondo Intendencia de Llanquihue, Solicitudes Varias 1912–13, vol. 227, AHN.

85. Prislei, "Imaginar la nación, modelar el desierto" 80–81; Bandieri, *Historia de la Patagonia*, 178; Vallmitjana, *Periodismo y otros medios en el pueblo*; Ruffini, *La Patagonia mirada desde arriba*, 78.

86. Velázquez to Ramón Gómez (minister of the interior), telegram, "Referencias con motivo de denuncias del ciudadano chileno Don Luis Oyarzun contra autoridades policiales de Cholila," March 19, 1920, Fondo Ministerio del Interior, Expedientes Generales, 1920, box 2, AGNI.

87. Ministerio de Relaciones Exteriores [Argentina], "Expediente 311 reservado—referencias con motivo de denuncias del ciudadano chileno Don Luis Oyarzun contra autoridades policiales de Cholila," April 5, 1920, Ministerio del Interior, Expedientes Generales, 1920, box 2, AGNI.

88. Ministerio del Interior [Argentina], *Memoria del ministro [. . .] 1912–1913*, 154.

89. Examples include Gobernación de Río Negro, "Expediente 4485R—licencia al Comisario Don Tomás Torres Ardiles," September 27, 1910, Fondo Ministerio del Interior, Expedientes Generales, 1910, box 15, AGNI; Gobernación de Neuquén, "Expediente 11940N—licencia al Subcomisario Don Pablo Valle," November 15, 1913, Fondo Ministerio del Interior, Expedientes Generales, 1913, box 61, AGNI; and Gobernación de Río Negro, "Expediente 14558R—licencia al Comisario D. C. Jardell," July 16, 1925, Fondo Ministerio del Interior, Expedientes Generales, 1926, box 34, AGNI.

90. The Border Police was created simultaneously to act in western Chubut and Río Negro and in Formosa and western Chaco, two areas that governors felt did not have a good grip on law enforcement. For Chaco, see Beck, *La vida en las fronteras interiores*.

91. For a legal genealogy of law enforcement in the national territories of Argentina, see Maggiori, *La cruzada patagónica*, 10–15.

92. For the US-Mexico border, see Hernandez, *Migra*; George Díaz, *Border Contraband*; and Dupree, "Roots of the Border Patrol Line Riders." In most of the world, border patrols were created in the second half of the twentieth century with the formation of new nation-states, e.g., in West Germany (1951), Israel (1949), and India (1965). In Latin America, border patrols only emerged toward the end of the century as countries tried to prevent civil wars from spilling into their territories.

93. Ministro del Interior [Argentina], *Memoria del ministro [. . .] 1912–1913*, 158–59.

94. "El bandolerismo en el territorio: Caso de conciencia," *La Nueva Era*, August 27, 1911.

95. Ministro del Interior [Argentina], *Memoria del ministro [. . .] 1910–1911*, 60.

96. N. R. Amuchástegui to Isidoro Ruiz Moreno, April 12, 1913, Fondo Ruiz Moreno, box 3090, AGNE; Pomar, *La concesión del Aysén y Valle Simpson*, 66; Maggiori, *Donde los lagos no tienen nombre*, 164; Vallmitjana, *Cruzando la cordillera*, 8; Pilar Pérez, "Las primeras policías fronterizas," 29; Cikota, "Frontier Justice," 100. I could not confirm his origins with the Austrian State Archives, but it is clear that Gebhard had military experience and spoke German as a first language.

97. Ministro del Interior [Argentina], *Memoria del ministro [. . .] 1912–1913*, 158.

98. Gobernación de Chubut, "Expediente 13639—nota pasada por el inspector general de la Policia Fronteriza en la que solicita la suspensión del Comisario Caminos del Castillo y la encarcelación del Sub-Comisario Maximiliano Montero," December 18, 1913, Fondo Ministerio del Interior, Expedientes Generales, 1913, box 70, AGNI.

99. "Policía fronteriza del territorio: Siguen los clamores," *La Nueva Era*, October 8, 1911.

100. Gebhard to Isidoro Ruiz Moreno, May 7, 1913, Fondo Ruiz Moreno, box 3091, AGNE.

101. "La cuestión del bandalaje en Bariloche," *El Llanquihue*, October 24, 1911.

102. "Atropellos inauditos de que son víctimas los colonos chilenos," *El Llanquihue*, October 24, 1911.

103. "Los bandidos del sur," *La Nueva Era*, October 22, 1911.

104. "Bandoleros yankees," *El Llanquihue*, November 22, 1911.

105. Ministro del Interior [Argentina], *Memoria del ministro [. . .] 1912–1913*, 156.

106. "Crímenes en Bariloche: Denuncia grave," *La Nueva Era*, February 2, 1913.

107. "Un reportaje," *La Nueva Era*, May 5, 1918. See also "Reorganización policial," *La Nueva Era*, February 6, 1921.

108. "Policías fronterizas," *La Nueva Era*, March 31, 1918.

109. José Vereertbrugghen, "De la zona cordillerana: Las garantías en Bariloche," *La Nueva Era*, December 21, 1919.

110. "Los bandoleros en plena fuga," *La Nueva Era*, March 6, 1921.

111. "Defensa de los territorios," *La Nueva Era*, January 30, 1921.

112. Ministro del Interior [Argentina], *Memoria del ministro* [...] *1912–1913*, 157–59.

113. Vereertbrugghen, "De la zona cordillerana."

114. Rawson, "Neuquén," 459–60.

115. "Gobierno de Chubut: Inmigración, ganadería, tierra fiscal," *La Nación* (Buenos Aires), May 17, 1908.

116. "El personal subalterno de policía," *La Nación* (Buenos Aires), October 28, 1907, Territorios Nacionales.

117. "Atropellos inauditos de que son víctimas los colonos chilenos."

118. "El personal subalterno de policía."

119. Bohoslavsky, "Modernización estatal y coerción."

120. Giménez to F. R. Beiró (minister of the interior), September 7, 1922, Fondo Ministerio del Interior, Expedientes Generales, 1922, box 41, AGNE.

121. Francisco Denis (governor of Neuquén) to F. R. Beiró (minister of the interior), 1922; and Giménez to Beiró, telegram, September 11, 1922, both in Fondo Ministerio del Interior, Expedientes Generales, 1922, box 41, AGNE.

122. Vereertbrugghen, "De la zona cordillerana."

123. Arturo Ríos, "Desde Bariloche," *La Nueva Era*, February 1, 1920.

124. Ríos, "Desde Bariloche."

125. Saldívar Arellano, "Etnografía de la nostalgia"; Bona and Vilaboa, "Las relaciones argentino-chilenas"; Gobantes et al., "Migraciones laborales entre la Isla de Chiloé (Chile) y Patagonia Austral"; Saldívar Arellano, "'Chilote tenía que ser.'"

126. Luis González, "Desde Bariloche," *La Nueva Era*, February 8, 1920.

127. Bertolotto, "Gaucho, atlético y nacional."

128. Elias, *Quest for Excitement*.

129. Romero Brest, "Organización general de la educación física," 6.

130. Cornelis, "Reflexiones sobre la trayectoria de Enrique Romero Brest," 122.

131. República Argentina, "Fomento del scoutismo argentino," 755; República Argentina, *Ley de presupuesto general*, 827.

132. Chiocconi, Chiappe, and Podlubne, "¡Todo por la patria!," 196–201.

133. Elias, *Quest for Excitement*.

134. See, e.g., Reyna, "Aproximaciones en torno al proceso de surgimiento y estructuración del fútbol"; Roldán, "Circulación, difusión y masificación"; and Le Bail, "Clubes sociales y deportivos."

135. Obligatory military service in Argentina was instituted in 1902, partly as a response to the border conflict with Chile. For debates from that time, see Manzoni, "Contra los arrastra sables." For the pedagogical role of obligatory military service, see Sillitti, "El servicio militar obligatorio y la 'cuestión social.'"

136. "Tiro Federal Bariloche—acta de constitución," April 30, 1915, 1. See also Emilio Frey to Serrano, July 22, 1915, both in Colección Frey, binder 8, ADMP.

137. Frey to Serrano, July 22, 1915, ADMP.

138. Raiter, "Ciudadanos y soldados," 50–51.

139. "Liga Patriótica Argentina: Constitución de la C.E.V. local," *La Nueva Era*, August 24, 1919.

140. Prislei, "Imaginar la nación, modelar el desierto." Other Tiro Federal chapters were founded around this time: Viedma (1907, then again 1914), Zapala (1916), and Neuquén City (1916).

141. Biedma, *Crónica histórica del lago Nahuel Huapi*, 273.

142. Pastor quoted in Méndez, "'El león de la cordillera,'" 38–39.

143. Lulio De Lermo, "Hacia arriba," *La Nueva Era*, January 14, 1928.

144. Bohoslavsky, *El complot patagónico*, chaps. 3–4.

Chapter Five

1. Lolich, "Bariloche y su Centro Cívico," 68.

2. Most of the DPN's public works in Nahuel Huapi are synthetized in Dirección de Parques Nacionales, *Obra pública, cultural y turística*.

3. Lekan, *Imagining the Nation in Nature*.

4. Navarro Floria, "La 'Suiza argentina.'"

5. Piglia, "El despertar del turismo"; Freitas, *Nationalizing Nature*, 37–58.

6. Confino, *Nation as a Local Metaphor*. For examples on the creation of national aesthetics, see Olsen, *Artifacts of Revolution*; and Rico, *Heritage State*. I have pulled much of the debates on architecture, nation-making, and the invention of tradition from Kusno, *Behind the Postcolonial*, chap. 3; and Rajagopalan and Desai, "Architectural Modernities."

7. For more on the Argentine Switzerland region as a way to bestow value on geographical space, see Navarro Floria, "La 'Suiza argentina.'"

8. Bustillo, *El despertar de Bariloche*; Navarro Floria, "El proceso de construcción social de la región de Nahuel Huapi."

9. Piglia, "En torno a los parques nacionales."

10. Urry, *Tourist Gaze*.

11. Nishimura, Waryszak, and King, "Guidebook Use by Japanese Tourists."

12. Tuan, *Space and Place*, 159.

13. *Boletín de la Sociedad de Fomento Fabril*.

14. Intendencia de Llanquihue, "Libro copiador—memoria de la Intendencia de Llanquihue," 1907, folio 9r, Fondo Intendencia de Llanquihue, AHN.

15. Méndez, *Estado, frontera y turismo*, 118.

16. Vallmitjana, *Sociedad Comercial y Ganadera Chile-Argentina*, 13.

17. Méndez, *Estado, frontera y turismo*, 181–82.

18. See Kirchmayr, *Plano general de la región del Nahuel Huapí*. For more on the Chile-Argentina Company's acquisiton of landed property, see Muñoz Sougarret, "Empresariado y política," 208–13; and Graciela Blanco, "Las sociedades anónimas cruzan los Andes," 118–21.

19. Méndez and Muñoz Sougarret, "Economías cordilleranas e intereses nacionales," 176; Harambour, *Soberanías fronterizas*, chap. 3.

20. For more on these features, see Prado et al., "Traces of Construction Following Migration." Unless otherwise noted, all buildings I discuss in this section appear in Carlos

Foresti, *Album de la Compañía Comercial y Ganadera Chile-Arentina*, 1900, Landesbibliothek Mecklenburg-Vorpommern Günther Uecker, Schwerin, Germany.

21. Prado, D'Alençon Castrillón, and Kramm, "Arquitectura alemana en el sur de Chile." For the specifics of massive timber techniques in southern Chile, see D'Alençon Castrillón and Prado García, "Construcción en madera maciza en el sur de Chile."

22. Tillería González and Vela Cossío, "Cuando habitábamos lo elemental," 862.

23. "La casa comercial Huber y Achelis," *El Llanquihue*, June 14, 1901.

24. Méndez, *Estado, frontera y turismo*, 186.

25. Two recent examples in the local news include "Mirá el video de la nueva atracción turística de la Suiza argentina: 'Laguna Neneo,'" *Diario El Cordillerano*, April 6, 2023, www.elcordillerano.com.ar/noticias/2023/06/04/162703-mira-el-video-de-la-nueva-atraccion-turistica-de-la-suiza-argentina-laguna-neneo; and "Bariloche, la Suiza que no fue," Bariloche 2000, June 25, 2023, www.bariloche2000.com/noticias/leer/bariloche-la-suiza-que-no-fue/147650.

26. See, e.g., "Escursión Criolla," *El Libre Pensador*, November 19, 1882. The author described travels to Chascomús, Dolores, Maipú, Ayacucho, and Tandil.

27. Francisco P. Moreno, *Apuntes preliminares*, 18. Holdich also used the term for Lake Nahuel Huapi. See Holdich, *Countries of the King's Award*, 316.

28. "Exposición británico-argentina," *La Nación* (Buenos Aires), November 30, 1905. See also "En la Suiza argentina: Locomoción automóvil," *La Nación* (Buenos Aires), June 25, 1908.

29. There is an early use of "Suiza chilena" to refer to an area in the Central Andes, near Santiago. See Vicuña Mackenna, "Elisa Bravo."

30. Booth, "De la selva araucana a la Suiza chilena," 15.

31. *Chile y Arjentina*, 1.

32. *Chile y Arjentina*, 4.

33. "La Suiza argentina en la región de los lagos," *La Nación* (Buenos Aires), December 26, 1913.

34. "La Suiza argentina en la región de los lagos."

35. Gerike, Manriquez, and Thies, *Turismo en las provincias australes de Chile*.

36. Wiederhold, *Turismo en la provincia de Llanquihue*, 199.

37. For advertisements, see "La concurrencia de nuestros agricultores e industriales a la Exposición Industrial y Agrícola," *El Llanquihue*, March 10, 1922; "El próximo viaje de turismo a la Suiza Chilena," *El Faro de Llanquihue*, November 15, 1922; and Wiederhold, *Turismo en la provincia de Llanquihue*, sec. Avisos. For other references to "Chilean Switzerland," see Santiago Marín Vicuña, "Por los canales (notas de viaje): Chile país de turismo - De Santiago a Puerto Montt - La Suiza chilena - De Puerto Montt a Punta Arenas - La Noruega chilena - El Canal de Moraleda - El Itsmo de Ofqui - Paisajes de Smith - En Punta Arenas," *El Mercurio*, January 26, 1917; H. Rosales Aranguiz, "Impresiones de viaje: A través de la Suiza chilena - De Villarrica a Riñihue II," *El Mercurio*, July 20, 1917; and H. Rosales Aranguiz, "Impresiones de viaje: A través de la Suiza chilena - De Villarrica a Riñihue V," *El Mercurio*, August 17, 1917. For travelers' impressions, see "Desde Puerto Varas: Impresiones de viaje," *El Llanquihue*, January 18, 1922.

38. Elflein, *Paisajes cordilleranos*, 89.

39. For more on Elflein's writing and her contribution to women's literature in twentieth-century Argentina, see Arambel-Güiñazú, *Las mujeres toman la palabra*.

40. Elflein, *Paisajes cordilleranos*, 129.

41. Vallmitjana, *San Carlos*, 10.

42. For examples of art and music, see López, *Crafting Mexico*, chap. 1; and McCann, *Hello, Hello Brazil*.

43. Itzigsohn and vom Hau, "Unfinished Imagined Communities." For a comparative study on modernist architecture in Latin America, see Guillén, "Modernism without Modernity."

44. Navarro Floria and Núñez, "Un territorio posible en la república imposible."

45. Romero, *History of Argentina*, 60. For a longer discussion of the coup, see Devoto, *Nacionalismo, fascismo y tradicionalismo*, 235–37.

46. Quiroga, "Notas sobre la historia de la democracia," 23.

47. Navarro Floria, "El proceso de construcción social de la región de Nahuel Huapi"; Freitas, *Nationalizing Nature*.

48. Jacoby, *Crimes against Nature*, 83.

49. Wakild, *Revolutionary Parks*; Freitas, *Nationalizing Nature*.

50. Cafaro, "Patriotism as an Environmental Virtue," 193. See also Bergman, *Exhibiting Patriotism*, 152.

51. Dirección de Parques Nacionales, *Parque Nacional de Nahuel Huapi*, 37–38.

52. Bustillo, *El despertar de Bariloche*, 10. See also Navarro Floria, "El proceso de construcción social de la región de Nahuel Huapi," 4.

53. Fortunato, "El territorio y sus representaciones como fuente de recursos turísticos," 316. For a succinct history of Nahuel Huapi National Park in English, see Kaltmeier, *National Parks from North to South*.

54. Gissibl, Höhler, and Kupper, "Towards a Global History," 4.

55. Dirección de Parques Nacionales y Turismo, "Memoria extraordinaria," 1944, 8, Fondo Bustillo, box 3343, AGNE. See also Bustillo, *El despertar de Bariloche*, 304; and M. Bierman to Bustillo, October 2, 1942, Fondo Bustillo, box 3343, folder 11, AGNE.

56. Karen Jones, "Unpacking Yellowstone," 40. An example of how the DPN embraced American and other countries' national parks is in Dirección de Parques Nacionales y Turismo, "Memoria extraordinaria," 5–6.

57. Dlamini, *Safari Nation*.

58. Ley de Parques Nacionales, Pub. L. No. 12.103 (1934), arts. 1, 5, Ministerio de Justicia de la Nación, InfoLEG, accessed June 5, 2024, http://servicios.infoleg.gob.ar/infolegInternet/anexos/195000-199999/196777/norma.htm.

59. E.g., Intendente del Parque Nacional "Los Alerces" to Exequiel Bustillo, November 10, 1938, Fondo Bustillo, box 3345, AGNE; Alexis Christensen to Bustillo, December 30, 1940; and Dalma Celia Lehner de Bresler to Bustillo, July 22, 1940, both in Fondo Bustillo, box 3346, AGNE.

60. Dante Roca to Bustillo, August 1939, Fondo Bustillo, box 3346, AGNE. Bustillo lay out this vision at the end of his term in Dirección de Parques Nacionales y Turismo, "Memoria extraordinaria."

61. Bohoslavsky, *El complot patagónico*, 234–37.

62. Bustillo, *El despertar de Bariloche*, 75.

63. Bustillo to Toribio Ayerza, July 17, 24, 1939; and Ayerza to Bustillo, July 19, 1939, all in Fondo Bustillo, box 3346, AGNE.

64. Bustillo to Sr. Varela, December 13, 1934, folio 141r; and to José Barbagelata, December 31, 1934, folio 152r–v, both in Fondo Bustillo, box 3351 "Correspondencia remitida," copybook 1, AGNE.

65. Emilio Frey to Bustillo, March 11, 1935, box 3344; and [illegible] to Bustillo, August 18, 1938, box 3345, both in Fondo Bustillo, AGNE.

66. Bustillo, *El despertar de Bariloche*, 281.

67. This sentiment appears clearly on a map pinpointing possible battlegrounds along the cordillera, "Probable concentración de beligerantes," in Informe que contribuye al planteo y desarrollo del problema operativo III, anexo 7, 1936, Servicio Histórico del Ejército, Buenos Aires.

68. Bustillo, *El despertar de Bariloche*, 26, 30.

69. Bustillo, *El despertar de Bariloche*, 34.

70. Bustillo, *El despertar de Bariloche*, 35.

71. Bustillo, *El despertar de Bariloche*, 36.

72. Bustillo, *El despertar de Bariloche*, 36.

73. Bustillo, *El despertar de Bariloche*, 56.

74. Bustillo to Toribio Ayerza. Bustillo to Toribio Ayerza, July 17, 24, 1939; and Ayerza to Bustillo, July 19, 1939, all in Fondo Bustillo, box 3346, AGNE.

75. Navarro Floria, "El proceso de construcción social de la región de Nahuel Huapi," 4–5.

76. Navarro Floria, "El proceso de construcción social de la región de Nahuel Huapi," 3–6.

77. Horacio Fernández Beschtedt to Bustillo, June 15, 1944, Fondo Bustillo, box 3343, AGNE.

78. Sarobe, *La Patagonia y sus problemas*. For more on Sarobe, see Navarro Floria and Núñez, "Un territorio posible en la república imposible."

79. Sarobe, *La Patagonia y sus problemas*, 134.

80. Sarobe, *La Patagonia y sus problemas*, 135–36.

81. Sarobe, *La Patagonia y sus problemas*, chaps. 22–23.

82. Sarobe, *La Patagonia y sus problemas*, 35. The 1930 Chilean census reported 419,614 people living in the provinces of Chiloé and Valdivia. Dirección General de Estadística, "Resultados del X Censo." In 1928, the government had divided the province of Llanquihue and assigned the northern part, Osorno, to Valdivia and the southern part—Puerto Montt, Chiloé, Palena, and farther south—to a new jurisdiction based in Chiloé. The following year, the southern part of this new province was transferred to the new Aysén Province. From 1940 through 1976, the province of Osorno existed as a jurisdiction separate from Valdivia, Llanquihue, and Chiloé. See Ministerio del Interior [Chile], "Decreto que fija"; Ministerio del Interior [Chile], Decreto 8583; and Ministerio del Interior [Chile], "Crea la provincia."

83. In fact, the Argentine military theorized that a war could occur against a Chile-Brazil alliance. See "Probable orientación y desarrollo inicial que asumirán las operaciones, en el supuesto de una próxima guerra de la Argentina contra la alianza de Chile-Brasil," 1936, Servicio Histórico del Ejército, Buenos Aires; and "Informe que contribuye al planteo y desarrollo del problema operativo III," 1936, Servicio Histórico del Ejército, Buenos Aires.

84. "Resumen del trabajo del Coronel José Ma. Sarobe de 'La Patagonia y sus problemas,'" 10, Ejército Argentino, 1934, Servicio Histórico del Ejército, Buenos Aires. Emphasis in the original.

85. "Resumen del trabajo del Coronel José Ma. Sarobe," 28.

86. "Resumen del trabajo del Coronel José Ma. Sarobe," 28.

87. Bustillo, *El despertar de Bariloche*, 195–96.

88. Bustillo, *El despertar de Bariloche*, 42.

89. For some examples of the centennial cultural movement in different arts, see Altamirano and Sarlo, "La Argentina del Centenario"; Gorelik and Silvestri, "Lo nacional en la historiografía de la arquitectura en la Argentina"; Goodrich, "La construcción de los mitos nacionales en la Argentina del centenario"; Delaney, "Imagining 'El Ser Argentino'"; and Cuarterolo, "El arte de 'instruir deleitando.'"

90. The symbolic opportunity to literally build the nation around centennial celebrations in Latin America can also be found in the following comparative studies: Fernández Bravo, "Celebraciones centenarias"; and Romero Vázquez and Betancourt Mendieta, "Emblemas del progreso."

91. Altamirano and Sarlo, "La Argentina del centenario"; Devoto, *Nacionalismo, fascismo y tradicionalismo*, chap. 2; Elbirt, "Representaciones de la nación en el ensayo argentino," 43.

92. For more on Catholic nationalism in Argentina, which provided a context for the DPN's choices, see McGee Deutsch, *Counterrevolution in Argentina*, 194–97; and Zanatta, *Del estado liberal a la nación católica*.

93. Prasad, "'Time-Sense.'"

94. Alexis Christensen to Exequiel Bustillo, November 30, 1940, Fondo Bustillo, box 3346, AGNE.

95. Lolich, "Bariloche y su centro cívico," 67.

96. Gorelik and Silvestri, "Past as the Future," 427.

97. Prefect of the Seine Georges-Eugène de Haussmann (1853–70) led an urban renovation in nineteenth-century Paris guided by the need for better circulation, better hygiene, and the erection of monuments. The old city gave way to parks, boulevards, and public works. For Haussmann's influence in Latin America, see Almandoz, "Urbanization and Urbanism in Latin America."

98. Even after Exequiel Bustillo left the DPN, the Bariloche style prevailed in new architecture. See Piantoni, Barrios García Moar, and Pierucci, "Las bellezas panorámicas argentinas," 243.

99. Barnitz and Frank, *Twentieth-Century Art of Latin America*, 181–207; Holston, *Modernist City*, 31–58; Gorelik, *La grilla y el parque*, 364–400. As a forum for discussion, CIAM did not have a unitary voice. For more on the internal history of CIAM toward a consensus, see Mumford, *CIAM Discourse on Urbanism*.

100. For more on the uneven agreement of the Charter of Athens and Le Corbusier's influence, see Gold, "Creating the Charter of Athens."

101. Bustillo, *El despertar de Bariloche*, 213.

102. Gorelik and Silvestri, "Past as the Future," 432–33.

103. For more on the urban transformation of Buenos Aires in the 1930s, see Ballent and Gorelik, "País urbano o país rural," 171–76.

104. Dirección de Parques Nacionales, *Parque Nacional de Nahuel Huapi*, 33–35.

105. Dirección de Parques Nacionales, *Obra pública, cultural y turística*.

106. For more on the creation of historical memory around the figure of Julio A. Roca, see Cersósimo and Lopes, "Julio A. Roca y la 'conquista del desierto.'"

107. Bustillo, *El despertar de Bariloche*, 221–22.

108. Bustillo, *El despertar de Bariloche*, 221–22.

109. See, e.g., Valdivieso Valenzuela and Coll-Hurtado, "La construcción y evolución del espacio turístico de Acapulco"; Flores, Gamarra, and Florián, *La ciudad y el mar*; and Ribeiro, "Rio de Janeiro e a Avenida Beira Mar."

110. Comisión Nacional de Museos, Monumentos y Lugares Históritocs et al., *Patrimonio Arquitectónico y Urbano de San Carlos de Bariloche*, vol. 1, file E21; vol. 2, files S and E72.

111. Comezaña to Vicente San Martín, May 29, 1938, Fondo Bustillo, box 3345, AGNE.

112. Bustillo to Alexis Christensen, July 12, 1939, Fondo Bustillo, box 3346, AGNE.

113. "Discurso de un poblador de Bariloche," ca. 1944, Fondo Bustillo, box 3343, AGNE. See also C. Müller, "Discurso de Ing. C. Müller," November 1, 1942, Fondo Bustillo, box 3343, AGNE; and *Vista parcial del Hotel Italia del Sr. Andrés Festa y primera sucursal del Banco Nación | Bariloche*, photograph, ca. 1942, Archivo Visual Patagónico.

114. Emilio Frey to Exequiel Bustillo, July 3, 1938, box 3345; and Pascual Gambino to Bustillo, May 5, 1940, box 3346, both in Fondo Bustillo, AGNE.

115. Gustavo Eppens to Exequiel Bustillo, February 4, 1935, Fondo Bustillo, box 3344, AGNE.

116. Juan Bautista Barbieri to Exequiel Bustillo, April 2 (two letters), May 16, 1940, Fondo Bustillo, box 3346, AGNE.

117. Bustillo, *El despertar de Bariloche*, 243.

118. Piantoni, "Objetos cotidianos," 117.

119. See, e.g., Artayeta to Susana O'Donnell, June 1940, Colección Artayeta, Cartas Recibidas, binder 1, BMCE 1O, folio 12r, ADMP.

120. Ramos Mexía, "Prólogo," 23.

121. Artayeta to Teodoro Aramendía, November 1940, Colección Artayeta, Cartas Recibidas, binder 1, BMCE 1A, folios 19r–20r, ADMP.

122. Bustillo to Pablo Nogués, October 13, 1939, Fondo Bustillo, box 3346, AGNE.

123. Bustillo, *El despertar de Bariloche*, 237.

124. Bustillo, *El despertar de Bariloche*, 236.

125. Biedma, *Crónica histórica del lago Nahuel Huapi*, 223–24.

126. Bustillo to Christensen, July 12, 1939, AGNE; Intendencia del Parque Nacional Nahuel Huapi to Bustillo, August 3, 1939, Fondo Bustillo, box 3346, AGNE.

127. Bustillo, *El despertar de Bariloche*, 239.

128. Stained-glass images can be found at "Iglesia Catedral Nuestra Señora del Nahuel Huapi | Vitraux," Iglesia Catedral de Bariloche (website), accessed August 1, 2022, https://iglesiacatedralbariloche.com/vitrales.html.

129. For a summary of the garden city diffusion in Latin America, see Almandoz, "Garden City in Early Twentieth-Century Latin America." For the movement's influence in Argentina, see Gutiérrez, "Los inicios del urbanismo en la Argentina."

130. Wakeman, *Practicing Utopia*, 2.

131. For Iguazú National Park, see Freitas, *Nationalizing Nature*, chap. 2. For Nahuel Huapi National Park, see Navarro Floria, "El proceso de construcción social de la región de Nahuel Huapi."

132. Bustillo, *El despertar de Bariloche*, 281.

133. "Solicitudes presentadas en el ofrecimiento de tierras en San Carlos de Bariloche," ca. 1938, Fondo Bustillo, box 3345, AGNE.

134. A. C. Giovaneli Perdiel to Bustillo, November 14, 1938, Fondo Bustillo, box 3345, AGNE.

135. A. Bättig to Bustillo, September 21, 1938; and Julio Peña to Bustillo, October 5, 1938, both in Fondo Bustillo, box 3345, AGNE.

136. Vargas and Klier, "Representaciones de Naturaleza en Isla Victoria." See also Núñez and Núñez, "Naturaleza construida."

137. Piantoni, "Un laboratorio a cielo abierto," 64–65.

138. Intendente del Parque Nacional "Los Alerces" to Bustillo, November 10, 1938, AGNE.

139. Ministerio de Agricultura, "Decreto declarando caduca la concesión de un lote en la Colonia Nahuel Huapi," 354; "Desde Bariloche," *La Nueva Era*, March 3, 1907.

140. Dirección de Parques Nacionales, *Parque Nacional Nahuel Huapi: Guía*, 64.

141. Briones and Cardin, *Informe preliminar sobre ocupación tradicional*.

142. "Registro de la Escuela Infantil no 16 de Río Negro," ca. 1910, Colección Frey, binder 6, ADMP.

143. Navarro Floria, "El proceso de construcción social de la región de Nahuel Huapi," 8.

144. Dirección de Parques Nacionales, "Permiso precario de ocupación y pastaje no 0020," October 1, 1937, Colección General, Permisos, ADMP.

145. Bustillo, *El despertar de Bariloche*, 288.

146. Bustillo, *El despertar de Bariloche*, 289.

147. Vallmitjana, *A cien años de la colonia agrícola Nahuel Huapi*, 17.

148. García and Valverde, "Políticas estatales y procesos de etnogénesis."

149. Hans Nöbl to Bustillo, March 2, 1939, Fondo Bustillo, box 3346, AGNE; Club Andino Bariloche, *Memoria, 1936*, 33.

150. Hans Nöbl to Bustillo, August 10, 1937; and Ehans Voehl to Bustillo, August 21, 1937, both in Fondo Bustillo, box 3345, AGNE.

151. Lolich, "Ernesto de Estrada como urbanista pionero," 51.

152. Dirección de Parques Nacionales, *Memoria correspondiente al año 1937*, 58.

153. Lolich et al., "Estado y paisaje," 68.

154. Directorio de la Dirección de Parques Nacionales, "Acta de sesión no 173," September 15, 1942, 5, Fondo Bustillo, box 3343, AGNE.

155. Adalberto Pagano, "Discurso del señor gobernador de Río Negro, Ing. Adalberto Pagano, Inauguración Hotel Llao-Llao," January 8, 1938, 1, Fondo Bustillo, box 3343, AGNE.

156. Anonymous to Bustillo, November 18, 1937, Fondo Bustillo, box 3345, AGNE.

Chapter Six

1. Carlota Andrée, "Los Bellos Paisajes de América," *En Viaje*, June 1937.

2. Urry and Larsen, *Tourist Gaze 3.0*, 119.

3. Pritchard and Morgan, "Privileging the Male Gaze."

4. Paula Gabriela Núñez, "'She-Land.'"

5. Ecofeminist scholars and activists have long challenged the portrayal of forests and women in terms of their virgin passivity. Some readings I found both useful and inspirational for this analysis include Plant, "Learning to Live with Differences"; and Simard, *Finding the Mother Tree*.

6. Andrée, "Los Bellos Paisajes de América," 37.

7. Wenceslao Landaeta, "Turismos: Las bellezas naturales del pais acrecentarin el Turismo entre el elemento extranjero y practicado entre conacionales, la riqueza pública." *En Viaje*, December 1933, 51; and "Viajar es educarse," *En Viaje*, December 1933.

8. Walsh, *Religion of Life*, chap. 4.

9. Collier and Sater, *Historia de Chile*, chap. 8.

10. For a detailed account on the making of this alliance, see Milos Hurtado, *Frente Popular en Chile*.

11. Booth, "Turismo y representación del paisaje," para. 7.

12. Barr-Melej, *Reforming Chile*, chaps. 3–4; Silva, "La espacialidad y el paisaje."

13. While Landaeta Sepúlveda did not appear on the editorial masthead of *En Viaje*, it is safe to assume that the transition to the next director occurred when his name began to appear in each issue. See García Matus de la Parra and Valdivia Garrido, "La Empresa de los Ferrocarriles del estado de Chile"; "Una oportunidad única para el turismo chileno," *En Viaje*, October 1939; and "Una sugerencia a los esquiadores," *En Viaje*, June 1943.

14. Booth, "Turismo y representación del paisaje," para. 17.

15. Dlamini, *Safari Nation*, 145.

16. Juan Esteban Iriarte, "Juan Antonio Ríos: Recuerdos juveniles," *En Viaje*, April 1942, 70.

17. *Guía del Veraneante*, 1941, 138. See also Schmohl, F. "El turismo deportivo en la impresionante región de Lago Ranco y Río Bueno," *Turismo Austral*, Agosto 1934, 31; Carlos Silva Vildosola, "Turismo estrangulado," *Turismo Austral*, January 1935, 46; *Guía del Veraneante*, 1940, 121, 144; *Guía del Veraneante*, 1943, 129; and "Excursiones en Chiloé y Puerto Montt," *Turismo Austral*, February 1935, 43.

18. "El llamado del sur," *En Viaje*, December 1939, 9.

19. Stella Corvalán, "Paisajes de Chile," *En Viaje*, February 1942, n.p.

20. "Cosas de Hollywood: El viaje de Errol Flynn por las Antillas," *En Viaje*, July 1938, n.p.

21. Some pieces in which this contrast appears are "Lo que debe saber una mujer," *En Viaje*, March 1935, n.p.; and Yvonne Burgalat, "A la luz de inmensos incendios surgió el cuerpo de bomberos," *En Viaje*, December 1943, n.p.

22. "Periodistas extranjeros opinan sobre el porvenir de Chile como país de turismo," *En Viaje*, February 1935, n.p.

23. Juvenal Guerra, "Guía de turismo de ferrocarriles del estado," *En Viaje*, January 1935, 25.

24. Servicio Nacional de Geología y Minería, *Chile*, 40–41.

25. E.g., see the images accompanying descriptions of Lake Puyehue and Pilmaiquen Falls in *Guía del Veraneante*, 1936, 148–51.

26. "Cautín," *En Viaje*, December 1945.

27. A few examples include "Jardinería, hortalizas, siembras," *En Viaje*, January 1935; "Turismo de invierno," *En Viaje*, May 1936; "Turismo de invierno," *En Viaje*, June 1941; and "Sociedad Fábrica de Cemento de 'El Melón,'" advertisement, *En Viaje*, June 1941.

28. "Santiago y el invierno," *En Viaje*, July 1942.

29. See, e.g., *Guía del Veraneante*, 1942, 149–43, 159, 161–73, 213–23.

30. "Febrero y el turismo," *En Viaje*, February 1942, n.p.

31. "Week end," *En Viaje*, October 1941, n.p.

32. "Lagos de Chile," *En Viaje*, February 1943.

33. For reproductive concerns in Chile, see Walsh, *Religion of Life*, 22–36. For that in other countries in Latin America, see Drinot, *Sexual Question*, chap. 1; and Hershfield, *Imagining La Chica Moderna*.

34. "Se inauguro el camino de P. Varas a Ensenada," *En Viaje*, January 1942, n.p.

35. Purcell, "Una mercancía irresistible."

36. "Insert [images of southern Chile]," *En Viaje*, March 1935; "Tres estrellas del cine: Elisa Landi, Jean Harlow y Dorothy Jordan, consideradas unas de las más hermosas de Hollywood," *En Viaje*, March 1935. See also "Tres de las más populares figuras femeninas del cine," *En Viaje*, February 1935, n.p.

37. E.g., "Damas de Antofagasta," *En Viaje*, March 1942, n.p.; "Damas de Osorno," *En Viaje*, January 1944, n.p.; and "Damas de Punta Arenas," *En Viaje*, February 1944, n.p.

38. "Una revista para los que viajan," *En Viaje*, November 1933.

39. "Boletos de invierno," *En Viaje*, June 1936, n.p.

40. E.g., "Para la mujer hacendosa: Cómo pueden emplearse diversos retazos," *En Viaje*, January 1936; "Página femenina: La mujer ante el matrimonio," *En Viaje*, March 1940; and André Maurois, "Los mandamientos matrimoniales," *En Viaje*, January 1943.

41. *Guía del Veraneante*, 1942, 139.

42. "Viaje usted con comodidad," advertisement, *En Viaje*, February 1938, n.p.; *Guía del Veraneante*, 1942, 250. For examples of itineraries, see "Resumen de los itinerarios entre Santiago y Puerto Montt y Ramales, Verano 1933-34," *En Viaje*, December 1933; and "Resumen de los itinerarios entre Santiago y Puerto Montt y Ramales, Invierno 1938," *En Viaje*, August 1938.

43. "El hotel de Puerto Varas ya está al servicio del turismo en la región del Lago Llanquihue," *En Viaje*, February 1938.

44. "El hotel de Puerto Varas."

45. Booth, "Turismo y representación del paisaje," para. 24.

46. "Ir al sur," *En Viaje*, February 1940. See also "El turismo de invierno," *En Viaje*, January 1935.

47. Alberto Carrasco Hermoza, "Chile, país de la belleza y de la democracia," *En Viaje*, March 1942, 43.

48. "La preparación de los viajes de vacaciones," *En Viaje*, November 1938.

49. Ballent, "Ingeniería y estado," 828.

50. Piglia, *Autos, rutas y turismo*, 187–88.

51. Gorelik, "La arquitectura de YPF," 194–95; Ballent and Gorelik, "País urbano o país rural," 189–91; Mallol i Moretti, "Institucionalización del imaginario moderno," 18.

52. Piglia, *Autos, rutas y turismo*, 27–37. For a succinct analysis on both clubs in the expansion of roads for leisure travel, see Ballent, "Kilómetro cero."

53. Mallol i Moretti, "Institucionalización del imaginario moderno," 11.

54. Jaime Gleg Bonorino, "Las bellezas del sur argentino: Ruta de turismo hacia la Patagonia," *Automovilismo*, April 1934; and Virginia Grego, "Las etapas patagónicas del Gran Premio Internacional," *Automovilismo*, January 1936. See also Ballent, "Kilómetro cero," 120–21.

55. Piglia, *Autos, rutas y turismo*, 77.

56. Fermín A. Blanco, "Excursiones a la región de los lagos," *Automovilismo*, October 1936.

57. *Automovilismo*, January 1934, cover; *Automovilismo*, June 1941, cover. For a deep analysis on the relationship between YPF and ACA in 1936–55, see Piglia, *Autos, rutas y turismo*, chap. 6.

58. See, e.g., Dirección de Parques Nacionales, *Obra pública, cultural y turística*, n.p.

59. "Avanza el desarrollo del plan A.C.A-Y.P.F.," *Automovilismo*, May 1940, n.p.

60. Dirección de Parques Nacionales, *Parque Nacional Nahuel Huapi: Guía*, 22–26.

61. Bustillo to Ricardo Silveyra, September 10, 1937, Fondo Bustillo, box 3351 "Correspondencia remitida," copybook 1, folio 14r-v, AGNE.

62. Dirección de Parques Nacionales, *Parque Nacional Nahuel Huapi: Guía*, 49–50. The old encampment, which included administrative offices, was located on land owned by the Crespo family, who made an appearance in earlier chapters of this book. For more on Sobral, including partial transcripts from his notebooks, see Ottone, "Sobral y la geología del Ñirihuau."

63. Agustín Rosas and Miguel Berro Madero to Exequiel Bustillo, January 1942; and Ricardo Silveyra to Bustillo, February 21, 1942, both in Colección Artayeta, Cartas Recibidas, binder 1, BMCE 1S, ADMP.

64. Club Andino Bariloche, "Libro de excursiones no 1," 1931–32, folios 38r–39r, Biblioteca del Club Andino Bariloche, Argentina.

65. Finó, *Andinismo en la Argentina*, n.p.

66. Carey, "Mountaineers and Engineers."

67. Finó, *Andinismo en la Argentina*. For the German expeditions, see Deutscher Wissenschaftlicher Verein, *Patagonia*.

68. Cornaglia, *Bariloche*, 47–49.

69. Club Andino Bariloche, *Memoria, 1942*, 74.

70. Club Andino Bariloche, *Memoria, 1941*, 33.

71. As Jacob Dlamini concludes, leisure provided disenfranchised groups a way to participate in society. Dlamini, *Safari Nation*, 95–101.

72. For examples of nation-making through sports in Latin America, see Louis A. Perez, "Between Baseball and Bullfighting"; Bocketti, *Invention of the Beautiful Game*; and Yoder, *Pitching Democracy*.

73. Colpan, Hachleitner, and Marschik, "Jewish Difference in the Context of Class, Profession and Urban Topography"; Chang, "Women in the Chase." Social class certainly played a crucial role in organizing sports in southern Chile and Argentina. In both countries, as in other parts of the world, sports clubs used other criteria to bring people together, such as race, ethnicity, and cultural background. See, e.g., the cases of the original Celtics in New York City or the foundation of a working-class football (soccer) club in Lima, to name a few. See Nelson, "Basketball as Cultural Capital"; and Abanto, *Una pelota de trapo*, chap. 3. See also Stein, "Miguel Rostaing."

74. Elsey, "Sport in Latin America," 362; Losada, "Sociabilidad, distinción y alta sociedad." For analyses on sports as social distinction, see Hernández Barral, "Polo"; and Dauncey, *French Cycling*, 24–28, 57–58.

75. "El turismo de invierno," January 1935. Among the governing elites of 1930s Chile and Argentina, the term "race" corresponded to a representation of a desirable inhabitant that could ethnically and culturally homogenize the nation. Argentines formulated their "race" as white Europeans and Chileans as a superior hybrid of immigrant and Native. For some analyses of both trajectories, see Halperín Donghi, *Una nación para el desierto argentino*; and Helg, "Race in Argentina and Cuba."

76. In some instances, the return to nature was framed as a modern practice, but in others it was a form of protesting modernism. See, e.g., Kaufmann, "'Naturalizing the Nation'"; and John Alexander Williams, *Turning to Nature in Germany*.

77. "Recado a los turistas rezagados," *En Viaje*, March 1940. See also "Otro Gran Hotel de Turismo," *En Viaje*, January 1938.

78. "Distribución de fondos acordados para fomento del turismo," *En Viaje*, February 1935.

79. Emilio Frey to Exequiel Bustillo, telegram, ca. 1939, Fondo Bustillo, box 3346, AGNE.

80. Paula Gabriela Núñez, "La dinámica de una localidad desde la articulación de sus instituciones."

81. Club Andino Bariloche, *Memoria, 1931–1932*, 7.

82. Kinzel Kahler and Horn Klenner, *Puerto Varas*, 333–34; López Cárdenas, *Osorno entre Julio Buschmann y René Soriano*, 9–10.

83. "Refugio Cerro López," *En Patagonia*, 2020. For a list of hiking refuges in the Andes known to the CAB, see Club Andino Bariloche, *Memoria, 1944*, 105.

84. "Club Andino Osorno: Antillanca," Antillanca Ski Chile, accessed July 27, 2022, https://antillanca.cl/quienes-somos-club-andino-osorno.

85. Paula Gabriela Núñez, "La región del Nahuel Huapi."

86. Paula Gabriela Núñez, "Memorias fragmentadas entre lo alpino y lo andino."

87. Club Andino Bariloche, "Libro de excursiones no 1," folios 1r–4r, Biblioteca del Club Andino Bariloche; Club Andino Bariloche, *Memoria, 1931–1932*, 4–6.

88. Club Andino Bariloche, *Memoria, 1931–1932*, 9.

89. Club Andino Bariloche, *Memoria, 1936*, 33; Dirección de Parques Nacionales, *Obra pública, cultural y turística*, n.p.

90. Club Andino Bariloche, *Memoria, 1931–1932*, 9–11, 19.

91. "Interesante programa de ski se organiza en la región de Nahuel Huapi," *La Nueva Era*, July 30, 1938.

92. Club Andino Bariloche, "Informe anual correspondiente al 13º ejercicio," 10.

93. Juan Carlos Vignale to Exequiel Bustillo, August 1942; and "Noticias Gráficas" director [name illegible] to Bustillo, July 7, 1941, both in Fondo Bustillo, box 3347, AGNE.

94. See, e.g., Atilano González to Bustillo, June 19, 1935, box 3344; Alfredo Ragusi to Bustillo, May 10, 1937, box 3345; and Pablo González Amorin to Bustillo, May 19, 1940, box 3346, all in Fondo Bustillo, AGNE.

95. Juan Neumeyer to Bustillo, August 20, 1937, Fondo Bustillo, box 3345, AGNE.

96. Bustillo to Mayor Famin, June 24, 1938, Fondo Bustillo, box 3351, libro copiador 1, folio 332r.

97. Nöbl to Bustillo, August 10, 1937, April 5, 1938, box 3345; and to Bustillo, July 8, 1939, box 3346, all in Fondo Bustillo, AGNE.

98. Bustillo to Mayor Famin, June 24, 1938.

99. Mársico, "Disputas por el sentido y el acceso a la práctica de esquí," 305; Navarro Floria, "El proceso de construcción social de la región de Nahuel Huapi," 9.

100. Bruno Ricardo Sálamon, *Playa Bonita i Nahuel Huapi*, photograph, ca. 1937, Archivo Visual Patagónico.

101. Dirección de Parques Nacionales, "Proyecto de ley de becas de especialización en silvicultura," September 15, 1942, Fondo Bustillo, box 3343, AGNE.

102. Bustillo to Carlos Hensel, May 8, 1939; and Hensel to Bustillo, June 27, 1939, both in Fondo Bustillo, box 3346, AGNE.

103. "Contrato entre Mercados de Haciendas y Carnes, la Dirección de Parques Nacionales y la Municipalidad de San Carlos de Bariloche," November 17, 1939, 1, Fondo Bustillo, box 3343, AGNE.

104. Club Andino Bariloche, "Libro de excursiones no 1," folio 11r, Biblioteca del Club Andino Bariloche.

105. Dirección de Parques Nacionales, "Permiso precario de ocupación y pastaje no 0047," November 8, 1938, Colección General, Permisos, ADMP. The Cayun family had resided on the northern side of Lake Nahuel Huapi, near the border with Chile. See Biedma, *Crónica histórica del lago Nahuel Huapi*, 152. Barbagelata's permit was for grazing in the Huemul district, an area across the lake from Bariloche where the family later had a small hotel. Dirección de Parques Nacionales, "Permiso precario de ocupación y pastaje no 0161," June 3, 1941, Colección General, Permisos, ADMP; Administración General de Parques Nacionales y Turismo, *Memoria correspondiente al año 1946*, 72.

106. *Guía del Veraneante*, 1936, 131.

107. *Guía del Veraneante*, 1936, 83.

108. Patroni, *Bellezas de los lagos argentinos-chilenos*. For a comparative analysis of Patroni's and the DPN's guidebooks, see Picone, "La idea de turismo."

109. Luis Durand, "Mercedes Urizar," *En Viaje*, February 1935, 58; *Guía del Veraneante*, 1936, 74–75; "El Gran Hotel Termas de Puyehue, una organización perfecta en servicio de la salud y el turismo," *En Viaje*, December 1944, 94.

110. *Guía del Veraneante*, 1942, 21–22, 176–78.

111. See, e.g., *Guía del Veraneante*, 1943, 197–203.

112. "Viajando por el sur," *En Viaje*, March 1943; "Hotel Bellavista," advertisement, *El Llanquihue*, January 1, 1937.

113. Pablo Cora to Frey, August 15, 1918, Colección Frey, binder 1, Turismo, document 5, ADMP. See also Navarro Floria, "La 'Suiza argentina,'" 13; Méndez, *Estado, frontera y turismo*, 186–87; and Montt de Etter, *Inmigracion suiza*, 148–53.

114. "Gran Hotel Suizo," advertisement, *Turismo Austral*, February 1935; "Hotel Italia," advertisement, *Turismo Austral*, February 1935; Sucesión Primo Capraro, "Navegación en el Lago Nahuel Huapi," *Turismo Austral*, February 1935.

115. "Entusiastas andinistas ascendieron hasta 1920 metros de altura en el volcán Osorno," *Turismo Austral*, September 1934.

116. Patroni, *Bellezas de los lagos argentinos-chilenos*, 62.

117. Finó, *Andinismo en la Argentina*, section 1.

Conclusion

1. For more on the cultural history of the Colón, see Benzecry, "Opera House."

2. This division was perhaps a bit different from the government's groupings (Northeast, Northwest, Center, Pampa, and Patagonia) but nonetheless applicable. For a discussion on Argentine regions, see Benedetti, "Los usos de la categoría región."

3. Mozzi et al., *Argentum*.

4. Ramifications of this persistence include the development of a "green economy," as examined by Marcos Mendoza. See Mendoza, *Patagonian Sublime*.

5. "Descansan desde ayer en la isla Centinela los restos del perito Francisco Moreno," *La Nación* (Buenos Aires), January 23, 1944.

6. Massey, *For Space*, 59.

Bibliography

Primary Sources
ARCHIVAL REPOSITORIES

Argentina
    Archivo Documental del Museo de la Patagonia (Bariloche)
    Archivo General de la Nación (Buenos Aires)
        Archivo Intermedio
        Documentos Escritos
    Archivo Regional Bariloche (Bariloche)
    Biblioteca Automóvil Club Argentino (Buenos Aires)
    Instituto Geográfico Nacional (Buenos Aires)
    Museo Emma Nozzi (Carmen de Patagones)
    Servicio Histórico del Ejército (Buenos Aires)

Chile
    Archivo Histórico del Ministerio de Relaciones Exteriores
        de la República de Chile (Santiago)
    Archivo Histórico de Puerto Montt
    Archivo Histórico Nacional (Santiago)
    Biblioteca Emilio Held (Santiago)
    Biblioteca Nacional (Santiago)
    Museo Histórico de Osorno

Digital Repositories
    Archivo Nacional de la Administración (Chile)—Conservador de Bienes Raíces
    Archivo Visual Patagónico (Argentina)
    Biblioteca Nacional (Argentina)—Colecciones Digitales
    Biblioteca Nacional Digital (Chile)

Ibero Amerikanisches Institut, Nachlass, Korrespondenz von
   Personen und Körperschaften A-Z (Germany)
Landesbibliothek Mecklenburg-Vorpommern Günther Uecker (Germany)

NEWSPAPERS AND PERIODICALS

Argentina
- *Automovilismo* (Buenos Aires)
- *El Libre Pensador* (Buenos Aires)
- *En Patagonia* (Buenos Aires)
- *La Nación* (Buenos Aires)
- *La Nueva Era* (Carmen de Patagones and Viedma)
- *La Prensa* (Buenos Aires)

Chile
- *Corre Vuela* (Santiago)
- *Deutsche Zeitung* (Valdivia)
- *El Faro de Llanquihue* (Puerto Varas)
- *El Llanquihue* (Puerto Montt)
- *El Mercurio* (Santiago)
- *El Reloncaví* (Puerto Montt)
- *En Viaje: Revista Mensual de los Ferrocarriles del Estado* (Santiago)
- *Guía del Veraneante*
- *La Nación* (Santiago)
- *Revista de Artes y Letras* (Santiago)
- *Turismo Austral: Revista Mensual Pro-fomento del Turismo en la Zona Austral de Chile* (Valdivia)

Rest of the World
- *Cardiff (UK) Times*
- *Chicago Daily Tribune*
- *Druid* (Scranton, PA)
- *El Comercio* (Lima)
- *Evening Express* (Cardiff, UK)
- *Los Angeles Times*
- *San Francisco Chronicle*
- *Weekly Mail* (Cardiff, UK)
- *Wide World Magazine* (London)
- *Y Rhedegydd* (Blaenau Ffestiniog, UK)

MAPS AND ATLASES

Dirección General de Correos y Telegráfos [Argentina]. "República Argentina [material cartográfico]: Carta de las comunicaciones postales y telegráficas / confeccionada según los datos, observaciones y exploraciones hechas por la Dirección General de Correos y Telégrafos; E. Escalante dibuj. cartógrafo Dn. Gl. de Cs. y Tl." 1:2,225,000 scale. Litografía "La Nueva Artística" de Alejandro Bianchi, 1904. Biblioteca Nacional (Argentina)—Colecciones Digitales.

Gutiérrez, Diego, and Hieronymus Cock. *Americae sive qvartae orbis partis nova et exactissima descriptio.* 83 × 86 cm, on sheet 100 × 102 cm. Antwerp, 1562. Library of Congress, Washington, DC. http://hdl.loc.gov/loc.gmd/g3290.ct000342.

Hondius, Hendrik, Jan Jansson, and Gerhard Mercator. *Americae pars meridionalis.* Ca. 1:13,800,000 scale. Amstelodami [Amsterdam]: Sumptibus Henrici Hondy, 1638. Norman B. Leventhal Map Center, Boston.

Instituto Geográfico Argentino. *Atlas de la República Argentina.* Kraft, 1898. Instituto Geográfico Nacional, Mapoteca.

Jansson, Jan. *Americae pars meridionalis.* Ca. 1:15,000,000 scale. Amsterdam: Jansson, 1644. Norman B. Leventhal Map Center, Boston.

Kirchmayr, Emilio von. *Plano general de la región del Nahuel Huapí con las tierras y propiedades de la Compañía Comercial y Ganadera "Chile-Argentina."* Ca. 1910. 1:370.000 scale. Biblioteca Nacional Mariano Moreno, Buenos Aires. https://catalogo.bn.gov.ar/F/?func=direct& doc_number=000648397&local_base=GENER.

L'Isle, Guillaume de, and Nicholas Guérard. *L'Amerique meridionale: Dressée sur les observations de Mrs. de L'Academie Royale des Sciences & quelques autres, & sur les memoires les plus recens.* 46 × 61 cm. Ca. 1:18,500,000 scale. Amsterdam: Chéz R. & J. Ottens, 1708. Geography and Map Division, Library of Congress, Washington, DC.

Moll, Herman. *Map of Chili, Patagonia, La Plata, Part of Brasil.* 1:18,000,000 scale. London: Thos, Bowles and John Bowles, 1736. List 5580.062, series 64, p. 62. David Rumsey Historical Map Collection.

Moreno, Francisco P. *Apuntes preliminares sobre una excursión a los territorios del Neuquén, Río Negro, Chubut y Santa Cruz.* La Plata, Argentina: Museo de La Plata, Taller de Publicaciones, 1897. Map at the end.

Olascoaga, Manuel J. *Plano del territorio de la Pampa y Río Negro.* 1:2,000,000 scale. Oficina Topográfica Militar, ca. 1880. Instituto Geográfico Nacional (Argentina)—Mapoteca.

Rohde, Jorge. *Descripción de las gobernaciones nacionales de la Pampa, del Río Negro y del Neuquén.* Buenos Aires: Compañía Sudamericana de Billetes de Banco, 1889. Map after p. 53.

Weber, Alfredo. "Mapa geográfico-comercial con la red completa de ferrocarriles de las repúblicas Argentina, Chile, Uruguay y Paraguay." Buenos Aires: Oficina Cartográfica, 1923.

PUBLISHED PRIMARY SOURCES

Administración General de Parques Nacionales y Turismo [Argentina]. *Memoria correspondiente al año 1946.* Buenos Aires: Administración General de Parques Nacionales y Turismo, 1947.

Amunátegui, Miguel Luis. *Títulos de la República de Chile a la soberanía i dominio de la estremidad austral del continente americano.* Santiago: Imprenta de Julio Belin i Ca., 1853.

Arata, Pedro. "Contribuciones al conocimiento higiénico de la ciudad de Buenos Aires: Las variaciones de nivel de las aguas subterráneas en sus relaciones con la presión atmosférica, lluvias, enfermedades infecciosas." *Anales de la Sociedad Científica Argentina,* no. 24 (1887): 101–19.

Astorga Pereira, Juan. "Contestacion a la circular número 5, de 1ro de junio de 1906." In Consulado Jeneral de Chile en la República Arjentina, *Informe correspondiente a 1906,* 5–25.

Barros Arana, Diego. *Esposicion de los derechos de Chile en el litijio de limites sometido al fallo arbitral de S. M. B.* Santiago: Imprenta Cervantes, 1899.

———. *Historia jeneral de Chile.* Vol. 1. Santiago: Rafael Jover, 1884.

———. *La cuestion de límites Chile i la República Arjentina.* Santiago: Imprenta Cervantes, 1895. https://catalog.hathitrust.org/Record/011682269.

Bello, Andrés. "Discurso pronunciado en la instalación de la Universidad de Chile el día 17 de septiembre de 1843." Speech, Santiago, Universidad de Chile, 1843.

Bernal, Liborio. "Memoria de la gobernación de Río Negro." In Ministerio del Interior, *Memoria del Ministro [. . .] 1895,* 494–520.

Bertrand, Alejandro. *Memoria sobre la rejión central de las tierras magallánicas*. Santiago: Imprenta Nacional, 1886.

Bilbao, Francisco. *La América en peligro: Las causas del peligro y charlatanismo del progreso*. 2nd ed. Buenos Aires: Imprenta y Litografía de Bernheim y Boneo, 1862.

Boletín de la Sociedad de Fomento Fabril. Vol. 22. Santiago: Imprenta Cervantes, 1905.

Bórquez, Felipe. "Memoria de la Prefectura Apostolica de Misioneros de Castro." In Ministerio de Relaciones Esteriores, Culto i Colonización, *Anexos a la memoria*, 293–97.

Briones Luco, Ramón. "Colonias estranjeras." In *Glosario de colonización [. . .] hasta el 1º de abril de 1905*, 113–26.

———. "Fueguinos." In *Glosario de colonización [. . .] hasta el 1º de abril de 1905*, 288–92.

———. *Glosario de colonización: Leyes, decretos y demás antecedentes relativos al despacho de colonización, hasta el 10 de julio de 1904, seguido de un apéndice hasta el 1º de abril de 1905*. Santiago: Imprenta Universitaria, 1905.

———. "Indíjenas." In *Glosario de colonización [. . .] hasta el 1º de abril de 1905*, 309–11.

———. "Ocupación de terrenos fiscales." In *Glosario de colonización [. . .] hasta el 1º de abril de 1905*, 511–29.

———. "Radicación de indíjenas." In *Glosario de colonización [. . .] hasta el 1º de abril de 1905*, 661–76.

Bustillo, Exequiel. *El despertar de Bariloche*. 2nd ed. Buenos Aires: Editorial y Librería Goncourt, 1988.

Carrasco (OFM), Domingo. "Memoria de la Prefectura Apostólica de Misioneros Franciscanos de Chillán." In Ministerio de Relaciones Esteriores, Culto i Colonización, *Memoria de Relaciones Exteriores*, 295–304.

Carrasco, Gabriel. "La provincia de Santa Fe y el Chaco." *Boletín del Instituto Geográfico Argentino* 8, no. 6 (June 1887): 125–45.

Cavada, J. Daniel. "Memoria del administrador de la colonia de Llanquihue y Chiloé." In Inspección Jeneral de Colonización e Inmigración, *Memoria*, 296–335.

Chile y Arjentina: De Puerto Montt al gran lago arjentino Nahuelhuapi; Obsequio de la Sociedad Chile y Arjentina a sus accionistas. Valparaíso, Chile: Imprenta Lit. Gustavo Weidman, 1904.

Christie, Roberto. *El camino de Vuriloche i su importancia para la ganadería de la rejion austral de Chile: Diario de viaje*. Santiago: Imprenta Cervantes, 1904.

Club Andino Bariloche. "Informe anual correspondiente al 13o ejercicio." In *Memoria, 1944*, 5–17.

———. *Memoria, 1931-1932*. Bariloche, Argentina: Imprenta Nahuel Huapi, 1932.

———. *Memoria, 1936: Quinto ejercicio*. Bariloche, Argentina: Imprenta Nahuel Huapi, 1936.

———. *Memoria, 1941: Décimo ejercicio*. Bariloche, Argentina: Imprenta Nahuel Huapi, 1941.

———. *Memoria, 1942: Undécimo ejercicio*. Bariloche, Argentina: Imprenta Nahuel Huapi, 1942.

———. *Memoria, 1944: Decimotercer ejercicio*. Bariloche, Argentina: Imprenta Nahuel Huapi, 1944.

Cobos, Norberto. "Expedición minera al territorio nacional del Chubut." In *Memoria del Departamento Nacional de Minas y Geología, 1893–1894*, 51–72. Buenos Aires: Imprenta de Obras, 1895.

Comercio é Industrias, Tierras y Colonias, Agricultura y Ganadería é Immigración y recopilación de mensajes al honorable Congreso, decretos, notas y otros documentos referentes a dichos ramos. Buenos Aires: Imprenta, Litografía y Encuadernación de J. Peuser, 1899.

Comisión Central del Censo [Chile]. *Memoria presentada al Supremo Gobierno*. Santiago: Comisión Central del Censo, 1908.

Comisión de Estudios Hidrológicos. *Northern Patagonia: Character and Resources*. Vol. 1. New York: Scribner, 1914.

Compañia del Gran Ferrocarril del Sud de Buenos Aires. *Ferrocarril del Sud: Inauguración oficial de la prolongación de Bahía Blanca al Neuquén*. Buenos Aires: Imprenta y Taller de Fotograbados de Fausto Ortega, 1899.

Congreso Nacional de la República Argentina. *Diario de sesiones de la Cámara de Diputados, año 1887*. Vol. 2. Buenos Aires: Imprenta La Universidad, 1888.

———. *Diario de sesiones de la Cámara de Senadores, año 1895*. Buenos Aires: Imprenta del Congreso, 1895.

Congreso Nacional de la República de Chile. "Colonias de naturales i estranjeros: Se autoriza al ejecutivo para establecerlas." Pub. L. No. s/n (1845). Ley Chile - Biblioteca del Congreso Nacional de Chile. www.bcn.cl/leychile/navegar?idNorma=1062510&buscar=ley%2Bde%2Bcolonias.

———. *Comisión Parlamentaria de Colonización: Informe, proyectos de ley, actas de sesiones y otros antecedentes*. Santiago: Sociedad Imprenta y Litografía Universo, 1912.

———. "Fundación de poblaciones en el territorio de los indíjenas, Pub. L. No. s/n (1866)." Ley Chile - Biblioteca del Congreso Nacional de la República de Chile. www.bcn.cl/leychile/navegar?idNorma=1045956.

———. "Ley 2757." Pub. L. No. 2757 (1913). Ley Chile - Biblioteca del Congreso Nacional de Chile. www.bcn.cl/leychile/navegar?idNorma=137950.

———. "Provincia de Arauco." Pub. L. No. s/n (1852). Ley Chile - Biblioteca del Congreso Nacional de Chile. https://bcn.cl/20q8m.

Consulado Jeneral de Chile en la República Arjentina. *Informe correspondiente a 1906*. Santiago: Imprenta Nacional, 1907.

de Angelis, Pedro. *Memoria historica sobre los derechos de soberania y dominio de la Confederacion Argentina a la parte austral del continente americano, comprendida entre las costas del oceano atlantico y la Gran Cordillera de los Andes, desde la boca del Rio de la Plata hasta el cabo de Hornos, inclusa la isla de los Estados*. Buenos Aires, 1852.

Departamento de Gobierno [Argentina]. *Memoria presentada por el ministro secretario en el Departamento de Gobierno a la Honorable Legislatura de la Provincia [de Buenos Aires], años 1881–1882*. Buenos Aires: Imprenta del Siglo, 1882.

Departamento de Interior [Argentina]. "Decreto no 15580 nombrando al agrimensor D. Alfredo Baigorri para mensurar y amojonar en el Territorio del Chubut una sección de terreno de un millón de hectáreas, 4 de diciembre de 1886." In República Argentina, *Registro nacional [...] desde 1810 hasta 1890*, 11:4–5. https://hdl.handle.net/2027/uiug.30112117712973.

———. "Decreto no 15581 nombrando al ingeniero D. Julián Romero para hacer mensura y amojonamiento en el Territorio del Chubut una sección de terreno de un millón de hectáreas, 4 de diciembre de 1886." In República Argentina, *Registro nacional [...] desde 1810 hasta 1890*, 11:5–6. https://hdl.handle.net/2027/uiug.30112117712973.

———. "Decreto no 17082 reconociendo a don Ildefonso Linares el derecho de adquirir en compra 4375 hectáreas de terreno en la Gobernación del Río Negro." In República Argentina, *Registro nacional [. . .] desde 1810 hasta 1890*, 8:679–80.

———. "Decreto no 17597 reconociendo al poblador D. Guillermo Iribarne, la propiedad del área que ocupa en el Territorio del Río Negro." In *Registro nacional de la República Arjentina, año 1888 (primer semestre)*, by República Argentina, 861. Buenos Aires: Imprenta La Universidad, 1889.

———. "Nombramiento de una comision encargada de formar la lista definitiva de los expedicionarios del Rio Negro, para la distribucion de las tierras que les corresponden, 13 de diciembre de 1890." In Ministerio del Interior, *Memoria del ministro*, 159–60.

———. "Resolución concediendo á D. José Maria Saavedra 416 hectáreas y 66 áreas de terreno sobre la márgen derecha del Río Negro, Mayo 9 1888." In *Registro nacional de la República Arjentina, año 1888 (primer semestre)*, by República Argentina, 591–92. Buenos Aires: Imprenta La Universidad, 1889.

Departamento de Tierras y Colonias [Argentina]. "Decreto aprobando un contrato de arrendamiento de tierras en el Chubut celebrado con don Bernardo Grange, 8 de enero de 1898." In *Registro nacional de la República Argentina, año 1898 (primer cuatrimestre)*, by República Argentina, 223. Buenos Aires: Taller Tipográfico de la Penitenciaría Nacional, 1898.

Departamento Nacional de Minas y Geología [Argentina]. "Autorizando a Paul Ahehelm a verificar la existencia de minerales auríferos en Chubut, 6 de julio de 1897." *Boletín Oficial de la República Argentina* 5, no. 1187 (July 22, 1897): 169.

Deutscher Wissenschaftlicher Verein, ed. *Patagonia: Resultados de las expediciones realizadas en 1910 a 1916*. Vol. 1. Buenos Aires: Compañía Sud-Americana de Billetes de Banco, 1917. https://catalog.hathitrust.org/Record/012224806.

Díaz, Gastón. "Uruguay." In Kuntz Ficker, *Historia mínima*, n.p.

Die Familie Achelis in Bremen, 1579–1921. Leipzig, Germany: n.p. 1921. http://archive.org/details/diefamilieacheliooleip.

Dirección de Parques Nacionales [Argentina]. *Memoria correspondiente al año 1937*. Buenos Aires: n.p., 1938.

———. *Obra pública, cultural y turística realizada en los parques nacionales*. Buenos Aires: Sección Propaganda y Turismo, 1939.

———. *Parque Nacional de Nahuel Huapi: Historia, tradiciones y etnología*. Buenos Aires: Ministerio de Agricultura, Dirección de Parques Nacionales, 1938.

———. *Parque Nacional Nahuel Huapi: Guía*. Buenos Aires: Ministerio de Agricultura, Dirección de Parques Nacionales, 1938.

Dirección General de Estadística [Chile]. *Censo jeneral de la poblacion de Chile levantado el 15 de diciembre de 1920*. Santiago: Sociedad Imprenta y Litografía Universo, 1925.

———. *Resultados del X Censo de Población efectuado el 20 de noviembre de 1930 y Estadísticas Comparativas con Censos Anteriores*. Vol. 1. Santiago: Universo, March 1931. www.memoriachilena.gob.cl/602/w3-article-86204.html.

Dirección General de Territorios Nacionales [Argentina]. *Censo de población de los territorios nacionales, República Argentina, 1912*. Buenos Aires: Imprenta G. Kraft, 1914.

Documentación de los contratos de colonización de la Compañia Comercial y Ganadera Chile-Argentina. Valparaíso, Chile: Sociedad Imprenta y Litográfia Universo, 1912.

Döll, Guillermo. "Exploracion del territorio de Osorno para segundo centro de la colonización actual." *Anales de la Universidad de Chile*, no. 15 (1858): 81–84.

Donoso, Ricardo, and Fanor Velasco. *Historia de la constitución de la propiedad austral.* Santiago: Imprenta Cervantes, 1928.

Drake, Francis, Francis Fletcher, W. S. W. [Williams Sandys Wright] Vaux, John Cooke, Francis Pretty, Nuno da Silva, Edward Cliffe, and Lopo Vaz. *The World Encompassed by Sir Francis Drake: Being His Next Voyage to That to Nombre de Dios; Collated with an Unpublished Manuscript of Francis Fletcher, Chaplain to the Expedition; With Appendices Illustrative of the Same Voyage, and Introduction.* London: Printed for the Hakluyt Society, 1854.

Elflein, Ada M. *Paisajes cordilleranos: Descripción de un viaje por los lagos andinos.* Buenos Aires: La Prensa, 1917. http://archive.org/details/paisajescordilleooelfl.

Escobedo, J. "El autóctono sud-americano." *Boletín del Instituto Geográfico Argentino* 3, no. 7 (1882): 129–32, 146–49.

"Expedición al Neuquén de los doctores Kurtz y Bodembender." *Boletín del Instituto Geográfico Argentino* 10, no. 10 (1889): 311–23.

Fernández, Alejandro. "La ley argentina de inmigración de 1876 y su contexto histórico." *Almanack*, no. 17 (December 2017): 51–85.

Fernández, Federico W., "Crónica geográfica." *Boletín del Instituto Geográfico Argentino* 9, no. 8 (August 1888): 213–17.

Finó, José Frederic. *Andinismo en la Argentina.* Bariloche, Argentina: Club Andido Bariloche, 1949.

Fonck, Francisco. *Exámen crítico de la obra del señor perito arjentino Francisco P. Moreno. Cuestión Chileno-Arjentina de Límites.* Valparaíso, Chile: Carlos F. Niemeyer, 1902.

———. *Introducción a la orografía y a la jeolojía de la rejión austral de Sud-América.* Valparaíso, Chile: Carlos F. Niemeyer, 1893.

———. *Libro de los diarios de Fray Francisco Menéndez.* Vol. 1. Valparaíso, Chile: Carlos F. Niemeyer, 1896.

Fonck, Francisco, and Fernando Hess. "Informe de los señores Francisco Fonk [*sic*] i Fernando Hess sobre la espedicion a Nahuelhuapi." *Anales de la Universidad de Chile* 15, no. 1 (1857): 1–11.

Fontana, Jorge Luis. "Expedición al Río Pilcomayo." *Boletín del Instituto Geográfico Argentino* 4 (1883): 25–40.

Furque, Hilarion. "Descripción del pueblo General Roca." *Boletín del Instituto Geográfico Argentino* 9, no 5 (1888): 124–32.

Gana, Augusto. "Conferencia hecha en la seccion topográfica, por el ex-comisionado para aplicar la lei núm. 2087, de 15 de febrero de 1908, el 7 de agosto de 1912." In Oficina de Mensura de Tierras, *Sexta memoria del director*, 136–60.

García, Jose. "Diario del viaje y navegación hechos por el padre José Garcia de la Compañia de Jesús: Desde su misión de Cailin, en Chiloe, hacia el sur en los años 1766 y 1767." In *Documentos para la historia de la náutica en Chile.* Vol. 14 of *Anuario Hidrográfico.* Santiago: Imprenta Nacional, 1889.

"Geographischer Monatsbericht: Amerika." *Dr. A. Petermann's Mittheilungen aus Justus Perthes' geographischer Anstalt* 40, no. 4 (1894): 94–95.

Gerike, Hugo, Ernesto Manriquez, and Rodolfo Thies. *Turismo en las provincias australes de Chile: Provincias Llanquihue y Valdivia, sus bellezas naturales, su historia, comercio e industria*. Valparaíso, Chile: South Pacific Mail, 1920.

Godio, Guillermo. "Tierra adentro." *Boletín del Instituto Geográfico Argentino* 18 (1897): 379–400.

Goodenough, William Howley, and James Cecil Dalton. *The Army Book for the British Empire: A Record of the Development and Present Composition of the Military Forces and Their Duties in Peace and War*. London: H. M. Stationery Office, 1893.

Guenem, Bemijio. "Manifestación de Alberto Wecker." In *Registro de conservatorio de descubrimientos*, 29:1–5v. Annotation 1. Puerto Montt: Conservador de Bienes Raíces de Llanquihue, 1894.

Hederra, Ramiro. "Francisco Huaiquipan i otro con Federico Hechenleitner, sobre restablecimiento, February 20, 1908. Corte de Apelaciones de Valdivia – sentencias civiles." *Gaceta de los Tribunales, 1908*, 1:366–67. Santiago: Imprenta Barcelona, 1908.

HLC. "Col. Sir Thomas Holdich, K.C.M.G., K.C.I.E." *Nature* 124, no. 3135 (1929): 847. https://doi.org/10.1038/124847a0.

Holdich, Thomas Hungerford. *The Countries of the King's Award*. London: Hurst and Blackett, 1904.

Huergo, Luis A. *Conversación exponiendo y aclarando los puntos principales de los informes producidos por los miembros de la delegación comercial enviada á Chile por el gobierno argentino, en abril de 1908*. Buenos Aires: Coni Hermanos, 1910.

Indice de decretos i leyes del Ministerio de Relaciones Esteriores Culto i Colonización: 1897 a 1903. Santiago: Taller Tipografico del Instituto de Sordo-Mudos, 1905.

Inspección Jeneral de Tierras i Colonias [Chile]. *Memoria de la inspección jeneral de tierras i colonización*. Edited by Temístocles Urrutia, Santiago: Imprenta Cervantes, 1906.

Inspección Jeneral de Tierras y Colonización [Chile]. *Memoria de la inspección jeneral de tierras i colonización*. Edited by Agustín Baeza Espiñeira. Santiago: Imprenta Moderna, 1902.

———. "Memoria de la inspeccion jeneral de tierras y colonizacion." In Ministerio de Relaciones Exteriores, Culto y Colonización, *Memoria de relaciones esteriores*, 5–25.

Iturbe, Miguel, and Marcial de Candioti. "Fábrica nacional de sombreros." *Anales de la Sociedad Científica Argentina* 31, no. 5 (1891): 271–80.

Jones, Luis. *Hanes y wladva Gymreig Tiriogaeth Chubut, yn y Weriniaeth Ariannin, De Amerig*. Caernarvon, UK: Gwmni'r Wasg Genedlaethol Gymreig, 1898.

Juárez Celman, Miguel Ángel. "Decreto nombrando a don Octavio Pico perito para la demarcación de límites con Chile, 20 agosto de 1888." In Zeballos, *Demarcación de límites*, 28.

Lista, Ramón. *Mis esploraciones y descubrimientos en la Patagonia 1877–1880*. Buenos Aires: Imprenta de M. Biedma, 1880. http://hdl.handle.net/2027/hvd.32044080471337.

Maldonado Coloma, Roberto. *Estudios geográficos é hidrográficos sobre Chiloé*. Santiago: Establecimiento Poligráfico Roma, 1897.

Martin, Carl. "Dr. Hans Steffens Reise im südlichen Patagonien." *Dr. A. Petermann's Mittheilungen aus Justus Perthes' geographischer Anstalt* 45, no. 5 (1899): 124–25.

Martínez de Ferrari, Marcial A. *Contribución al estudio del problema pendiente, sobre represión de la embriaguez*. Santiago: Imprenta Mejía, 1899.

Ministerio de Agricultura [Argentina]. "Decreto declarando caduca la concesión de un lote en la Colonia Nahuel Huapi, otorgada a don Eladio Lizasoain y condiéndosela a don Antonio

Buenuleo." In *Registro nacional de la República Argentina, 1911 (primer trimestre)*, by República Argentina, 354. Buenos Aires: Talleres Gráficos de la Penitenciaría Nacional, 1911.

———. "Decreto declarando en vigencia el de 29 de marzo de 1894, sobre concesión de tierras al Sr. Warren Lowe, 19 de diciembre de 1899." In *Registro nacional de la República Argentina, año 1899 (tercer cuatrimestre)*, by República Argentina, 733–36. Buenos Aires: Taller Tipográfico de la Penitenciaría Nacional, 1899.

———. "Decreto fundando colonias pastoriles en el Río Negro." In *Memorias de las Direcciones*, 166–67.

———. "Decreto fundando una colonia pastoril y reservando tierras en el Río Negro y en el Chubut, para el mismo objeto." In *Memorias de las Direcciones*, 170–71.

———. *Memoria presentada al Honorable Congreso, 1899–1900*. Buenos Aires: Imprenta de M. Biedma é Hijo, 1900.

———. *Memorias de las Direcciones de Comercio é Industrias, Tierras y Colonias, Agricultura y Ganadería é Immigración y recopilación de mensajes al Honorable Congreso, decretos, notas y otros documentos referentes a dichos ramos*. Buenos Aires: Imprenta, Litografía y Encuadernación de J. Peuser, 1899.

Ministerio de Guerra [Argentina]. "Ley 1628 mandando que el Poder Ejecutivo haga ubicar en los Territorios Nacionales del Sud, 5 de septiembre de 1895." In *Registro nacional de la República Argentina, 1885 (segundo semestre)*, by República Argentina, 28:250–52. Buenos Aires: Taller Tipográfico de la Penitenciaría, 1886.

Ministerio de Hacienda [Chile]. "Que suspende el impuesto que grava la internación de ganado vacuno i ovino i reduce en un 50 por ciento los derechos de aduana de varios artículos." Pub. L. No. 2060 (1907). Ley Chile - Biblioteca del Congreso Nacional de Chile. https://bcn.cl/35dos.

Ministerio de Justicia, Culto é Instrucción Pública. *Memoria presentada al Congreso Nacional de 1896*. Edited by Antonio Bermejo. Vol. 1, *Texto*. Buenos Aires: Taller Tipográfico de la Penitenciaría Nacional, 1896.

Ministerio del Interior [Argentina]. "Acuerdo autorizando á la gobernación del Neuquén para construir un camino carretero entre Chos-Malal y Pichachen." In *Registro nacional de la República Argentina, año 1895*, by República Argentina, 2:370–71. Buenos Aires: Taller Tipográfico de la Penitenciaría Nacional, 1895.

———. "Ley de Defensa Social." *Boletín Oficial de la República Argentina* 18, no. 4969 (July 8, 1910): 149–50. www.boletinoficial.gob.ar/detalleAviso/primera/7006913/19100708.

———. *Memoria del ministro del interior al Congreso Nacional, 1883*. Edited by Bernardo Irigoyen. Buenos Aires, 1884.

———. *Memoria del ministro del interior al Congreso Nacional, 1889*. Edited by Norberto Quirno Costa. Buenos Aires: Imprenta Sud-América, 1889.

———. *Memoria del ministro del interior al Congreso Nacional, 1895*. Edited by Norberto Quinor Costa. Vols. 1–2. Buenos Aires: Imprenta Tribuna, 1896.

———. *Memoria del ministro del interior al Congreso Nacional, 1891*. Edited by Julio A. Roca. Buenos Aires: Imprenta La Tribuna Nacional, 1892.

———. *Memoria del ministro del interior al Congreso Nacional, 1904–1905*. Edited by Rafael Castillo. Buenos Aires: Imprenta de V. Daroqui y Cía, 1905.

———. *Memoria del ministro del interior al Congreso Nacional, 1910–1911*. Edited by Indalecio Gómez. Buenos Aires: Imprenta y Casa Editora Juan A. Alsina, 1911.

———. *Memoria del ministro del interior al Congreso Nacional, 1912–1913*. Edited by Indalecio Gómez. Buenos Aires: Talleres Gráficos de la Penintenciaría Nacional, 1913.

———. *Memoria del Ministerio del Interior presentada al Congreso Nacional, 1893*. Edited by Wenceslao Escalante. Buenos Aires: Imprenta de la Tribuna, 1894. https://catalog.hathitrust.org/Record/100629856.

———. "Neuquén." In *Memoria del Ministerio*, 86–89.

Ministerio del Interior [Chile]. "Crea la provincia de Osorno, le fija sus departamentos y su capital." Pub. L. No. 6505 (1940). Ley Chile - Biblioteca del Congreso Nacional de Chile. https://bcn.cl/345ct.

———. "Crea y fija los límites de los departamentos de Palena, Aysén, Coyhaique y Chile Chico y la de sus respectivas comunas [...]." Pub. L. No. 13375 (1959). Ley Chile - Biblioteca del Congreso Nacional de Chile. www.bcn.cl/leychile/navegar?idNorma=27455&idVersion=1959-09-09.

———. Decreto 2335. Pub. L. No. 2335 (1929). Ley Chile - Biblioteca del Congreso Nacional de Chile. https://bcn.cl/3aezp.

———. Decreto 8583. Pub. L. No. 8583 (1927). Ley Chile - Biblioteca del Congreso Nacional de Chile. https://bcn.cl/2guyr.

———. "Decreto que fija la nueva división territorial de la república." Pub. L. No. 8582 (1927). Ley Chile - Biblioteca del Congreso Nacional de Chile. https://bcn.cl/2u4xz.

———. "Divide la actual provincia de Chiloé en las de Llanquihue y Chiloé y restablece la comuna de Puqueldon." Pub. L. No. 6027 (1937). Ley Chile - Biblioteca del Congreso Nacional de Chile. https://bcn.cl/3aezy.

———. *Rol de cartas de naturalización*. Santiago: Imprenta Nacional, 1927.

Ministerio de Obras Públicas [Argentina]. *Ley no. 5559 de fomento de los territorios nacionales*. Buenos Aires: Taller Tipográfico del Ministerio de Obras Públicas, 1909. http://hdl.handle.net/2027/umn.31951d01838556u.

Ministerio de Obras Públicas [Chile]. "Título XI. Obras públicas. Caminos, canales, puentes i calzadas." Ley s/n. (1842). Ley Chile - Biblioteca del Congreso Nacional de Chile. https://bcn.cl/2nn4h.

Ministerio de Relaciones Esteriores, Culto i Colonización [Chile]. *Anexos a la memoria del ministro de colonización i culto presentada al Congreso Nacional i correspondiente a 1896*. Vol. 2. Santiago: Imprenta Nacional, 1897.

———. *Memoria de relaciones exteriores y culto presentada al Honorable Congreso Nacional en 1899*. Santiago: Imprenta Nacional, 1899.

Ministerio de Relaciones Exteriores [Chile]. *Memoria del Ministerio de Relaciones Exteriores, Culto y Colonización presentada al Congreso Nacional en 1909*. Santiago: Imprenta Cervantes, 1910.

Ministerio de Relaciones Exteriores, Culto y Colonización [Chile]. "Memoria del ajente jeneral de colonización de Chile en Europa." In *Memoria de relaciones esteriores*, 135–279.

———. *Memoria de relaciones esteriores, culto, y colonización, 1899*. Vol. 3. Santiago: Imprenta Nacional, 1899.

Ministerio de Relaciones Exteriores y Colonización [Chile]. *Recopilacion de leyes i decretos supremos sobre colonizacion: 1810–1896*. 2 vols. Edited by Julio Zenteno Barros. Santiago: Imprenta Nacional, 1892–96.

Ministerio de Relaciones Exteriores y Culto [Argentina]. *La frontera argentino-chilena: Documentos de la demarcación*. Compiled by Zacarías Sánchez, Vol. 2. Buenos Aires: Talleres Gráficos de la Penitenciaría Nacional, 1908.

———. *Memoria de relaciones exteriores y culto presentada al Honorable Congreso Nacional en 1899*. Buenos Aires: Taller Tipográfico de la Penitenciaría Nacional, 1899.

Montes de Oca, Manuel Augusto. *Límites argentino-chilenos: El divortium aquarum continental ante el tratado de 1893*. Buenos Aires: Imprenta de M. Biedma é Hijo, 1899.

Montt, Jorge. "Introduccion e instalacion de cinco mil familias de colonos—mensaje del presidente de la República por el que pide la aprobacion del contrato celebrado con Don A. Charles Colson." In Ministerio de Relaciones Exteriores y Colonización, *Recopilacion de leyes i decretos supremos*, 2:1432–37.

Moreno, Eduardo, ed. *Reminiscencias de Francisco P. Moreno: Versión propia documentada*. 2nd ed. Lucha de fronteras con el indio. Buenos Aires: EUDEBA, 1979.

Moreno, Francisco P. *Apuntes preliminares sobre una excursión a los territorios del Neuquén, Río Negro, Chubut y Santa Cruz*. La Plata, Argentina: Museo de La Plata, Taller de Publicaciones, 1897.

———. "Dr. Steffen's Exploration in South America." *Geographical Journal* 14, no. 2 (August 1899): 219–20.

———. "El Museo de La Plata: Rápida ojeada sobre su fundación y desarrollo." *Revista del Museo de La Plata* 1 (1890): 27–55.

———. "Explorations in Patagonia." *Geographical Journal* 14, no. 3 (September 1899): 241–69.

Munizaga, Enrique. "Informe el injeniero don Enrique Munizaga, a cuyo cargo está la entrega de los terrenos subastados, 15 de maro de 1895." In Zenteno Barros, *Recopilacion de leyes i decretos supremos*, 2:1294. Santiago: Imprenta Nacional, 1896.

Museo Nacional de Historia Natural [Chile]. *Guía del museo nacional de Chile en septiembre de 1878*. Santiago: Imprenta de los Avisos, 1878.

"New Maps." *Geographical Journal* 25, no. 1 (1905): 123–28.

Oficina Central de Estadística [Chile]. *Sétimo censo jeneral de la poblacion de Chile levantado el 28 de noviembre de 1895*. Vol. 1. Valparaíso, Chile, 1900.

Oficina de Estadística de la Superintendencia de Aduanas [Chile]. *Estadística comercial de la república de Chile, 1913*. Santiago: Compañía Inglesa de Imprenta y Litografía, 1914.

———. *Estadística comercial de la república de Chile, 1914*. Santiago: Sociedad Imprenta y Litografía Universo, 1915.

———. *Estadística comercial de la república de Chile, 1915*. Santiago: Sociedad Imprenta y Litografía Universo, 1916.

Oficina de Estadística e Informaciones Agrícolas [Chile]. *Indice de propietarios rurales y valor de la propiedad rural según los roles de avalúos comunales*. Santiago: Sociedad Imprenta y Litografía Universo, 1908.

Oficina de Límites [Chile]. *La linea de frontera con la República Arjentina entre las latitudes 35 i 46 S*. Edited by Luis Riso Patrón. Santiago: Imprenta i Encuadernación Universitaria, 1907.

Oficina de Mensura de Tierras [Chile]. "Concesion Woodhouse (compañia colonizadora de Villarrica)." In *Primera memoria del director*, 106–7.

———. *Primera memoria del director de la Oficina de Mensura de Tierras*. Santiago: Imprenta y Encuadernación Universitaria, 1908.

———. *Sexta memoria del director de la Oficina de Mensura de Tierras*. Santiago: Imprenta Universitaria, 1913.

Olascoaga, Manuel J. *Estudio topográfico de la Pampa y Río Negro*. Buenos Aires: Editorial Universitaria de Buenos Aires, 1974.

———. *La conquista del desierto proyectada y llevada a cabo por Exmo Señor Ministro de la Guerra y Marina General D. Julio A. Roca*. Buenos Aires: Ostwald y Martinez, 1881.

———. "Memoria del gobernador de Neuquén." In *Memoria del ministro del interior al Congreso Nacional, 1887*, edited by Eduardo Wilde, 567–74. Buenos Aires: Imprenta de Sud-America, 1888.

———. *Regiones australes: Topografía andina, ferrocarril paralelo á los Andes como fomento de población y seguridad de la frontera; Complemento indispensable de la campaña de 1879*. Buenos Aires: Imprenta, Litografía y Encuadernación de J. Peuser, 1901.

Ortiz, Roberto M. *Memoria presentada al Honorable Congreso por el Ministerio de Obras Públicas*. Buenos Aires: Talleres Gráficos del Ministerio de Obras Públicas, 1926.

Patroni, Adrián. *Bellezas de los lagos argentinos-chilenos*. Buenos Aires: Lotito Hnos y Cía, 1938.

Philippi, Rodolfo. "Espedición al volcán de Osorno." *Anales de la Universidad de Chile* (1853): 107–10.

Pomar, José. *La concesión del Aysén y Valle Simpson*. Santiago: Imprenta Cervantes, 1923.

"Presentacion de algunos habitantes del departamento de Osorno en que piden indemnizacion a consecuencia del remate de tierras de esa rejion, c. 1895." In Zenteno Barros, *Recopilacion de leyes i decretos supremos*, 3:1290–93.

Quintana, Manuel. *Memoria del ministro del interior al Congreso Nacional, 1893*. Buenos Aires: Imprenta de la Tribuna, 1894.

Quirno Costa, Norberto. "Decreto comisionando al agrimensor González para efectuar el deslinde de las concesiones para colonizar y en propiedad sobre la márgen derecha de los Ríos Negro y Limay, 22 de septiembre de 1888." In *Memoria del ministro [...] 1889*, 284–85.

Ramos Mexía, Ezequiel. *Memoria presentada al Honorable Congreso por el Ministerio de Obras Públicas*. Buenos Aires: Compañía Sud-Americana de Billetes de Banco, 1910.

———. "Mensaje y projecto." In *Veinte meses de administración*, 19–30.

———. "Prólogo." In *La Patagonia y sus problemas; estudio geográfico, económico, político y social de los territorios nacionales del sur*, by José María Sarobe, 23–29. Buenos Aires: Editorial Centro de Estudios Unión para la Nueva Mayoría, 1999.

———. *Veinte meses de administración en el Ministerio de Agricultura*. Buenos Aires: Imprenta de la Agricultura Nacional, 1908. https://catalog.hathitrust.org/Record/012480395.

Rawson, Franklin. "Neuquén: Memoria de la gobernación." In Ministerio del Interior, *Memoria del ministro [...] 1895*, 1:450–89.

"Record of Geographical Progress." *Journal of the American Geographical Society of New York* 29, no. 1 (1897): 68–82.

Rehren, Otto. "Anexo no 5." In *Memoria de inspección jeneral de tierras i colonización*, by Ministerio de Relaciones Exteriores de Chile, 93–144. Santiago: Imprenta Moderna, 1902.

———. "Memoria correspondiente al año 1905 de la sub-inspeccion de Tierras y Colonizacion de Llanquihue y Chiloe." In Urrutia, *Memoria*, 228–42.

República Argentina. *Argentine-Chilian Boundary: Report Presented to the Tribunal Appointed by Her Britannic Majesty's Government "to Consider and Report upon the Differences Which*

Have Arisen with Regard to the Frontier between the Argentine and Chilian Republics" to Justify the Argentine Claims for the Boundary in the Summit of the Cordillera de Los Andes, According to the Treaties of 1881 and 1893; Printed in Compliance with the Request of the Tribunal, Dated December 21, 1899. London: Printed for the government of the Argentine Republic by W. Clowes and Sons, 1900. http://archive.org/details/argentinechilianooargerich.

———. *Código rural de los territorios nacionales en vigencia desde el 10 de octubre de 1894.* Códigos y leyes usuales de la República Argentina. Buenos Aires: J. Lajouane y Cía, 1909.

———. "Decreto reservando para fundación de pueblos, cuatrocientas hectáreas en cada uno de los puntos que se citan del terri torio del Neuquén y Rio Negro." In *Registro nacional [...] 1902,* 2:134.

———. "Fomento del scoutismo argentino." In *Sesiones ordinarias,* 754–57. Vol. 2 of *Diario de sesiones de la Cámara de Diputados, año 1919.* Buenos Aires: Talleres Gráficos de L. J. Rosso y Cía, 1919. https://catalog.hathitrust.org/Record/011441654.

———. *Leyes nacionales sancionadas por el Congreso argentino durante el período legislativo de 1916.* Colección Legislative de la República Argentina. Buenos Aires: J. Lajouane y Cía, 1917. https://catalog.hathitrust.org/Record/100667434.

———. "Ley no 1501 del Hogar (2 de octubre de 1884)." In *Registro nacional de la República Arjentina, año 1884 (segundo semestre),* 354–55. Buenos Aires: Taller Tipográfico de la Penitenciaría, 1885.

———. *Primer censo de la República Argentina.* Buenos Aires: Imprenta El Porvenir, 1872.

———. *Registro nacional de la República Argentina, año 1902.* Buenos Aires: Taller Tipográfico de la Penitenciaría Nacional, 1901, 3 vols.

———. *Registro nacional de la República Argentina que comprende los documentos espedidos desde 1810 hasta 1890.* 14 vols. Buenos Aires: La República, 1879–1911.

———. *Segundo censo de la República Argentina, Mayo 10, 1895.* Vol. 2, *Población.* Buenos Aires: Taller Tipográfico de la Penitenciaría Nacional, 1898.

———. *Tercer censo nacional levantado el 1ro de Junio de 1914.* Vol. 1, *Antecedenes y comentarios.* Buenos Aires: Talleres Gráficos de L. J. Rosso y Cía, 1916.

República de Chile. *Anuario estadístico de la República de Chile, año 1910.* Vol. 4, *Movimiento marítimo.* Santiago: Sociedad Imprenta y Litografía Universo, 1911.

———. *Anuario estadístico de la República de Chile, año 1916.* Vol. 12, *Comunicaciones.* Santiago: Sociedad Imprenta y Litografía Universo, 1918.

———. "Contardi Juan B.—contrato sobre colonizacion celebrado con dicho señor." *Boletín de las Leyes i Decretos del Gobierno* 73, no. 3 (1903): 309–11.

———. "Contrato entre la República de Chile y Charles Colson." In Zenteno Barros, *Recopilacion de leyes i decretos supremos,* 3:1437–39.

"Resumen del trabajo del Coronel José Ma. Sarobe de 'La Patagonia y sus problemas.'" Ejército Argentino, 1934. Servicio Histórico del Ejército.

Riso Patrón, Luis. "Tierras de colonización." In Oficina de Mensura de Tierras, *Primera memoria del director,* 33–65.

Roca, Julio A. "Proyecto de creación de un museo nacional, 12 de septiembre de 1881." In *Memoria presentada al Congreso Nacional de 1882 por el Ministro de Justicia, Culto é Instrucción Pública,* edited by Eduardo Wilde, 483–84. Buenos Aires: Imprenta de la Penitenciaría, 1882.

Rohde, Jorge. *Descripción de las gobernaciones nacionales de la Pampa, del Río Negro y del Neuquén*. Buenos Aires: Compañía Sudamericana de Billetes de Banco, 1889.

———. "El Paso de Bariloche: Conferencia leída en los salones del Instituto Geográfico Argentino." *Boletín del Instituto Geográfico Argentino* 4 (1883): 161–79.

Romero Brest, Enrique. "Organización general de la educación física en la enseñanza secundaria: Conferencia del doctor E. Romero Brest." *Anales de la Sociedad Científica Argentina, 1904 (segundo semestre)* 58 (1904): 5–16.

Royal Geographical Society [London]. "Meetings of the Royal Geographical Society, Session 1898–1899: Eleventh Ordinary Meeting, May 29, 1899." *Geographical Journal* 14, no. 1 (July 1899): 102–9.

Sáenz Peña, Roque. "Decreto aceptando renuncias." *Boletín Oficial de la República Argentina* 21, no. 5844 (June 25, 1913): 1061.

Sarmiento, Domingo F. *Life in the Argentine Republic in the Days of the Tyrants; or, Civilization and Barbarism*. Translated by Mary Tyler Peabody Mann. New York: Hurd and Houghton, 1868.

Sarobe, José María. *La Patagonia y sus problemas: Estudio geográfico, económico, político y social de los territorios nacionales del sur*. Buenos Aires: Editorial Centro de Estudios Unión para la Nueva Mayoría, 1999.

Seguí, Francisco. "Expedición al Chaco (I)." *Boletín del Instituto Geográfico Argentino* 6 (1885): 53–54.

———. "Expedición del Bermejo." *Boletín del Instituto Geográfico Argentino* 6 (1885): 26–30.

Serrano Montaner, Ramón. *Derrotero del Estrecho de Magallanes, Tierra del Fuego i canales de la Patagonia: Desde el Canal de Chacao hasta el Cabo de Hornos*. Santiago: Imprenta Nacional, 1891.

———. "Discusión sobre los Andes australes." *Boletín del Instituto Geográfico Argentino* 9, no. 8 (August 1888): 195–97.

———. "Reconocimiento del Río Buta-Palena i del Canal Fallos por el vapor de la República 'Toro' bajo la dirección del capitán graduado de Fragata Ramón Serrano M." *Anuario hidrográfico de la marina de Chile*, no. 11 (1886): 73–176.

Silva Narro, Domingo. *Guía administrativa, industrial y comercial de las provincias de Tacna, Tarapacá y Antofagasta*. Santiago: Imprenta y Encuadernación Chile, 1913.

Sociedad Agrícola i Frigorífica de Cochamó. *Prospecto de la Sociedad Agrícola i Frigorífica de Cochamó: Capital; $3.000,000*. Santiago: Imprenta Enc. i Litograf. Esmeralda, 1908.

Sociedad Científica Argentina. *Anales de la Sociedad Científica Argentina*. Vols. 2–3. Buenos Aires: Imprenta de Pablo E. Coni, 1876–77.

———. "Estatutos fundamentales." In *Anales de la Sociedad Científica Argentina*, 1:7–8. Buenos Aires: Imprenta de Pablo E. Coni, 1876.

Steffen, Hans. "Memoria jeneral sobre la espedicion esploradora del Rio Palena, diciembre 1893–marzo 1894 (Capítulos V y VI)." *Anales de la Universidad de Chile* 88 (1894): 137–240.

———. "On Recent Explorations in the Patagonian Andes, South of 41° S. Lat." *Scottish Geographical Magazine* 13, no. 2 (1897): 57–71.

———. *Patagonia occidental: Las cordilleras patagónicas y sus regiones circundantes*. 2 vols. Reprint. Santiago: Ediciones de la Universidad de Chile, 2008.

———. *Problemas limítrofes y viajes de exploración en la Patagonia: Recuerdos de la época del conflicto entre Chile y Argentina*. Santiago: Dirección de Bibliotecas, Archivos, y Museos, 2015.

———. *Viajes de exploración y estudio en la Patagonia occidental 1892–1902*. Edited by Rafael Sagredo Baeza. Vol. 2. 1909. Reprint, Santiago: Cámara Chilena de la Construcción; Pontificia Universidad Católica de Chile; Dirección de Bibliotecas, Archivos y Museos, 2010.

Superintendencia de Aduanas [Chile]. *Memoria del superintendente de aduanas sobre la renta y el comercio exterior en 1902*. Santiago: Imprenta del Universo, 1903.

Tagle y Jordan, Enrique. *El tratado de comercio entre Chile y la República Argentina*. Buenos Aires: Est. Tip. A. Cantiello, 1909.

ten Kate, Herman. "Matériaux pour servir à l'anthropologie des Indiens de la République Argentine." *Revista del Museo de La Plata* 12 (1906): 31–57.

"Tratado de 1881." In *Tratados de límites entre Chile y la República Argentina*, 3–8. Buenos Aires: Imprenta de J. Peuser, 1898. Microform. http://archive.org/details/tratadosdelimiteoochil.

United Nations. *The Cordillera of the Andes Boundary Case (Argentina, Chile)*. UN Reports of International Arbitral Awards, November 20, 1902.

Uriburu, José Evaristo. "Decreto no 300: Creación de la colonia pastoril 'Coronel Barcala.'" Oficina de Tierras y Colonias - Actos Dispositivos, March 6, 1896.

———. "Decreto nombrando perito para la demarcación de límites con Chile al Dr. D. Francisco P. Moreno." Registro Nacional de la República Argentina, año 1896 (second semester), September 21, 1896.

Uriburu, José Félix. "Decreto nombrando y confirmando personal." *Boletín Oficial de la República Argentina* 40, no. 11.305 (December 1931): 902.

US House of Representatives. *Message of the President of the United States Respecting the Relations with Chile, Together with the Diplomatic Correspondence; the Correspondence with the Naval Officials; the Inquiry into the Attack on the Seamen of the U.S.S. Baltimore in the Streets of Valparaiso; and the Evidence of the Officers and Crew of the Steamer Keweenaw Respecting the Ill-Treatment of Patrick Shields by the Chilean Police*. Washington, DC: Government Printing Office, 1892.

Varas, José Antonio. *Colonización de Llanquihue, Valdivia i Arauco, o sea coleccion de las leyes i decretos supremos concernientes a esta materia: Desde 1823 a 1871 inclusive*. Santiago: La Republica de J. Nuñez, 1872.

Vega, Nicolás. "Memoria del ajente jeneral de colonizacion en Europa." In Ministerio de Relaciones Esteriores, Culto i Colonización, *Anexos a la memoria*, 76–173.

"Viaje de Enrique Brouwer." In *Anuario hidrográfico de la marina de Chile*, no. 16 (1892), 3–88.

"Viajes del padre Francisco Menéndez al lago Nahelguapi en 1791–1794." In *Anuario hidrográfico de la marina de Chile*, no. 15 (1890), 3–72.

Vicuña Mackenna, Benjamín. "Elisa Bravo, o sea el misterio de su vida, de su cautividad y de su muerte, con las consecuencias políticas i públicas que la última tuvo para Chile." *Revista de Artes y Letras* 1, no. 1 (October 1884): 493–529.

———. *La conquista de Arauco: Discurso pronunciado en la Cámara de Diputados en su sesión de 10 de agosto*. Santiago: Imprenta del Ferrocarril, 1868.

———. *La Patagonia: Estudios jeográficos i políticos dirijidos a esclarecer la "cuestión Patagonia" con motivo de las amenazas recíprocas de guerra entre Chile i la República Arjentina*. Santiago: Imprenta del Centro Editorial, 1880.

Villegas, Conrado. "Diario general de las operaciones de la Segunda División del Ejército en la Expedición al Gran Lago Nahuel-Huapi y batida general en el cuadrilátero." *Boletín del Instituto Geográfico Argentino* 4 (1883): 49–61.

Weber, Alfredo. *Chiloé: Su estado actual—su colonizacion. Su porvenir.* Santiago: Imprenta Mejía, 1903.

———. "Inspeccion de colonizacion de Llanquihue i Chiloé." In Ministerio de Relaciones Esteriores, Culto i Colonización, *Anexos a la memoria*, 189–258.

———. "Memoria de la inspección de Llanquihue i Chiloé." In Ministerio de Relaciones Exteriores, Culto y Colonización, *Memoria de relaciones esteriores*, 41–50.

Wiederhold, Germán. *Turismo en la provincia de Llanquihue a través de la Suiza chilena y argentina con datos de los canales de Chiloé.* 2nd ed. Santiago: Soc. Imprenta y Litografía Universo, 1921.

Willis, Bailey, and Comisión de Estudios Hidrológicos. *El norte de la Patagonia.* Vol. 2, *Estrategias y proyectos.* Edited by Gerardo Mario De Jong, Marcos Damián Mare, and Eduardo Miguel Bessera. Translated by Diana Raimondo, Alba Mora, and Romina Sánchez. EDUCO, 2018. http://rdi.uncoma.edu.ar:8080/handle/123456789/6826.

Whiteside Toro, Arturo. "Chiloé i sus colonias, 2da parte." *La Revista de Chile*, no. 5 (1900): 11–19.

Yofré, Felipe. *Memoria del Departamento del Interior correspondiente al año 1899.* Buenos Aires, 1900.

Zeballos, Estanislao, ed. *Demarcación de límites entre la República Argentina y Chile: Extracto de la memoria presentada al Congreso de la Nación.* Buenos Aires: Empresa La Nueva Universidad, 1892.

———. *Descripción amena de la República Argentina.* Vol. 1. Buenos Aires: Imprenta de Jacobo Peuser, 1881.

———. *La conquista de quince mil leguas: Estudio sobre la traslacion de la frontera sud de la república al Rio Negro, dedicado á los gefes y oficiales del ejército expedicionario.* Buenos Aires: La Prensa, 1878.

Secondary Sources

Abanto, Carlos Martín Benavides. *Una pelota de trapo, un corazón blanquiazul: Tradición e identidad en Alianza Lima, 1901–1996.* Lima: Fondo Editorial PUCP, 2000.

Ablard, Jonathan, María Silvia Di Liscia, and Ernesto Lázaro Bohoslavsky, eds. *Instituciones y formas de control social en América Latina, 1840–1940: Una revisión.* Buenos Aires: Universidad Nacional de La Pampa; Universidad Nacional de General Sarmiento; Prometeo Libros, 2005.

Aguiar, Marian. *Tracking Modernity: India's Railway and the Culture of Mobility.* Minneapolis: University of Minnesota Press, 2011.

Ainsa, Fernando. *Historia, utopía y ficción de la Ciudad de los Césares: Metamorfosis de un mito.* Madrid: Alianza Editorial, 1992.

———. "The Myth, Marvel, and Adventure of El Dorado: Semantic Mutations of a Legend." *Diógenes* 4¼, no. 164 (1993): 13–26.

Allen, Stephen D. *A History of Boxing in Mexico: Masculinity, Modernity, and Nationalism.* Albuquerque: University of New Mexico Press, 2017.

Allevi, José Ignacio. "La creación clínica de normas sexuales: Nosología, patologización y contramodelos sexuales en la Penitenciaría Nacional de Buenos Aires (Argentina, 1901–1904)." *Sexualidad, Salud y Sociedad* 26 (August 2017): 126–47.

Almandoz, Arturo. "The Garden City in Early Twentieth-Century Latin America." *Urban History* 31, no. 3 (2004): 437–52.

———. "Urbanization and Urbanism in Latin America: From Haussmann to CIAM." In *Planning Latin America's Capital Cities 1850–1950*, edited by Arturo Almandoz, 27–58. London: Routledge, 2002.

Almonacid Zapata, Fabián. "Comercio entre Chile y Argentina en la zona sur, en el contexto de una economía regional agropecuaria (1930–1960)." In *Cultura y espacio: Araucanía-Norpatagonia*, edited by Pedro Navarro Floria and Walter Delrio, 182–99. Bariloche, Argentina: Universidad Nacional de Río Negro—Instituto de Investigaciones en Diversidad Cultural y Procesos de Cambio, 2011.

———. "El desarrollo de la propiedad rural en las provincias de Valdivia y Llanquihue, 1850–1920." *Revista Austral de Ciencias Sociales* 2 (1998): 27–36.

———. "El mercado de tierras en el departamento de Valdivia, 1859–1877." *Revista de Historia de la Universidad de Concepción* 1, no. 8 (1998): 195–206.

———. "El problema de la propiedad de la tierra en el sur de Chile (1850–1930)." *Historia* 42, no. 1 (2009): 5–56.

Altamirano, Carlos, and Beatriz Sarlo. "La Argentina del centenario: Campo intelectual, vida literaria y temas ideológicos." *Hispamérica* 9, no. 25/26 (1980): 33–59.

Álvarez, Gonzalo Pérez. "Industrias y proyectos de desarrollo en Chubut antes de la implantación de los polos industriales subsidiados." *H-industria: Revista de historia de la industria y el desarrollo en América Latina* 15, no. 29 (2021): 1–22.

Álvarez Correa, Lily. "Cartografía y geodesia: Las innovaciones de la Oficina de Mensura de Tierras de Chile a principios del siglo XX (1907–1914)." *Scripta Nova. Revista Electrónica de Geografía y Ciencias Sociales* 12, no. 69 (2000). www.ub.edu/geocrit/sn-69-12.htm.

Andermann, Jens. "Reshaping the Creole Past: History Exhibitions in Late Nineteenth-Century Argentina." *Journal of the History of Collections* 13, no. 2 (2001): 145–62.

Appelbaum, Nancy P. *Mapping the Country of Regions: The Chorographic Commission of Nineteenth-Century Colombia*. Chapel Hill: University of North Carolina Press, 2016.

Arambel-Güiñazú, María Cristina. *Las mujeres toman la palabra: Escritura femenina del siglo XIX*. Madrid: Iberoamericana, 2001.

Ardüser, Jorge. *Un suizo en la Patagonia: El diario de Leonhard Ardüser; Su trabajo y sus vivencias a la par de la construcción del ferrocarril, en la hoy llamada "Línea Sur" desde San Antonio al lago Nahuel Huapi en 1911–1912*. Bariloche, Argentina: Self-published, 2004.

Argeri, María. *De guerreros a delincuentes: De la desarticulacion de las jefaturas indigenas y el poder judicial; Norpatagonia, 1880–1930*. Madrid: Consejo Superior de Investigaciones Científicas, 2005.

Armus, Diego. *The Ailing City: Health, Tuberculosis, and Culture in Buenos Aires, 1870–1950*. Durham, NC: Duke University Press, 2011.

Arneil, Barbara. *Domestic Colonies: The Turn Inward to Colony*. Oxford: Oxford University Press, 2017.

Augé, Marc. *Non-places: Introduction to an Anthropology of Supermodernity*. Translated by John Howe. London: Verson, 1995.

Autrique Escobar, Cecilia. "Los orígenes de los movimientos prohibicionistas del alcohol y las drogas: El caso de México (1917–1928)." *Historia y grafía*, no. 53 (2019): 145–83.

Aylwin, José. *Estudio sobre tierras indígenas de la Araucanía: Antecedentes históricos legislativos (1850–1920)*. Temuco, Chile: Universidad de la Frontera, Instituto de Estudios Indígenas, 1995.

Azar, Pedro, Gabriela Nacach, and Pedro Navarro Floria. "Antropología, genocidio, y olvido en la representación del Otro étnico a partir de la Conquista." In *Paisajes del progreso: La resignificación de la Patagonia Norte, 1880–1916*, edited by Pedro Navarro Floria, 79–106. Neuquén, Argentina: Educo, 2007.

Babini, José. *Historia de la ciencia en la Argentina*. Buenos Aires: Ediciones del Solar, 1986.

Bade, Klaus J. "From Emigration to Immigration: The German Experience in the Nineteenth and Twentieth Centuries." *Central European History* 28, no. 4 (1995): 507–35.

Baeza, Manuel Antonio, and Grace Silva. "Imaginarios sociales del Otro: El personaje del forastero en Chile (de 1845 a nuestros días)." *Sociedad Hoy*, no. 17 (2009): 29–38.

Ballent, Anahí. "Ingeniería y estado: La red nacional de caminos y las obras públicas en la Argentina, 1930–1943." *História, Ciências, Saúde-Manguinhos* 15, no. 3 (2008): 827–47.

———. "Kilómetro cero: La construcción del universo simbólico del camino en la Argentina de los años treinta." *Boletín del Instituto de Historia Argentina y Americana Dr. Emilio Ravignani*, no. 27 (June 2005): 107–36.

Ballent, Anahí, and Adrián Gorelik. "País urbano o país rural: La modernización territorial y su crisis." In *Nueva historia argentina: Crisis económica, avance del estado e incertidumbre política (1930–1943)*, edited by Alejandro Cattaruzza, 143–200. Buenos Aires: Sudamericana, 2001.

Bandieri, Susana. "Condicionantes históricos del asentamiento humano después de la ocupación militar del espacio." In *Historia de Neuquén*, edited by Susana Bandieri, Orietta Favaro, and Marta Morinelli, 109–46. Buenos Aires: Plus Ultra, 1993.

———. "Del discurso poblador a la praxis latifundista: La distribución de la tierra pública en la Patagonia." *Mundo agrario* 6, no. 11 (2005). http://ref.scielo.org/g4b97n.

———. *Historia de la Patagonia*. 2nd ed. Buenos Aires: Sudamericana, 2011.

Bandieri, Susana, Sonia Fernández, Graciela Blanco, and Laura Fontana. "Los propietarios de la nueva frontera: Tendencia de la tierra y estructuras del poder en el área andina de Neuquén; Primeros avances." *Revista de Historia*, no. 5 (1995): 133–52.

Barbería, Elsa Mabel. *Los dueños de la tierra en la Patagonia austral, 1880–1920*. 3rd ed. Río Gallegos, Argentina: Universidad Nacional de la Patagonia Austral, 2001.

Barnitz, Jacqueline, and Patrick Frank. *Twentieth-Century Art of Latin America*. Rev. and expanded ed. William and Bettye Nowlin Series in Art, History, and Culture of the Western Hemisphere. Austin: University of Texas Press, 2015.

Baron, Nick. "New Spatial Histories of Twentieth Century Russia and the Soviet Union: Surveying the Landscape." *Jahrbücher Für Geschichte Osteuropas* 55, no. 3 (January 1, 2007): 374–400.

Barr-Melej, Patrick. *Reforming Chile: Cultural Politics, Nationalism, and the Rise of the Middle Class*. Chapel Hill: University of North Carolina Press, 2001.

Bassin, Mark, Christopher David Ely, and Melissa Kirschke Stockdale. *Space, Place, and Power in Modern Russia: Essays in the New Spatial History*. DeKalb: Northern Illinois University Press, 2010.

Bayer, Osvaldo. "Proyecto de ley." In *Historia de la crueldad argentina: Julio Argentino Roca y el genocidio de los Pueblos Originarios*, edited by Osvaldo Bayer and Diana Lenton, 11–28. Buenos Aires: Red de Investigadores en Genocidio y Política Indígena, 2010.

Beck, Hugo Humberto. *La vida en las fronteras interiores del territorio formoseño: La naturaleza hostil del último baluarte aborigen*. Tucumán, Argentina: Facultad de Filosofía y Letras, Universidad Nacional de Tucumán, 2007. https://cdsa.aacademica.org/000-108/948.pdf.

Belini, Claudio. *Historia económica de la Argentina en el siglo XX*. Buenos Aires: Siglo Veintiuno Editores, Fundación OSDE, 2012.

Bello, Álvaro. "Exploración, conocimiento geográfico y nación: La 'creación' de la Patagonia Occidental y Aysén a fines del siglo XIX." In *Imaginarios geográficos, prácticas y discursos de frontera*, edited by Andrés Núñez, Enrique Aliste, Álvaro Bello, and Mauricio Osorio, 61–86. Santiago: Instituto de Geografía, Facultad de Historia, Geografía y Ciencia Política - Ñire Negro Ediciones, 2017.

Benedetti, Alejandro. "Los usos de la categoría región en el pensamiento geográfico argentino." *Scripta Nova. Revista Electrónica de Geografía y Ciencias Sociales* 13, no. 286 (2009). www.ub.edu/geocrit/sn/sn-286.htm.

Bengoa, José. Historia del pueblo mapuche (siglo XIX y XX). 7th ed. Santiago: LOM Ediciones, 2008.

———. *Mapuche, colonos y el estado nacional*. Santiago: Catalonia, 2014.

Benzecry, Claudio. "An Opera House for the 'Paris of South America': Pathways to the Institutionalization of High Culture." *Theory and Society* 43, no. 2 (2014): 169–96.

Bergman, Teresa. *Exhibiting Patriotism: Creating and Contesting Interpretations of American Historic Sites*. Walnut Creek, CA: Left Coast, 2013.

Bernal, Azuela, and Luz Fernanda. "La Sociedad Mexicana de Geografía y Estadística, la organización de la ciencia, la institucionalización de la geografía y la construcción del país en el siglo XIX." *Investigaciones geográficas*, no. 52 (2003): 153–66.

Bertolotto, Antonella. "Gaucho, atlético y nacional: Imágenes de la famosa pelea del siglo XX." *Claves* 7, no. 12 (2021): 131–52.

Bess, Michael Kirkland. *Routes of Compromise: Building Roads and Shaping the Nation in Mexico, 1917–1952*. Lincoln: University of Nebraska Press, 2017.

Biedma, Juan Martín. *Crónica histórica del lago Nahuel Huapi*. 4th ed. Buenos Aires: Del Nuevo Extremo, Caleuche, 2003.

Blackwelder, Eliot. "Bailey Willis, 1857–1949." In *Biographical Memoirs*, edited by National Academy of Science, 35:333–50. Washington, DC: National Academies Press, 1961.

Blakeslee, Brandon Todd. "Foot-Ball! Turning Colombian Boys into Patriotic Men: How Sport and Education Developed with Early Twentieth-Century Colombian Nationalism." PhD diss., University of Texas at Arlington, 2021. www.proquest.com/docview/2597501436/abstract/45DB62D0C6A34FBAPQ/1.

Blanc, Jacob. *Before the Flood: The Itaipu Dam and the Visibility of Rural Brazil*. Durham, NC: Duke University Press, 2019.

Blanco, Graciela. "Las sociedades anónimas cruzan los Andes: Los inversores chilenos en Neuquén al comenzar el siglo XX." *América Latina en la Historia Económica* 19, no. 2 (2012): 107–30.

———. "Neuquén en el espacio patagónico: Tierra, ganado y comercio al comenzar el siglo XX." In *Tierra adentro: Instituciones económicas y sociales en los territorios nacionales (1884–1951)*, edited by Andrea Lluch and Marisa Moroni, 43–62. Rosario, Argentina: Prohistoria Ediciones, 2010. https://ri.conicet.gov.ar/handle/11336/137070.

Blanco, María del Pilar, and Joanna Page, eds. *Geopolitics, Culture, and the Scientific Imaginary in Latin America*. Gainesville: University Press of Florida, 2020.

Bocketti, Gregg. *The Invention of the Beautiful Game: Football and the Making of Modern Brazil*. Gainesville: University Press of Florida, 2019.

Bohoslavsky, Ernesto L. *El complot patagónico: Nación, conspiracionismo y violencia en el sur de Argentina y Chile (siglos XIX y XX)*. Buenos Aires: Prometeo, 2009.

———. "Modernización estatal y coerción: El lugar de la policía en el avance del estado argentino en la frontera (1880–1946)." In *La policía en perspectiva histórica: Argentina y Brasil (del siglo XIX a la actualidad)*, edited by Ernesto Bohoslavsky, Lila Caimari, and Cristina Schettini, 250–77. Buenos Aires: Crimen y Sociedad, 2009. CD-ROM. www.crimenysociedad.com.ar (site discontinued).

Booth, Rodrigo. "De la selva araucana a la Suiza chilena." *Ciudad y Arquitectura*, no. 143 (2011): 10–32.

———. "Turismo y representación del paisaje: La invención del sur de Chile en la mirada de la 'Guía del Veraneante' (1932–1962)." *Nuevo Mundo Mundos Nuevos* (February 2008). https://doi.org/10.4000/nuevomundo.25052.

Bona, Aixa, and Juan Vilaboa. "Las relaciones argentino-chilenas en el extremo austral, 1930–1955." *Magallania: Punta Arena* 32 (2004): 15–21.

Borneman, John. "Grenzregime (Border Regime): The Wall and Its Aftermath." In *Border Identities: Nation and State at International Frontiers*, edited by Thomas M. Wilson and Hastings Donnan, 162–90. Cambridge: Cambridge University Press, 1998.

Bouret, Daniela. "Lo sano y lo enfermo: El consumo de vinos y los problemas sociales del alcoholismo en el Montevideo del novecientos." *Boletín Americanista*, no. 65 (2012): 167–90.

Briones, Claudia, and Lorena Cardin. *Informe preliminar sobre ocupación tradicional del Lof Buenuleo*. Bariloche, Argentina: Universidad Nacional de Río Negro, 2020. http://rid.unrn.edu.ar/handle/20.500.12049/8213.

Briones, Claudia, and Walter Delrio. "La 'Conquista del Desierto' desde perspectivas hegemónicas y subalternas." *Runa: Archivo para las Ciencias del Hombre* 27 (2007): 23–48.

Buck, Daniel, and Anne Meadows. "Neighbors on the Hot Seat: Revelations from the Long-Lost Argentine Police File." *WOLA Journal* 5, no. 2 (1996): 6–15, 59–60.

Bueno, Christina. *The Pursuit of Ruins: Archaeology, History, and the Making of Modern Mexico*. Albuquerque: University of New Mexico Press, 2016.

Cafaro, Philip. "Patriotism as an Environmental Virtue." *Journal of Agricultural and Environmental Ethics* 23, no. 1 (March 2010): 185–206.

Caimari, Lila. *Apenas un delincuente*. Buenos Aires: Siglo XXI Editores, 2002.

Campos, Amie. "Territorial Conflicts, Bureaucracy, and State Formation in Chile's Southern Frontera 1866–1912." PhD diss., University of California, San Diego, 2022.

Cañizares-Esguerra, Jorge. *Nature, Empire, and Nation: Explorations of the History of Science in the Iberian World*. Stanford, CA: Stanford University Press, 2006.

Cánovas, Rodrigo. "A cien años de la matanza de Forrahue: Digna sepultura para los comuneros mapuches." *Cuadernos de Literatura* 22, no. 44 (2018): 251–75.

Cárdenas Palma, Mauricio. "El conflicto por la tierra: Repatriación de chilenos del Neuquén a las provincias de Malleco y Cautín (1896–1923)." In *Sujetos sin voz en la región sur y austral de Chile y Argentina: Frontera, colonización, marginalidad y organización popular chileno-mestiza en los siglos XIX y XX*, edited by Mathías Órdenes Delgado and Pablo Marimán Quemenado, 191–247. Santiago: LOM Ediciones, 2022.

Carey, Mark. "Mountaineers and Engineers: The Politics of International Science, Recreation, and Environmental Change in Twentieth-Century Peru." *Hispanic American Historical Review* 92, no. 1 (February 2012): 107–41.

Cariola, Carmen, and Osvaldo Sunkel. *Un siglo de historia económica de Chile, 1830–1930*. Santiago: Editorial Universitaria, 1991.

Carrasco, Gabriel. "La provincia de Santa Fe y el Chaco." *Boletín del Instituto Geográfico Argentino* 8, no. 6 (June 1887): 125–45.

Carreras, Sandra. "¿Un mismo origen con diferente destino? Los científicos alemanes en Argentina y Chile entre mediados del siglo XIX y comienzos del siglo XX." In Sanhueza Cerda, *La movilidad del saber*, 127–48.

Carter, Paul. *The Road to Botany Bay: An Exploration of Landscape and History*. Minneapolis: University of Minnesota Press, 2010.

Castellanos, M. Bianet. "Introduction: Settler Colonialism in Latin America." *American Quarterly* 69, no. 4 (December 2017): 777–81.

Castro, Juan, and Manuela Lavinas Picq. "Stateness as Landgrab: A Political History of Maya Dispossession in Guatemala." *American Quarterly* 69, no. 4 (2017): 791–99.

Castro, Sergio A., Ariel Camousseight, Mélica Muñoz-Schick, and Fabián M. Jaksic. "Rodulfo Amando Philippi, el naturalista de mayor aporte al conocimiento taxonómico de la diversidad biológica de Chile." *Revista Chilena de Historia Natural*, no. 79 (2006): 133–43.

Cersósimo, Facundo, and Maíne B. Lopes. "Julio A. Roca y la 'conquista del desierto': Monumentalización, patrimonio y usos del pasado durante las décadas de 1930 y 1940." *Quinto Sol* 23, no. 1 (2019): 1–19. https://doi.org/10.19137/qs.v23i1.2510.

Chalier, Gustavo. "El puerto comercial de Punta Alta: El capital francés y la pugna por el espacio económico de la Bahía Blanca (1900–1930)." *Cuadernos del Sur: Historia*, no. 34 (2005): 299–317.

Chang, Ning Jennifer. "Women in the Chase: Sports, Empire, and Gender in Shanghai, 1860–1945." *Chinese Studies in History* 54, no. 2 (2021): 130–48.

Chapman, Roger. "Throwing the Explorer out with the Fountain: American History Textbooks and Juan Ponce de León." *Florida Historical Quarterly* 94, no. 1 (2015): 92–107.

Chasteen, John Charles, and Sara Castro-Klarén. *Beyond Imagined Communities: Reading and Writing the Nation in Nineteenth-Century Latin America*. Washington, DC: Woodrow Wilson Center Press, 2003.

Cheater, A. P. "Transcending the State? Gender and Borderline Constructions of Citizenship in Zimbabwe." In *Border Identities: Nation and State at International Frontiers*, edited by Thomas M. Wilson and Hastings Donnan, 191–214. Cambridge: Cambridge University Press, 1998.

Chiocconi, María, Mariano Chiappe, and Adriana Podlubne. "¡Todo por la patria! Nacionalismo, practicas corporales y tiempo libre en asociaciones civiles region del Nahuel Huapi; Primera mitad del siglo XX." In *Historias en movimiento: Cuerpo, educación y tiempo libre en la Norpatagonia, 1884–1945*, edited by Laura Méndez, 181–254. Rosario, Argentina: Prohistoria Ediciones, 2011.

Cikota, Javier. "Frontier Justice: State, Law, and Society in Patagonia, 1880–1940." PhD diss., University of California, Berkeley, 2017.

Clark, A. Kim. *The Redemptive Work: Railway and Nation in Ecuador, 1895–1930*. Lanham, MD: Rowman and Littlefield, 1998.

Coatsworth, John H. *Growth against Development: The Economic Impact of Railroads in Porfirian Mexico*. DeKalb: Northern Illinois University Press, 1981.

Collier, Simon, and William F. Sater. *Historia de Chile, 1808–2017*. Madrid: Akal, 2019.

———. *A History of Chile, 1808–2002*. Cambridge Latin American Studies 82. 2nd ed. New York: Cambridge University Press, 2004.

Collinao, Francisco, Lorenzo Loncón, Damian Olivero, Laura Subiri, Soledad Tropan, Verónica Márquez, Florentino Nahuel, et al. *Lof Paichil Antreao: Comunidad mapuche ancestral de la región de Villa la Angostura*. Buenos Aires: Universidad de Buenos Aires, Facultad de Filosofía y Letras, 2019. https://ri.conicet.gov.ar/handle/11336/138250.

Colpan, Sema, Bernhard Hachleitner, and Matthias Marschik. "Jewish Difference in the Context of Class, Profession and Urban Topography: Studies of Jewish Sports Officials in Interwar Vienna." *Austrian Studies*, no. 24 (2016): 140–55.

Comisión Nacional de Museos, Monumentos y Lugares Históricos (Administración de Parques Nacionales), Universidad Nacional del Comahue, Municipalidad de San Carlos de Bariloche, and Asociación Amigos del Museo de la Patagonia Francisco P. Moreno. *Patrimonio arquitectónico y urbano de San Carlos de Bariloche: Inventario de edificios y sitios del centro de la ciudad*. 2 vols. Bariloche, Argentina: Municipalidad de San Carlos de Bariloche, 1995.

Confino, Alon. *The Nation as a Local Metaphor: Württemberg, Imperial Germany, and National Memory, 1871–1918*. Chapel Hill: University of North Carolina Press, 1997.

Cornaglia, Miguel Ángel. *Bariloche: Su pasado y su gente*. Buenos Aires: Plus Ultra, 1983.

Cornelis, Stella M. "Reflexiones sobre la trayectoria de Enrique Romero Brest: Un profesional al servicio del estado (primeras décadas del siglo XX)." *Res Gesta*, no. 57 (2021): 112–35.

Correa, María José. "'¿Quiénes son los profesionales?' Justicia, profesionalización y ejercicio médico en el Chile urbano de la segunda mitad del siglo XIX." *Dynamis* 37, no. 2 (2017): 273–93.

Craib, Raymond B. *Cartographic Mexico: A History of State Fixations and Fugitive Landscapes*. Durham, NC: Duke University Press, 2004.

———. "Cartography and Decolonization." In *Decolonizing the Map: Cartography from Colony to Nation*, edited by James R. Akerman, 11–71. Chicago: University of Chicago Press, 2017.

———. *The Cry of the Renegade: Politics and Poetry in Interwar Chile*. New York: Oxford University Press, 2016.

Cresswell, Tim. *Place: An Introduction*. 2nd ed. Chichester, UK: Wiley Blackwell 2015.

Crow, Joanna. *The Mapuche in Modern Chile: A Cultural History*. Gainesville: University Press of Florida, 2013.

Cuarterolo, Andrea. "El arte de 'instruir deleitando': Discursos positivistas y nacionalistas en el cine argentino del primer centenario." *Iberoamericana* 10, no. 39 (2010): 197–210.

Dalakoglou, Dimitris, and Penny Harvey. "Roads and Anthropology: Ethnographic Perspectives on Space, Time and (Im)Mobility." *Mobilities* 7, no. 4 (November 2012): 459–65.

D'Alençon Castrillón, Renato, and Francisco Prado García. "Construcción en madera maciza en el sur de Chile: Un sistema constructivo excepcional en peligro de extinción." In *Actas del noveno congreso nacional y primer congreso internacional hispanoamericano de historia de la construcción: Segovia, 13 a 17 de octubre de 2015*, edited by Instituto Juan de Herrera, 521–30. Madrid: Instituto Juan de Herrera, 2015. https://dialnet.unirioja.es/servlet/articulo?codigo=5576445.

Dauncey, Hugh. *French Cycling: A Social and Cultural History*. Liverpool, UK: Liverpool University Press, 2012. https://directory.doabooks.org/handle/20.500.12854/30754.

de Laforcade, Geoffroy. "Memories and Temporalities of Anarchist Resistance: Community Traditions, Labor Insurgencies, and Argentine Shipyard Workers, Early 1900s to Late 1950s." In *In Defiance of Boundaries: Anarchism in Latin American History*, edited by Geoffroy de Laforcade and Kirwin R. Shaffer, 185–215. Gainesville: University Press of Florida, 2015.

Delaney, Jean H. "Imagining 'El Ser Argentino': Cultural Nationalism and Romantic Concepts of Nationhood in Early Twentieth-Century Argentina." *Journal of Latin American Studies* 34, no. 3 (2002): 625–58.

De la Torre, Oscar. *The People of the River: Nature and Identity in Black Amazonia, 1835–1945*. North Carolina Scholarship Online. Chapel Hill: University of North Carolina Press, Project MUSE, 2018.

del Castillo, Lina. "Cartography in the Production (and Silencing) of Colombian Independence History, 1807–1827." In *Decolonizing the Map: Cartography from Colony to Nation*, edited by James R. Akerman, 110–60. Chicago: University of Chicago Press, 2017.

Delrio, Walter. "De 'salvajes' a 'indios nacionales': Interpelaciones hegemónicas y campañas militares en Norpatagonia y la Araucanía (1879–1885)." *Mundo de Antes*, no. 3 (2002): 189–207.

———. *Memorias de expropriación: Sometimiento e incorporación indígena en la Patagonia, 1872–1943*. Bernal, Argentina: Universidad Nacional de Quilmes, 2005.

Delrio, Walter, Diana Lenton, Marcelo Musante, Mariano Nagy, Alexis Papazian, and Pilar Pérez. "Discussing Indigenous Genocide in Argentina: Past, Present, and Consequences of Argentinean State Policies toward Native Peoples." *Genocide Studies and Prevention* 5, no. 2 (2010): 138–59.

Delrio, Walter, and Pilar Pérez. "Territorializaciones y prácticas estatales: Percepciones del espacio social luego de la Conquista del Desierto." In *Cultura y espacio: Araucanía-Norpatagonia*, edited by Pedro Navarro Floria and Walter Delrio, 237–52. Bariloche, Argentina: Universidad Nacional de Río Negro, Instituto de Investigaciones en Diversidad Cultural y Procesos de Cambio, 2011.

Devoto, Fernando. *Nacionalismo, fascismo y tradicionalismo en la Argentina moderna: Una historia*. Buenos Aires: Siglo XXI Editores, 2002.

Diacon, Todd A. *Millenarian Vision, Capitalist Reality: Brazil's Contestado Rebellion, 1912–1916*. Durham, NC: Duke University Press, 1991.

Díaz, George. *Border Contraband: A History of Smuggling across the Rio Grande*. Austin: University of Texas Press, 2015.

Dlamini, Jacob S. T. *Safari Nation: A Social History of the Kruger National Park*. Athens: Ohio University Press, 2020.

Dodds, Klaus-John. "Geography, Identity and the Creation of the Argentine State." *Bulletin of Latin American Research* 12, no. 3 (1993): 311–31.

Donoghue, Michael. "Roberto Durán, Omar Torrijos, and the Rise of Isthmian Machismo." In *Sports Culture in Latin American History*, edited by David M. K. Sheinin, 17–38. Pittsburgh, PA: University of Pittsburgh Press, 2015.

Doura, Miguel Armando. "Acerca del topónimo 'Patagonia,' una nueva hipótesis de su génesis." *Nueva Revista de Filología Hispánica* 59, no. 1 (2011): 37–78.

Drinot, Paulo. *The Sexual Question: A History of Prostitution in Peru, 1850s–1950s*. Cambridge: Cambridge University Press, 2020.

Dupree, James. "The Roots of the Border Patrol Line Riders and the Bureaucratization of US-Mexican Border Policing, 1894–1924." In *Border Policing: A History of Enforcement and Evasion in North America*, edited by Holly M. Karibo and George T. Díaz, 115–28. Austin: University of Texas Press, 2020.

Edney, Matthew H. *Mapping an Empire: The Geographical Construction of British India, 1765–1843*. Chicago: University of Chicago Press, 1997.

Edwards, Ryan C. *A Carceral Ecology: Ushuaia and the History of Landscape and Punishment in Argentina*. Oakland: University of California Press, 2022.

Elbirt, Ana Laura. "Representaciones de la nación en el ensayo argentino: Una lectura crítica a la centralidad pampeana." *Latinoamérica: Revista de Estudios Latinoamericanos*, no. 65 (2017): 35–56.

Elias, Norbert. *The Quest for Excitement: Sport and Leisure in the Civilizing Process*. Oxford, UK: B. Blackwell, 1986.

Elsey, Brenda. *Citizens and Sportsmen: Fútbol and Politics in Twentieth-Century Chile*. Austin: University of Texas Press, 2011.

———. "Sport in Latin America." In *The Oxford Handbook of Sports History*, edited by Robert Edelman and Wayne Wilson, 361–76. Oxford: Oxford University Press, 2017.

Erbig, Jeffrey Alan. *Where Caciques and Mapmakers Met: Border Making in Eighteenth-Century South America*. Chapel Hill: University of North Carolina Press, 2020.

Escolar, Diego, and Leticia Saldi. "Apropiación y destino de los niños indígenas capturados en la campaña del desierto: Mendoza, 1878–1889." *Nuevo Mundo Mundos Nuevos* (December 2018). https://doi.org/10.4000/nuevomundo.74602.

———. "Castas invisibles de la nueva nación: Los prisioneros indígenas de la Campaña del Desierto en el registro parroquial de Mendoza." In *En el país de nomeacuerdo: Archivos y memorias del genocidio del estado argentino sobre los pueblos originarios, 1870–1950*, edited by Walter Delrio, Diana Lenton, and Marisa Malvestitti, 99–136. Viedma, Argentina: Editorial UNRN, 2018.

Escolar, Diego, Celia Claudia Salomon Tarquini, and Julio Esteban Vezub. "La 'Campaña del Desierto' (1870–1890): Notas para una crítica historiográfica." In *Guerras de la historia argentina*, edited by Federico Lorenz, 223–47. Buenos Aires: Ariel, 2015.

Essinger, Bent. "La emigración danesa." In *La emigración europea a la América Latina: Fuentes y estado de investigación; Informes presentados a la IV. Reunión de Historiadores*

Latinoamericanistas Europeos, edited by Wilhelm Stegmann, 85–99. Berlin: Colloquium-Verlag, 1979. https://publications.iai.spk-berlin.de/receive/riai_mods_00002430.

Estefante, Andrés. "Estado y ordenamiento territorial en Chile, 1810-2016." In *Historia política de Chile, 1810–2010*, edited by Francisca Rengifo, 2:87–138. Santiago: FCE-UAI, 2018.

Favaro, Orietta, and Marta Morinelli. "Los reformistas de la clase dominante (1890–1916)." *Revista de Historia*, no. 1 (1990): 59–81.

Fernández Bravo, Álvaro. "Celebraciones centenarias: Nacionalismo y cosmopolitismo en las conmemoraciones de la independencia; Buenos Aires, 1910—Río de Janeiro, 1922." In *Galerías del progreso: Museos, exposiciones y cultura visual en América Latina*, edited by Beatriz González Stephan and Jens Andermann, 331–72. Rosario, Argentina: Beatriz Viterbo Editora, 2006.

Fernández Lobbé, Marcos. "La virtud como militancia: Las organizaciones temperantes y la lucha anti-alcohólica en Chile, 1870–1930." *Cuadernos de Historia (Departamento de Ciencias Históricas, Universidad de Chile)*, no. 27 (2007): 125–58.

Fiquepron, Maximiliano. *Morir en las grandes pestes: Las epidemias de cólera y fiebre amarilla en la Buenos Aires del siglo XIX*. Buenos Aires: Siglo XXI Editores, Asociación Argentina de Investigadores en Historia, 2020.

Flores, Daniel, Marco Gamarra, and Juan Florián. *La ciudad y el mar: La Avenida Costanera entre la Lima aristocrática y la Lima del oncenio (1917–1930)*. Research report. Lima: Programa de Apoyo a la Iniciación en la Investigación, 2015. https://repositorio.pucp.edu.pe/index/handle/123456789/136397.

Fortunato, Norberto. "El territorio y sus representaciones como fuente de recursos turísticos: Valores fundacionales del concepto de 'parque nacional.'" *Estudios y Perspectivas en Turismo* 14, no. 4 (2005): 314–48.

Foucault, Michel. *Abnormal: Lectures at the Collège de France, 1974–1975*. New York: Picador, 2003.

Freeman, Cordélia. "Identity and the Militarized Border: Mi mejor enemigo (Chile, 2005)." *Espaço e Cultura*, no. 33 (2013): 65–86.

Freitas, Frederico. *Nationalizing Nature: Iguazu Falls and National Parks at the Brazil-Argentina Border*. Cambridge: Cambridge University Press, 2021.

Gallucci, Lisandro Juan. "Nación, república y constitución: La Liga Patriótica Argentina y su Congreso General de Territorios Nacionales." *Jahrbuch für Geschichte Lateinamerikas*, no. 54 (2017): 306–37.

García, Analía, and Sebastián Valverde. "Políticas estatales y procesos de etnogénesis en el caso de poblaciones mapuche de Villa La Angostura, provincia de Neuquén, Argentina." *Cuadernos de Antropología Social*, no. 25 (2007): 111–32.

García Matus de la Parra, María, and Ingrid Valdivia Garrido. "La Empresa de los Ferrocarriles del Estado de Chile y el despertar del turismo nacional: Rutas y paisajes." *Estudios Hemisféricos y Polares* 3, no. 2 (Second Quarter 2012): 88–101.

Gattás Vargas, Maia, Paula Gabriela Núñez, and Carolina Lema. "La monstruosa cartografía patagónica o los mapas como discursos retóricos." *Bitácora arquitectura*, no. 36 (2017): 122–29.

Gersdorf, Catrin. *The Poetics and Politics of the Desert: Landscape and the Construction of America*. Amsterdam: BRILL, 2009.

Gissibl, Bernhard, Sabine Höhler, and Patrick Kupper. "Towards a Global History of National

Parks." In *Civilizing Nature: National Parks in Global Historical Perspective*, edited by Bernhard Gissibl, Sabine Höhler, and Patrick Kupper, 1:1–27. New York: Berghahn Books, 2012.

Gobantes, Catalina, Jonathan Barton, Álvaro Román, and Alejandro Salazar. "Migraciones laborales entre la Isla de Chiloé (Chile) y Patagonia austral: Relaciones históricas y cambios recientes en un espacio transnacional." In *Miradas transcordilleranas: Selección de trabajos del IX Congreso Argentino-Chileno de Estudios Históricos e Integración Cultural*, edited by Paula Gabriela Núñez, 20–30. Bariloche, Argentina: Instituto de Investigaciones en Diversidad Cultural y Procesos de Cambio-Universidad Nacional de Río Negro-CONICET, 2011.

Goebel, Michael. "Settler Colonialism in Postcolonial Latin America." In *The Routledge Handbook of the History of Settler Colonialism*, edited by Edward Cavanagh and Lorenzo Veracini, 139–51. Abingdon, UK: Routledge, 2016. Gold, John R. "Creating the Charter of Athens: CIAM and the Functional City, 1933–43." *Town Planning Review* 69, no. 3 (1998): 225–47.

González-Caniulef, Elsa Gabriela. "Mujeres mapuche en manos de primitivos dueños: Orden y control sociosexual en Chile a mediados del siglo XIX." *Revista Española de Antropología Americana*, no. 48 (2018): 105–20.

González Leiva, José Ignacio, and Belisario Andrade Johnson. "Geografía física de la República de Chile por Pedro José Amado Pissis Marín, 1812–1889." In *Geografía física de la República de Chile*, by Pedro José Amado Pissis and edited by Rafael Sagredo Baeza, ix–xlv. Santiago: Cámara Chilena de la Construcción; Pontificia Universidad Católica de Chile; Dirección de Bibliotecas, Archivos y Museos, 2011.

González Leiva, José Ignacio, and Patricio Bernedo Pinto. "Cartografía de la transformación de un territorio: La Araucanía 1852–1887." *Revista de Geografía Norte Grande*, no. 54 (May 2013): 179–98.

Goodrich, Diana Sorensen. "La construcción de los mitos nacionales en la Argentina del centenario." *Revista de Crítica Literaria Latinoamericana* 24, no. 47 (1998): 147–66.

Gordillo, Gastón. *Landscapes of Devils: Tensions of Place and Memory in the Argentinean Chaco*. Durham, NC: Duke University Press, 2004.

Gorelik, Adrián. "La arquitectura de YPF: 1934–1943; Notas para una interpretación de las relaciones entre estado, modernidad e identidad en la arquitectura argentina de los años 30." *Anales del Instituto de Arte Americano e Investigaciones Estéticas Mario Buschiazzo*, no. 25 (1987): 178–201.

———. *La grilla y el parque: Espacio público y cultura urbana en Buenos Aires, 1887–1936*. Buenos Aires: Universidad Nacional de Quilmes, 1998.

Gorelik, Adrián, and Graciela Silvestri. "Lo nacional en la historiografía de la arquitectura en la Argentina: El peso de la tradición." In *Historiografía argentina, 1958–1988: Una evaluación crítica de la producción histórica nacional*, 174–87. Buenos Aires: Comité Internacional de Ciencias Históricas, Comité Argentino, 1990.

———. "The Past as the Future: A Reactive Utopia in Buenos Aires." In *The Latin American Cultural Studies Reader*, edited by Ana Del Sarto, Alicia Ríos, and Abril Trigo, 427–40. Durham, NC: Duke University Press, 2004.

Greenberg, Dolores. "Reassessing the Power Patterns of the Industrial Revolution: An Anglo-American Comparison." *American Historical Review* 87, no. 5 (1982): 1237–61.

Greenwald, Hannah. "'Improve Their Condition While Making Them Useful': Colonia General Conesa and the Dynamics of Settler Colonialism in Nineteenth-Century Argentina." *Hispanic American Historical Review* 103, no. 1 (2023): 101–37.

———. "Now I Walk on Foreign Soil: Settler Colonialism in Argentina's Southern Borderlands, 1867–1899." PhD diss., Yale University, 2022. https://elischolar.library.yale.edu/gsas_dissertations/475.

Gregory, Ian N. *Troubled Geographies: A Spatial History of Religion and Society in Ireland*. Bloomington: Indiana University Press, 2013.

Greider, Thomas, and Lorraine Garkovich. "Landscapes: The Social Construction of Nature and the Environment." *Rural Sociology* 59, no. 1 (1994): 1–24.

Guajardo Soto, Guillermo. "Infraestructura y movilidad: Una reflexión histórica comparativa sobre Chile y México, 1840–1980." *Revista de Historia y Geografía*, no. 30 (2014): 155–65.

———. *Tecnología, estado y ferrocarriles en Chile, 1850–1950*. Mexico City: Fundación Ferrocarriles Españoles/UNAM, 2007.

Guillén, Mauro F. "Modernism without Modernity: The Rise of Modernist Architecture in Mexico, Brazil, and Argentina, 1890–1940." *Latin American Research Review* 39, no. 2 (2004): 6–34.

Gundermann Kröll, Hans. "Los pueblos originarios del norte de Chile y el estado." *Diálogo Andino*, no. 55 (2018): 93–109.

Gutiérrez, Ramón. "Los inicios del urbanismo en la Argentina: El aporte francés y la acción de Ernesto de Estrada." In *Ernesto de Estrada: El arquitecto frente al paisaje*, edited by Ramón Gutiérrez, 23–48. Buenos Aires: Centro de Documentación de Arte y Arquitectura Latinoamericana, 2007.

Hajduk, Adán, Ana María Albornoz, Maximiliano J. Lezcano, and Graciela Montero. "De Chiloé al Nahuel Huapi: Nuevas evidencias materiales del accionar jesuítico en el gran lago (siglos XVII y XVIII)." In *Araucanía-Norpatagonia: La territorialidad en debate; Perspectivas ambientales, culturales, sociales, políticas y económicas*, edited by María Andrea Nicoletti and Paula Gabriela Núñez, 243–79. Bariloche, Argentina: Instituto de Investigaciones en Diversidad Cultural y Procesos de Cambio, 2013.

Halperín Donghi, Tulio. *Una nación para el desierto argentino*. Buenos Aires: Centro Editor de América Latina, 1982.

Harambour, Alberto, ed. "Fronteras nacionales, Estados colonials. ¿Para una historia plurinacional de América Latina?" In "Fronteras, empresas de colonización y pueblos indígenas, siglos XIX y XX," special issue, *Historia crítica* 82 (2021): 3–27.

———. *Soberanías fronterizas: Estados y capital en la colonización de Patagonia (Argentina y Chile, 1830–1922)*. Valdivia: Ediciones Universidad Austral de Chile, 2019.

———. "Soberanía y corrupción: La construcción del estado y la propiedad en Patagonia austral (Argentina y Chile, 1840–1920)." *Historia (Santiago)* 50, no. 2 (2017): 555–96.

Hartshorne, Richard. *Perspective on the Nature of Geography*. Monograph Series of the Association of American Geographers 1. Chicago: Rand McNally/Association of American Geographers, 1959.

Harvey, Kyle E. "Engineering Value: The Transandine Railway and the 'Techno-capital' State in Chile at the End of the Nineteenth Century." *Journal of Latin American Studies* 52, no. 4 (November 2020): 711–33.

Healey, Mark A. *The Ruins of the New Argentina: Peronism and the Remaking of San Juan after the 1944 Earthquake*. Durham, NC: Duke University Press, 2011.

Hecht, Susanna. *The Scramble for the Amazon and the "Lost Paradise" of Euclides Da Cunha*. Chicago: University of Chicago Press, 2013.

Helg, Aline. "Race in Argentina and Cuba, 1880–1930: Theory, Policies, and Popular Reaction." In *The Idea of Race in Latin America, 1870–1940*, edited by Richard Graham, 37–69. Austin: University of Texas Press, 1990.

Herda, Phyllis S. "Ethnology in the Enlightenment: The Voyage of Alejandro Malaspina in the Pacific." In *Enlightenment and Exploration in the North Pacific, 1741–1805*, edited by Stephen Haycox, James K. Barnett, and Caedmon A. Liburd, 65–76. Seattle: University of Washington Press, 1997.

Hernandez, Kelly Lytle. *Migra: A History of the U.S. Border Patrol*. Berkeley: University of California Press, 2010.

Hernández Barral, José Miguel. "Polo: Social Distinction and Sports in Spain, 1900–1950." *International Journal of the History of Sport* 36, no. 2–3 (2019): 149–68.

Herner, María Teresa. "La invisibilización del otro indígena en el proceso de construcción nacional: El caso de la Colonia Emilio Mitre, La Pampa." *Huellas*, no. 18 (2014): 118–31.

Herr, Pilar. *Contested Nation: The Mapuche, Bandits, and State Formation in Nineteenth-Century Chile*. Albuquerque: University of New Mexico Press, 2019.

Hershfield, Joanne. *Imagining La Chica Moderna: Women, Nation, and Visual Culture in Mexico, 1917–1936*. Durham, NC: Duke University Press, 2008.

Hilgartner, Stephen. *Science on Stage: Expert Advice as Public Drama*. Stanford, CA: Stanford University Press, 2000.

Holston, James. *The Modernist City: An Anthropological Critique of Brasilia*. Chicago: University of Chicago Press, 1989.

Hunt, John Dixon. *Genius Loci: An Essay on the Meanings of Place*. London: Reaktion Books, 2022.

Irarrázaval Larraín, José Miguel. *La Patagonia: Errores geográficos y diplomáticos*. Santiago: Andrés Bello, 1966.

Itzigsohn, José, and Matthias vom Hau. "Unfinished Imagined Communities: States, Social Movements, and Nationalism in Latin America." *Theory and Society* 35, no. 2 (2006): 193–212.

Jackson, Shona N. "Subjection and Resistance in the Transformation of Guyana's Mythocolonial Landscape." In *Caribbean Literature and the Environment: Between Nature and Culture*, edited by Elizabeth M. DeLoughrey, Renée K. Gosson, and George B. Handley, 85–98. Charlottesville: University of Virginia Press, 2005.

Jacoby, Karl. *Crimes against Nature: Squatters, Poachers, Thieves, and the Hidden History of American Conservation*. Berkeley: University of California Press, 2003.

Jameson, W. C. *Butch Cassidy: Beyond the Grave*. Lanham, MD: Taylor Trade, 2012.

Jones, Karen. "Unpacking Yellowstone: The American National Park in Global Perspective." In *Civilizing Nature: National Parks in Global Historical Perspective*, edited by Bernhard Gissibl, Sabine Höhler, and Patrick Kupper, 1:31–49. New York: Berghahn Books, 2012.

Kaltmeier, Olaf. *National Parks from North to South: An Entangled History of Conservation and Colonization in Argentina*. New Orleans, LA: University of New Orleans Press, 2021.

Kaufmann, Eric. "'Naturalizing the Nation': The Rise of Naturalistic Nationalism in the United States and Canada." *Comparative Studies in Society and History* 40, no. 4 (October 1998): 666–95.

Kerr, Ashley Elizabeth. *Sex, Skulls, and Citizens: Gender and Racial Science in Argentina (1860–1910)*. Nashville, TN: Vanderbilt University Press, 2020.

Kinzel Kahler, Enrique, and Bernardo Horn Klenner. *Puerto Varas: 130 años de historia 1852–1983*. Puerto Varas, Chile: Imprenta Horn, 1983.

Korol, Juan Carlos. "La economía." In *Nueva historia argentina: Crisis económica, avance del estado e incertidumbre política (1930–1943)*, edited by Alejandro Cattaruzza, 17–48. Buenos Aires: Sudamericana, 2001.

Kuntz Ficker, Sandra, ed. *Historia mínima de la expansión ferroviaria en América Latina*. Mexico City: El Colegio de México, 2017. Ebook.

Kusno, Abidin. *Behind the Postcolonial: Architecture, Urban Space, and Political Cultures in Indonesia*. London: Routledge, 2000.

Lacoste, Pablo. *La imagen del otro en las relaciones de la Argentina y Chile (1534–2000)*. Buenos Aires: Fondo de Cultura Económica, 2003.

———. "Las propuestas de integración económica sudamericana: De Diego Portales a Alfredo Palacios (1830–1939)." *Historia* 32, no. 1 (1999): 103–29.

———. "Vinos, carnes, ferrocarriles y el tratado de libre comercio entre Argentina y Chile (1905–1910)." *Historia (Santiago)* 37, no. 1 (June 2004): 97–127.

Lafuente, Antonio, and Leoncio López-Ocón. "Bosquejos de la ciencia nacional en la América Latina del siglo XIX." *Asclepio: Revista de Historia de la Medicina y de la Ciencia* 50, no. 2 (1998): 5–10.

Lagos Carmona, Guillermo. *Historia de las fronteras de Chile: Los tratados de límites con Argentina*. Santiago: Andrés Bello, 1966.

Langer, Erick D. *Expecting Pears from an Elm Tree: Franciscan Missions on the Chiriguano Frontier in the Heart of South America, 1830–1949*. Durham, NC: Duke University Press, 2009.

Larson, Carolyne R., ed. *The Conquest of the Desert: Argentina's Indigenous Peoples and the Battle for History*. Albuquerque: University of New Mexico Press, 2020.

Latour, Bruno. "Drawing Things Together." In *Representation in Scientific Practice*, edited by Michael Lynch and Steve Woolgar, 19–68. Cambridge: Massachusetts Institute of Technology Press, 1990.

Lavandaio, Eddy Omar Luis, and Edmundo Catalano. *Historia de la minería argentina*. Vol. 2. Buenos Aires: Servicio Geológico Minero Argentino, 2004.

Lavrín, Asunción. *Women, Feminism, and Social Change in Argentina, Chile, and Uruguay, 1890–1940*. Lincoln: University of Nebraska Press, 1995.

Lazzari, Axel, and Diana Lenton. "Araucanization and Nation, or How to Inscribe Modern Indians upon the Pampas during the Last Century." In *Living on Edge: Contemporary Perspectives on the Native Peoples of Pampa, Patagonia, and Tierra del Fuego*, edited by Claudia Briones and José Luis Lanata, 33–46. Westport, CT: Bergin and Garvey, 2002.

Le Bail, Etienne. "Clubes sociales y deportivos en ingenios azucareros: Tucumán, 1875–1930." *Travesía* 22, no. 2 (2020): 59–94.

Lefebvre, Henri. *The Production of Space*. Oxford, UK: Blackwell, 1991.

Lekan, Thomas M. *Imagining the Nation in Nature: Landscape Preservation and German Identity, 1885–1945*. Cambridge, MA: Harvard University Press, 2004.

Lenton, Diana. "La 'cuestión de los indios' y el genocidio en los tiempos de Roca: Sus repercusiones en la prensa y la política." In *Historia de la crueldad argentina: Julio Argentino Roca y el genocidio de los Pueblos Originarios*, edited by Osvaldo Bayer and Diana Lenton, 29–49. Buenos Aires: Red de Investigadores en Genocidio y Política Indígena, 2010.

———. "Relaciones interétnicas: Derechos humanos y autocrítica en la Generación del 80." In *La problemática indígena: Estudios antropológicos sobre pueblos indígenas de la Argentina*, edited by Juan Carlos Radovich, 27–65. Buenos Aires: Centro Editor de América Latina, 1992.

Lespai Silva, Joel. "Consolidación del capitalismo agrario en la región austral y propiedad indígena en Osorno (1883–1931)." In *Amotinados, abigeos y usurpadores: Una mirada regional acerca de las formas de violencia en Osorno (1821–1931)*, edited by Jorge Ernesto Muñoz Sougarret and Raúl Nuñez Muñoz, 101–45. Osorno, Chile: Editorial Universidad de Los Lagos, 2007.

Lesser, Jeffrey. *Immigration, Ethnicity, and National Identity in Brazil, 1808 to the Present*. New York: Cambridge University Press, 2013.

Livon-Grosman, Ernesto. *Geografías imaginarias: El relato de viaje y la construcción del espacio patagónico*. Rosario, Argentina: Beatriz Viterbo Editora, 2003.

Lois, Carla. "La invención del desierto chaqueño: Una aproximación a las formas de apropiación simbólica de los territorios del Chaco en los tiempos de formación y consolidación del estado nación Argentino." *Scripta Nova. Revista Electrónica de Geografía y Ciencias Sociales*, no. 38 (1999). www.ub.edu/geocrit/sn-38.htm.

———. "La Patagonia en el mapa de la Argentina moderna: Política y 'deseo territorial' en la cartografía oficial argentina en la segunda mitad del siglo XIX." In *Paisajes del progeso: La resignificación de la Patagonia Norte, 1880–1916*, edited by Pedro Navarro Floria, 107–34. Neuquén, Argentina: Educo, 2007.

———. *Mapas para la nación: Episodios en la historia de la cartografía argentina*. Buenos Aires: Biblos, 2014.

Lolich, Liliana. "Bariloche y su centro cívico." In *Ernesto de Estrada: El arquitecto frente al paisaje*, edited by Ramón Gutiérrez, 64–72. Buenos Aires: Centro de Documentación de Arte y Arquitectura Latinoamericana, 2007.

———. "Ernesto de Estrada como urbanista pionero en la Patagonia." In *Ernesto de Estrada: El arquitecto frente al paisaje*, edited by Ramón Gutiérrez, 49–60. Buenos Aires: Centro de Documentación de Arte y Arquitectura Latinoamericana, 2007.

Lolich, Liliana, Laila Vejsbjerg, Hugo Weibel, and Gian Piero Cherubini Zanetel. "Estado y paisaje: Estudio comparativo de la arquitectura hotelera desde una perspectiva binacional." In *Araucanía-Norpatagonia: La territorialidad en debate; Perspectivas ambientales, culturales, sociales, políticas y económicas*, edited by María Andrea Nicoletti and Paula Gabriela Núñez, 55–78. Bariloche, Argentina: Instituto de Investigaciones en Diversidad Cultural y Procesos de Cambio, 2013.

López, Rick Anthony. *Crafting Mexico: Intellectuals, Artisans, and the State after the Revolution*. Durham, NC: Duke University Press, 2010.

López Cárdenas, Patricio. *Osorno entre Julio Buschmann y René Soriano*. Osorno, Chile: Dokumenta Comunicaciones, 2008.

López-Ocón, Leoncio. "La Sociedad Geográfica de Lima y la formación de una ciencia nacional en el Perú republicano." *Terra Brasilis: Revista da Rede Brasileira de História da Geografia e Geografia Histórica*, no. 3 (2001). https://doi.org/10.4000/terrabrasilis.330.

Losada, Leandro. "Sociabilidad, distinción y alta sociedad en Buenos Aires: Los clubes sociales de la elite porteña (1880–1930)." *Desarrollo Económico* 45, no. 180 (2006): 547–72.

Macor, Darío. "Partidos, coaliciones y sistemas de poder." In *Nueva historia argentina: Crisis económica, avance del estado e incertidumbre política (1930–1943)*, edited by Alejandro Cattaruzza, 49–96. Buenos Aires: Sudamericana, 2001.

Maggiori, Ernesto. *Donde los lagos no tienen nombre: La historia de Río Pico, sus pobladores, sus alrededores y la colonia alemana "Friedland."* Comodoro Rivadavia, Argentina: Editorial Universitaria de la Patagonia, 2001.

———. *La cruzada patagónica de la Policía Fronteriza*. Gaiman, Argentina: Del Cedro, 2012.

Malkiel, María Rosa Lida de. "Para la toponimia argentina: Patagonia." *Hispanic Review* 20, no. 4 (1952): 321–23.

Mallol i Moretti, Adrián. "Institucionalización del imaginario moderno en las estaciones de servicio del Plan ACA • YPF: Los concursos de anteproyectos de 1936 y 1937." *Registros* 14, no. 2 (2018): 4–27.

Manara, Carla G. "La disputa por un territorio indígena: Argentina y Chile tras Varvarco (siglo XIX)." *Tefros* 11, no. 1 (2013): 7–37.

Manzoni, Gisela. "Contra los arrastra sables... militarismo y antimilitarismo en los comienzos de la Argentina moderna." *Avances del Cesor* 15, no. 19 (2018): 77–100.

Marchant, Anyda. *Viscount Maua and the Empire of Brazil: A Biography of Irineu Evangelista De Sousa (1813–1889)*. Berkeley: University of California Press, 2021.

Mársico, Leonardo Daniel. "Disputas por el sentido y el acceso a la práctica de esquí en Bariloche (1999–2013)." *Cuadernos del Claeh* 40, no. 114 (2021): 299–315.

Martinic Beros, Mateo. "Ferrocarriles en la zona austral de Chile, 1869–1973." *Historia (Santiago)* 38, no. 2 (2005): 367–95.

Massey, Doreen. *For Space*. Thousand Oaks, CA: SAGE, 2005.

Matthews, Michael. *The Civilizing Machine: A Cultural History of Mexican Railroads, 1876–1910*. Lincoln: University of Nebraska Press, 2013.

Mazzitelli Mastricchio, Malena. *Imaginar, medir, representar y reproducir el territorio: Una historia de las prácticas y las políticas cartográficas del estado argentino 1904–1941*. Buenos Aires: Universidad de Buenos Aires, Facultad de Filosofía y Letras, 2017. https://ri.conicet.gov.ar/handle/11336/112399.

McCann, Bryan. *Hello, Hello Brazil: Popular Music in the Making of Modern Brazil*. Durham, NC: Duke University Press, 2004.

McCook, Stuart. "Global Currents in National Histories of Science: The 'Global Turn' and the History of Science in Latin America." *Isis* 104, no. 4 (2013): 773–76.

———. *States of Nature: Science, Agriculture, and Environment in the Spanish Caribbean, 1760–1940*. Austin: University of Texas Press, 2002.

McGee Deutsch, Sandra. *Counterrevolution in Argentina, 1900–1932: The Argentine Patriotic League*. Lincoln: University of Nebraska Press, 1986.

Méndez, Laura. "'El león de la cordillera': Primo capraro y el desempeño empresario de la región de Nahuel Huapi, 1902–1932." *Boletín Americanista*, no. 59 (2009): 29–46.

———. *Estado, frontera y turismo: Historia de San Carlos de Bariloche*. Buenos Aires: Prometeo, 2010.

Méndez, Laura, and Jorge Ernesto Muñoz Sougarret. "Alianzas sectoriales en clave regional: La Norpatagonia argentino-chilena entre 1895 y 1920." In *Araucanía-Norpatagonia: La territorialidad en debate; Perspectivas ambientales, culturales, sociales, políticas y económicas*, edited by María Andrea Nicoletti and Paula Gabriela Núñez, 149–64. Bariloche, Argentina: Instituto de Investigaciones en Diversidad Cultural y Procesos de Cambio, 2013.

———. "Economías cordilleranas e intereses nacionales: Genealogía de una relación; El caso de la Compañía Comercial y Ganadera Chile-Argentina (1895–1920)." In *Fronteras en movimiento e imaginarios geográficos: La cordillera de los Andes como espacialidad sociocultural*, edited by Andrés Núñez, Rafael Sánchez, and Federico Arenas 163–87. Santiago: RIL Editores, 2013.

Mendoza, Marcos. *The Patagonian Sublime: The Green Economy and Post-neoliberal Politics*. New Brunswick, NJ: Rutgers University Press, 2018.

Meza Bazán, Mario. "El enfoque médico social sobre el uso y consumo de la coca y la cocaína en Perú en la primera mitad del siglo XX." *Nueva Corónica*, no. 2 (2013): 487–503.

Milos Hurtado, Pedro. *Frente Popular en Chile: Su configuración; 1935–1938*. Santiago: LOM Ediciones, 2008.

Montt de Etter, Rosario. *Inmigración suiza en Chile en el siglo XIX por su propia fuerza: El pionero Ricardo Roth*. Santiago: Centro de Estudios del Bicentenario, 2009.

Mosse, George L. *The Image of Man: The Creation of Modern Masculinity*. Oxford, NY: Oxford University Press, 1998.

Mozzi, Gustavo, Nicolás Sorín, Nicolás Guerschber, and Ricky Pashkus. "Argentum, el espectáculo que emocionó al G20." Buenos Aires, Televisión Pública Argentina, 2018. Youtube video, 36:49. www.youtube.com/watch?v=goHxpNWthbk.

Mumford, Eric. *The CIAM Discourse on Urbanism, 1928–1960*. Cambridge: Massachusetts Institute of Technology Press, 2002.

Muñoz Sougarret, Jorge E. "Apropiación pública y privada del valle central de la provincia de Llanquihue (1893–1910)." *Cuadernos de Historia*, no. 58 (2023): 229–53.

———. "Empresariado y política: Aproximación histórica a las relaciones políticas de los empresarios germanos de la provincia de Llanquihue (1891–1914)." PhD diss., Pontificia Universidad Católica de Chile, 2016.

Navarro Floria, Pedro. "El desierto y la cuestión del territorio en el discurso político argentino sobre la frontera Sur." *Revista Complutense de Historia de América*, no. 28 (2002): 139–68.

———. "El proceso de construcción social de la región de Nahuel Huapi en la práctica simbólica y material de Exequiel Bustillo (1934–1944)." *Revista Pilquen* 9, no. 9 (2008): 1–14.

———. "La 'Suiza argentina,' de utopía agraria a postal turística: La resignificación de un espacio entre los siglos XIX y XX." Paper presented at III Jornadas de Historia de la Patagonia. Bariloche, Argentina, 2008.

Navarro Floria, Pedro, and Paula Gabriela Núñez. "Un territorio posible en la república imposible: El coronel Sarobe y los problemas de la Patagonia argentina." *Andes* 23, no. 2 (2012). http://ref.scielo.org/663trx.

Nelson, Murry R. "Basketball as Cultural Capital: The Original Celtics in Early Twentieth-Century New York City." In *Sporting Nationalisms: Identity, Ethnicity, Immigration, and*

Assimilation, edited by Mike Cronin and David Mayall, 67–81. London: Taylor and Francis, 1998.

Newkirk, Pamela. *Spectacle: The Astonishing Life of Ota Benga*. New York: Amistad, 2015.

Nicoletti, María Andrea, and Pedro Navarro Floria. *Confluencias: Una breve historia del Neuquén*. Buenos Aires: Dunken, 2000.

Nishimura, Sachiko, Robert Waryszak, and Brian King. "Guidebook Use by Japanese Tourists: A Qualitative Study of Australia Inbound Travellers." *International Journal of Tourism Research* 2, no. 15 (2006): 13–26.

Nobbs-Thiessen, Ben. *Landscape of Migration: Mobility and Environmental Change on Bolivia's Tropical Frontier, 1952 to the Present*. Chapel Hill: University of North Carolina Press, 2020.

Núñez, Andrés. "El país de las cuencas: Fronteras en movimiento e imaginarios territoriales en la construcción de la nación; Chile siglos XVIII–XIX." *Scripta Nova. Revista Electrónica de Geografía y Ciencias* 16, no. 418 (2012). www.ub.edu/geocrit/sn/sn-418/sn-418-15.htm.

Núñez, Paula Gabriela. "La dinámica de una localidad desde la articulación de sus instituciones: El municipio de San Carlos de Bariloche, el Club Andino y Parques Nacionales (1931–1955)." In *Nuevos espacios, nuevos problemas: Los territorios nacionales*, edited by Graciela Iuorno and Edda Crepo, 173–93. Neuquén, Argentina: Universidad Nacional del Comahue, Universidad Nacional de la Patagonia, 2008.

———. "La región del Nahuel Huapi en el último siglo: Tensiones en un espacio de frontera." *Revista Pilquen* 17, no. 1 (2014). http://ref.scielo.org/y945g8.

———. "Memorias fragmentadas entre lo alpino y lo andino: El refugio Italia y las percepciones sobre el poblamiento en la región del Nahuel Huapi." *Estudios Transandinos* 18, no. 1 (2013): 101–20.

———. "Naturaleza ajena en un territorio a integrar: La región del Nahuel Huapi hasta 1955." In *Cultura y espacio: Araucanía-Norpatagonia*, edited by Pedro Navarro Floria and Walter Delrio, 126–40. Bariloche, Argentina: Universidad Nacional de Río Negro, Instituto de Investigaciones en Diversidad Cultural y Procesos de Cambio, 2011.

———. "The 'She-Land,' Social Consequences of the Sexualized Construction of Landscape in North Patagonia." *Gender, Place and Culture* 22, no. 10 (2015): 1445–62.

Núñez, Paula Gabriela, and Martín Núñez. "Naturaleza construida: Una revisión sobre la interpretación del paisaje en la zona del Nahuel Huapi." Paper presented at the Terceras Jornadas de Historia de la Patagonia, Bariloche, Argentina, November 6, 2008.

Nunn, Frederick M. *Yesterday's Soldiers: European Military Professionalism in South America, 1890–1940*. Lincoln: University of Nebraska Press, 1983.

Ogden, Laura A. *Loss and Wonder at the World's End*. Durham, NC: Duke University Press, 2021.

Oldani, Karina, Miguel Añon Suarez, and Fernando Miguel Pepe. "Las muertes invisibilizadas del Museo de La Plata." *Corpus* 1, no. 1 (2011). https://doi.org/10.4000/corpusarchivos.986.

Olea Rosenbluth, Catalina. *La mujer en la sociedad mapuche: Siglos XVI al XIX*. Santiago: Servicio Nacional de la Mujer, Gobierno de Chile, 2010.

Olsen, Patrice Elizabeth. *Artifacts of Revolution: Architecture, Society, and Politics in Mexico City, 1920–1940*. Lanham, MD: Rowman and Littlefield, 2008.

Órdenes Delgado, Mathías. "La experiencia de los sin voz: Una propuesta epistemológica para el abordaje transdisciplinar del sujeto profundo en la colonización de la región sur y austral." In *Sujetos sin voz en la región sur y austral de Chile y Argentina: Frontera, colonización, marginalidad y organización popular chileno-mestiza en los siglos XIX y XX*, edited by Mathías Órdenes Delgado and Pablo Marimán Quemenado, 17–62. Santiago: LOM Ediciones, 2022.

Ortega Martínez, Luis. "La crisis de 1914–1924 y el sector fabril en Chile." *Historia (Santiago)* 45, no. 2 (2012): 433–54.

Ortelli, Sara. "La 'araucanización' de las Pampas: ¿Realidad histórica o construcción de los etnólogos?" *Anuario del Instituto de Estudios Histórico-Sociales*, no. 11 (1996): 203–25.

Ospital, María Silvia. "Patrones e inmigrantes: Los planteos sobre inmigración de la Asociación del Trabajo; 1918–1930." In *Inmigración y nacionalismo: La Liga Patriótica y la Asociación del Trabajo (1910–1930)*, 12–25. Buenos Aires: Centro Editor de América Latina, 1994.

Ottone, Eduardo Guillermo. "Sobral y la geología del Ñirihuau." *Revista del Museo de La Plata* 1, no. 3 (2016): 195–204.

Palacio, Juan Manuel. "La antesala de lo peor: La economía argentina entre 1914–1930." In *Nueva historia argentina*, 101–50. Vol. 6 of *Democracia, conflicto social y renovación de ideas, 1916–1930*, edited by Ricardo Falcón. Buenos Aires: Sudamericana, 2000.

Papazian, Alexis, and Mariano Nagy. "La Isla Martín García como campo de concentración de indígenas hacia fines del siglo XIX." In *Historia de la crueldad argentina: Julio Argentino Roca y el genocidio de los Pueblos Originarios*, edited by Osvaldo Bayer and Diana Lenton, 77–104. Buenos Aires: Red de Investigadores en Genocidio y Política Indígena, 2010.

Pearson, Mike. "One Letter and 55 Footnotes: The Assassination of Llwyd Ap Iwan by the Outlaws Wilson and Evans." *Parallax* 19, no. 4 (2013): 63–73.

Perez, Louis A. "Between Baseball and Bullfighting: The Quest for Nationality in Cuba, 1868–1898." *Journal of American History* 81, no. 2 (1994): 493–517.

Pérez, Pilar. "Futuros y fuentes: Las listas de indígenas presos en el campo de concentración de Valcheta, Río Negro (1887)." *Nuevo Mundo Mundos Nuevos*, 2015. https://doi.org/10.4000/nuevomundo.68751.

———. "Historia y silencio: La Conquista del Desierto como genocidio no-narrado." *Corpus* 1, no. 2 (2011). https://doi.org/10.4000/corpusarchivos.1157.

———. "Las primeras policías fronterizas en Río Negro y Chubut (1911–1914): Creación, desarrollo y balance de una experiencia policial." *Cuadernos de Marte* 8, no. 13 (2017): 19–54.

Pesoa, Melisa, and Joaquín Sabate. "La Plata y la construcción de un país, del papel a la realidad." Paper presented at XIV Coloquio Internacional de Geocrítica, Barcelona, 2016.

Piantoni, Giulietta. "Objetos cotidianos: El tratamiento de las colecciones indígenas en el museo de la Patagonia, San Carlos de Bariloche (1938–1944)." *Estudios del ISHiR* 5, no. 11(2015): 114–31.

———. "Un laboratorio a cielo abierto." *Desde la Patagonia* 18, no. 31 (2021): 60–67.

Piantoni, Giulietta, Gonzalo Barrios García Moar, and Liliana Valeria Pierucci. "Las bellezas panorámicas argentinas: Una revisión histórica de las políticas públicas y el desarrollo del turismo en el Parque Nacional Nahuel Huapi durante el peronismo (1943–1955)." *Pasado Abierto*, no. 9 (2019): 236–55.

Piccato, Pablo. *City of Suspects: Crime in Mexico City, 1900–1931*. Durham, NC: Duke University Press, 2001.

Picone, María de los Ángeles. "La idea de turismo en San Carlos de Bariloche a través de dos guías (1938)." *Estudios y Perspectivas en Turismo* 22, no. 2 (2013): 198–215.

Piglia, Melina. *Autos, rutas y turismo: El Automóvil Club Argentino y el estado.* Buenos Aires: Siglo XXI Editores, 2014.

———. "The Awakening of Tourism: The Origins of Tourism Policy in Argentina, 1930–1943." *Journal of Tourism History* 3, no. 1 (2011): 57–74.

———. "En torno a los parques nacionales: Primeras experiencias de una política turística nacional centralizada en la Argentina (1934–1950)." *PASOS* 10, no. 1 (2012).

Pinto Rodríguez, Jorge. "Al final de un camino: El mundo fronterizo en Chile en tiempos de Balmaceda (1860–1900)." *Revista Complutense de Historia de América*, no. 22 (1996): 287–322.

———. "Bárbaros, demonios y bárbaros de nuevo: Estereotipos del Mapuche en Chile, 1550–1900." In *Cruzando la cordillera: La frontera argentino-chilena como espacio social*, edited by Susana Bandieri, 119–40. Neuquén, Argentina: Centro de Estudios de Historia Regional, Facultad de Humanidades, Universidad Nacional del Comahue, 2001.

———. *De la inclusión a la exclusión: La formación del estado, la nación y el pueblo Mapuche.* Santiago: Editorial de la Universidad de Santiago de Chile, 2000.

Plant, Judith. "Learning to Live with Differences: The Challenge of Ecofeminist Community." In *Ecofeminism: Women, Culture, Nature*, edited by Karen Warren, 120–39. Bloomington: Indiana University Press, 1997.

Podgorny, Irina. "De razón a facultad: Ideas acerca de las funciones del Museo de la Plata en el período 1890–1918." *Runa* 22, no 1 (1995): 89–104.

Podgorny, Irina, and Maria Margaret Lopes. *El desierto en una vitrina: Museos e historia natural en la Argentina, 1810–1890.* Mexico City: Limusa, 2008.

Popescu, Gabriel. *Bordering and Ordering the Twenty-First Century: Understanding Borders.* Lanham, MD: Rowman and Littlefield, 2012.

Prado, Francisco, Renato D'Alençon, Daniel Korwan, and Johanna Moser. "Traces of Construction Following Migration: Transverse Gable; Massive Timber and Carpenter's Marks in the Houses of the 19th Century German Settlers in Southern Chile." In *Nuts and Bolts of Construction History: Culture, Technology and Society*, edited by Robert Carvais, André Guillerme, Valérie Nègre, Joël Sakarovitch, 1:1–9. Paris: Picard, 2012.

Prado, Francisco, Renato D'Alençon Castrillón, and F. Kramm. "Arquitectura alemana en el sur de Chile: Importación y desarrollo de patrones tipológicos, espaciales y constructivos." *Revista de la Construcción* 10, no. 2 (2011): 104–21.

Prasad, Ritika. "'Time-Sense': Railways and Temporality in Colonial India." *Modern Asian Studies* 47, no. 4 (2013): 1252–82.

Prislei, Leticia. "Imaginar la nación, modelar el desierto: Los '20 en tierras del 'Neuquén.'" In *Pasiones sureñas: Prensa, cultura y política en la frontera norpatagónica, 1884–1946*, edited by Leticia Prislei, 79–100. Buenos Aires: Prometeo, 2001.

Pritchard, Annette, and Nigel J. Morgan. "Privileging the Male Gaze: Gendered Tourism Landscapes." *Annals of Tourism Research* 27, no. 4 (2000): 884–905.

Purcell, Fernando. "Una mercancía irresistible: El cine norteamericano y su impacto en Chile, 1910–1930." *Historia Crítica* 1, no. 38 (2017): 46–69. https://doi.org/10.7440/histcrit38.2009.04.

Quijada, Mónica. "La ciudadanización del 'indio bárbaro': Políticas oficiales y oficiosas hacia la población indígena de la Pampa y la Patagonia, 1870–1920." *Revista de Indias* 59, no. 217 (1999): 675–704.

Quiroga, Hugo. "Notas sobre la historia de la democracia en la Argentina." In *Estado y territorios nacionales: Política y ciudadanía en Río Negro 1912–1930*, edited by Hugo Quiroga and Martha Ruffini, 13–44. Neuquén, Argentina: Editorial de la Universidad Nacional del Comahue, 2011.

Radding, Cynthia. *Landscapes of Power and Identity: Comparative Histories in the Sonoran Desert and the Forests of Amazonia from Colony to Republic*. Durham, NC: Duke University Press, 2006.

Raiter, Bárbara. "Ciudadanos y soldados: El Tiro Federal Concordia de la República Argentina, 1898–1923." *Revista Universitaria de Historia Militar* 5, no. 9 (2016): 33–51.

Raj, Kapil. *Relocating Modern Science: Circulation and the Construction of Knowledge in South Asia and Europe, 1650–1900*. New York: Palgrave Macmillan, 2007.

Rajagopalan, Mrinalini, and Madhuri Desai. "Architectural Modernities of Imperial Pasts and Nationalist Presents." Introduction to *Colonial Frames, Nationalist Histories: Imperial Legacies, Architecture and Modernity*, edited by Mrinalini Rajagopalan and Madhuri Desai, 1–23. Burlington, VT: Ashgate, 2012.

Ramos, Ana. *Los pliegues del linaje: Memorias políticas mapuches-tehuelches en contextos de desplazamiento*. Buenos Aires: Editorial Universitaria de Buenos Aires, 2010.

Reeves, Rene. *Ladinos with Ladinos, Indians with Indians: Land, Labor, and Regional Ethnic Conflict in the Making of Guatemala*. Stanford, CA: Stanford University Press, 2006.

Rein, Raanan. *Argentine Jews or Jewish Argentines? Essays on Ethnicity, Identity, and Diaspora*. Leiden, Netherlands: Brill, 2010.

Rey, Héctor. "La economía del Nahuel Huapi." In *La cordillera rionegrina: Economía, estado y sociedad en la primera mitad del siglo XX*, 31–65. Viedma, Argentina: Editorial 2010 Bicentenario, 2005.

Reyna, Franco D. "Aproximaciones en torno al proceso de surgimiento y estructuración del fútbol en la ciudad de Córdoba (1890–1920)." *Anuario de la Escuela de Historia Virtual*, no. 1 (2010): 218–35.

Ribeiro, Rafael Winter. "Rio de Janeiro e a Avenida Beira Mar: Desejo de paisagem e cidade balneário nas primeiras décadas do século XX." *Confins*, no. 39 (March 2019). https://doi.org/10.4000/confins.18065.

Rico, Trinidad. *The Heritage State: Religion and Preservation in Contemporary Qatar*. Ithaca, NY: Cornell University Press, forthcoming.

Rocchi, Fernando. "El péndulo de la riqueza: La economía argentina en el período 1880–1916." In *El progreso, la modernización y sus límites*, edited by Mirta Zaida Lobato, 15–70. Vol. 5 of *Nueva Historia Argentina*, edited by Juan Suiano. Buenos Aires: Sudamericana, 2000.

Rodriguez, Julia. "Beyond Prejudice and Pride: The Human Sciences in Nineteenth- and Twentieth-Century Latin America." *Isis* 104, no. 4 (2013): 807–17.

———. *Civilizing Argentina: Science, Medicine, and the Modern State*. Chapel Hill, NC: University of North Carolina Press, 2006.

Rodriguez, María Laura, María Dolores Rivero, and Adrian Carbonetti. "Convicciones, saberes y prácticas higiénicas argentinas en la segunda mitad del siglo XIX: Sus condiciones de posibilidad en los estudios de las epidemias de cólera, 1868, 1871 y 1887." *Investigaciones y Ensayos*, no. 66 (October 2018): 75–110.

Rogers, Thomas D. *The Deepest Wounds: A Labor and Environmental History of Sugar in Northeast Brazil*. Chapel Hill: University of North Carolina Press, 2010.

Roig, Fidel, Gustavo Costa, Darío Trombotto, Lucas Ruiz, Ivanna Pecker Marcosig, Laura Zalazar, Gustavo Aloy, Juan Pablo Scarpa, and Lidia Ferri Hidalgo. *Informe de las subcuencas de los ríos Carrenleufú y Pico. Cuenca de los ríos Carrenleufú y Pico*. Buenos Aires: Inventario Nacional de Glaciares, Instituto Argentino de Nivología, Glaciología y Ciencias Ambientales (Consejo Nacional de Investigaciones Científicas y Técnicas - Universidad Nacional de Cuyo), 2018.

Roldán, Diego P. "Circulación, difusión y masificación: El futbol en Rosario (Argentina) 1900–1940." *Secuencia*, no. 93 (December 2015): 137–61.

Romero, Luis Alberto. *A History of Argentina in the Twentieth Century*. Updated and rev. ed. University Park: Pennsylvania State University Press, 2014.

Romero Vázquez, Jesús, and Alexander Betancourt Mendieta. "Emblemas del progreso: El Teatro Colón y el Palacio de Bellas Artes en la construcción de la nación, Argentina y México, 1880–1910." *Signos Históricos* 22, no. 44 (2020): 260–92.

Ruffini, Martha. "Ecos del centenario: La apertura de un espacio de deliberación para los territorios nacionales; La primera conferencia de gobernadores (1913)." *Revista Pilquen* 12, no. 12 (2010). http://ref.scielo.org/j2zz4q.

———. "La Patagonia en el pensamiento y la acción de un reformista liberal: Ezequiel Ramos Mexía (1852–1935)." *Quinto Sol*, no. 12 (2008): 127–50.

———. *La Patagonia mirada desde arriba: El Grupo Braun Menéndez Behety y la revista Argentina Austral (1929–1967)*. Rosario, Argentina: Prohistoria Ediciones, 2017.

Ruggiero, Kristin. *Modernity in the Flesh: Medicine, Law, and Society in Turn-of-the-Century Argentina*. Stanford, CA: Stanford University Press, 2003.

Safier, Neil. "Fugitive El Dorado: The Early History of an Amazonian Myth." In *Fugitive Knowledge: The Loss and Preservation of Knowledge in Cultural Contact Zones*, edited by Andreas Beer and Gesa Mackenthun, 51–62. Münster; NY: Waxmann, 2015.

Saldívar Arellano, Juan Manuel. "'Chilote tenía que ser': Vida migrante transnacional en territorios patagónicos de Chile y Argentina." *Cultura-Hombre-Sociedad* 27, no. 2 (2017): 175–200.

———. "Etnografía de la nostalgia: Migración transnacional de comunidades chilotas en Punta Arenas (Chile) y Río Gallegos (Argentina)." *Chungará* 50, no. 3 (2018): 501–12.

Saldivia Maldonado, Zenobio, and Griselda De la Jara Nova. "La Sociedad Nacional de Agricultura en el siglo XIX chileno: Su rol social y su aporte al desarrollo científico-tecnológico." *Scripta Nova. Revista Electrónica de Geografía y Ciencias Sociales* 5, no. 100 (2001). www.ub.edu/geocrit/sn-100.htm.

Salessi, Jorge. *Médicos, maleantes y maricas*. Rosario, Argentina: Beatriz Viterbo Editora, 1995.

Salvatore, Ricardo. "Burocracias expertas y exitosas en Argentina: Los casos de educación primaria y salud pública (1870–1930)." *Estudios Sociales del Estado* 2, no. 3 (2016): 22–64.

Sánchez Delgado, Marcelo Javier. "Chile y Argentina en el escenario eugénico de la primera mitad del siglo XX." PhD diss., Universidad de Chile, 2015.

Sanhueza Cerda, Carlos, ed. *La movilidad del saber científico en América Latina: Objetos, prácticas e instituciones*. Santiago: Editorial Universitaria, 2018.

Santacreu Soler, José Miguel. "Unidad monetaria, vertebración territorial y conformación nacional: El caso de la República Argentina." *Anales de Historia Contemporánea* 20 (2004): 439–62.

Saus, María Alejandra. "La 'britanización' de Bahía Blanca: Estado, capital global, ferrocarril y espacio local en perspectiva multiescalar." *Revista Universitaria de Geografía* 27, no. 2 (2018): 79–102.

Schell, Patience. *The Sociable Sciences: Darwin and His Contemporaries in Chile*. London: Palgrave Macmillan, 2013.

Schulten, Susan. *The Geographical Imagination in America, 1880–1950*. Chicago: University of Chicago Press, 2001.

Schvarzer, Jorge, Andrés Regalsky, and Teresita Gómez. *Estudios sobre la historia de los ferrocarriles argentinos (1857–1940)*. Buenos Aires: Universidad de Buenos Aires, 2007.

Scott, James C. *Seeing Like a State: How Certain Schemes to Improve the Human Condition Have Failed*. New Haven, CT: Yale University Press, 1998.

Servicio Nacional de Geología y Minería. *Chile: Territorio volcánico*. Santiago: SERNAGEOMIN, 2018.

Sevilla, Elisa, and Ana Sevilla. "Inserción y participación en las redes globales de producción de conocimiento: El caso del Ecuador del siglo XIX." *Historia Crítica*, no. 50 (2013): 79–103.

Seyferth, Giralda. "German Immigration and the Formation of German-Brazilian Ethnicity." *Anthropological Journal on European Cultures* 7, no. 2 (1998): 131–54.

Shoemaker, Nancy. "Settler Colonialism: Universal Theory or English Heritage?" *William and Mary Quarterly* 76, no. 3 (2019): 369–74.

Sillitti, Nicolas G. "El servicio militar obligatorio y la 'cuestión social': Apuntes para la construcción de un problema historiográfico." *Pasado Abierto* 4, no. 7 (2018): 265–75. http://fh.mdp.edu.ar/revistas/index.php/pasadoabierto/article/view/2570.

Silva, Bárbara. "La espacialidad y el paisaje en las representaciones nacionales durante el Frente Popular chileno: 1938–1941." *Revista de Historia Social y de las Mentalidades* 22, no. 1 (June 2018): 129–53.

Simard, Suzanne. *Finding the Mother Tree: Discovering the Wisdom of the Forest*. New York: Knopf, 2021.

Sivasundaram, Sujit. "Sciences and the Global: On Methods, Questions, and Theory." *Isis* 101, no. 1 (2010): 146–58.

Soluri, John. *Creatures of Fashion. Animals, Global Markets, and the Transformation of Patagonia*. Chapel Hill: University of North Carolina Press, 2024.

Stein, Steve. "Miguel Rostaing: Dodging Blows on and off the Soccer Field." In *The Human Tradition in Modern Latin America*, edited by William H. Beezley and Judith Ewell, 147–60. Lanham, MD: Rowman and Littlefield, 1987.

Stepan, Nancy. *The Hour of Eugenics: Race, Gender, and Nation in Latin America*. Ithaca, NY: Cornell University Press, 1991.

Stern, Claudia. "'Professionals, Merchants, and Industrialists Unite!': Middle-Class Masculinities, Subjectivities, and Nationhood in Chile, 1932–1952." *Men and Masculinities* 25, no. 2 (2022): 271–91.

Suriano, Juan. "El anarquismo." In *El progreso, la modernización y sus límites*, edited by Mirta Zaida Lobato, 291–326. Vol. 5 of *Nueva Historia Argentina*. Buenos Aires: Sudamericana, 2000.

Taranda, Demetrio. "Papel del estado y del capital británico en el proceso de constitución de la matriz productiva del Alto Valle de Rio Negro y Neuquén." *Revista de Historia*, no. 3 (November 2014): 181–206.

Taylor, Lucy. "The Welsh Way of Colonisation in Patagonia: The International Politics of Moral Superiority." *Journal of Imperial and Commonwealth History* 47, no. 6 (2019): 1073–99.

Torrano, Andrea. "Ontologías de la monstruosidad: El cyborg y el monstruo biopolítico." In *Actas del VI Encuentro Interdisciplinario de Ciencias Sociales y Humanas*. Córdoba, Argentina: Universidad Nacional de Córdoba, 2009.

Trejo Barajas, Deni, ed. *Los desiertos en la historia de América: Una mirada multidisciplinaria*. Morelia, Mexico: Instituto de Investigaciones Históricas de la Universidad Michoacana de San Nicolás de Hidalgo/Universidad Autónoma de Coahuila, 2011.

Tuan, Yi-fu. *Space and Place: The Perspective of Experience*. 6th ed. Minneapolis: University of Minnesota Press, 2008.

Turner, Billie L. "Contested Identities: Human-Environment Geography and Disciplinary Implications in a Restructuring Academy." *Annals of the Association of American Geographers* 92, no. 1 (2002): 52–74.

Tutino, John. "From Involution to Revolution in Mexico: Liberal Development, Patriarchy, and Social Violence in the Central Highlands, 1870–1915." *History Compass* 6, no. 3 (2008): 796–842.

Ugarte, Magdalena, Mauro Fontana, and Matthew Caulkins. "Urbanisation and Indigenous Dispossession: Rethinking the Spatio-legal Imaginary in Chile vis-à-vis the Mapuche Nation." *Settler Colonial Studies* 9, no. 2 (December 2017): 1–20.

Urbina Carrasco, María Ximena. *La frontera de arriba en Chile colonial: Interacción hispano-indígena en el territorio entre Valdivia y Chiloé e imaginario de sus bordes geográficos, 1600–1800*. Santiago: Centro de Investigaciones Diego Barros Arana, 2009.

Urry, John. *The Tourist Gaze*. 2nd ed. Thousand Oaks, CA: SAGE, 2002.

Urry, John, and Jonas Larsen. *The Tourist Gaze 3.0*. London: SAGE, 2011. https://doi.org/10.4135/9781446251904.

Valdivieso Valenzuela, Ernesto, and Atlántida Coll-Hurtado. "La construcción y evolución del espacio turístico de Acapulco (México)." *Anales de Geografía de la Universidad Complutense* 30, no. 1 (2010): 163–90.

Vallmitjana, Ricardo. *A cien años de la colonia agrícola Nahuel Huapi, 1902–2002*. Bariloche, Argentina: Self-published, n.d.

———. *Cruzando la cordillera*. Bariloche, Argentina: Self-published, n.d.

———. *Periodismo y otros medios en el pueblo*. Bariloche, Argentina: Self-published, n.d.

———. *San Carlos*. Bariloche, Argentina: Self-published, n.d.

———. *Sociedad Comercial y Ganadera Chile-Argentina, 1900–1916*. Bariloche, Argentina: Self-published, n.d.

Vandendriessche, Joris, Evert Peeters, and Kaat Wils. *Scientists' Expertise as Performance: Between State and Society, 1860–1960*. London: Routledge, 2015.

Van Hoy, Teresa Miriam. *A Social History of Mexico's Railroads: Peons, Prisoners, and Priests*. Lanham, MD: Rowman and Littlefield, 2008.

Varela, Gladys, and Carla Manara. "Tiempos de transición en las fronteras surandinas: De la colonia a la República." In *Cruzando la cordillera: La frontera argentino-chilena como espacio social*, edited by Susana Bandieri, 31–64. Neuquén, Argentina: Centro de Estudios de Historia Regional, Facultad de Humanidades, Universidad Nacional del Comahue, 2001.

Vargas, Maia, and Gabriela Klier. "Representaciones de Naturaleza en Isla Victoria." *Aisthesis*, no. 69 (2021): 259–80.

Vela Cossío, Fernando, and Jocelyn del Carmen Tillería González. "Cuando habitábamos lo elemental. Una mirada crítica sobre la vivienda tradicional en el Chile austral a través de la fotografía del siglo XIX." *RITA. Revista indexada de textos académicos*, no. 8 (2017): 118–25.

Vergara del Solar, Jorge. *La herencia colonial del Leviatán: El estado y los mapuche-huilliches, 1750–1881*. Iquique, Chile: Centro de Investigaciones del Hombre en el Desierto, Ediciones Instituto de Estudios Andinos, Universidad Arturo Prat, 2005.

Viotti da Costa, Emilia. *The Brazilian Empire: Myths and Histories*. Rev. ed. Chapel Hill: University of North Carolina Press, 2000.

Wakeman, Rosemary. *Practicing Utopia: An Intellectual History of the New Town Movement*. Chicago: University of Chicago Press, 2016.

Wakild, Emily. *Revolutionary Parks: Conservation, Social Justice, and Mexico's National Parks, 1910–1940*. Tucson, AZ: University of Arizona Press, 2011.

Walsh, Sarah. *The Religion of Life: Eugenics, Race, and Catholicism in Chile*. Pittsburgh, PA: University of Pittsburgh Press, 2022.

Weber, David. *Barbaros: Spaniards and Their Savages in the Age of Enlightenment*. New Haven, CT: Yale University Press, 2005.

Williams, Glyn. *Naturalists at Sea: Scientific Travellers from Dampier to Darwin*. New Haven, CT: Yale University Press, 2013.

Williams, John Alexander. *Turning to Nature in Germany: Hiking, Nudism, and Conservation, 1900–1940*. Stanford, CA: Stanford University Press, 2007.

Williams, R. Bryn. *Y Wladfa*. Caernarfon, UK: Gwasg Pantycelyn, 2001.

Yannakakis, Yanna. "Digital Resources: Power of Attorney, a Digital Spatial History of Indigenous Legal Culture in Colonial Oaxaca, Mexico." *Oxford Research Encyclopedia of Latin American History* (online), May 24, 2018.

Yoder, April. *Pitching Democracy: Baseball and Politics in the Dominican Republic*. Austin: University of Texas Press, 2023.

Zanatta, Loris. *Del estado liberal a la nación católica: Iglesia y ejército en los orígenes del peronismo, 1930–1943*. Buenos Aires: Universidad Nacional de Quilmes, 1996.

Zanetti Lecuona, Oscar. *Sugar and Railroads: A Cuban History, 1837–1959*. Chapel Hill: University of North Carolina Press, 1998.

Zartman, I. William. "Identity, Movement, and Response." Introduction to *Understanding Life in the Borderlands: Boundaries in Depth and in Motion*, edited by I. William Zartman, 1–18. Athens: University of Georgia Press, 2010.

Zimmermann, Eduardo A. *Los Liberales Reformistas: La cuestión social en la Argentina, 1890–1916*. Buenos Aires: Sudamericana, 1995.

Zusman, Perla, and Sandra Minvielle. "Sociedades geográficas y delimitación del territorio en la construcción del Estado-Nación argentino." Paper presented at V Encuentro de Geógrafos de América Latina, Cuba, 1995.

Index

Page numbers in italics refer to illustrations.

Abrazo del Estrecho, 25
ACA (Argentine Automobile Club), 191–92
Achelis, Adolfo, 101, 150
Adventure (ship), 38
aesthetics: Bariloche, 143; nationalist, 142–43, 174; Spanish Mission, 162–63; trans-Andean, 143, 152, 162. *See also* architectural aesthetics
AGI (Argentine Geographic Institute), 29–30
agriculture, 11, 100, 109, 202; and DPN support, 198–99; exclusion of, 181; and fertility trope, 183
Aguirre Campos, José Joaquín, 27
Aguirre Cerda, Pedro, 177
Ahehelm, Paul, 36
Alcoba Pitt, Pedro, 139
alcohol consumption, 115, 117–18
Alert (ship), 38
Alessandri Palma, Arturo, 13–14, 177
alpinism, 201
Alsina, Adolfo, 70
Aluminé River, 94
Álvarez, Ciriaco, 61
Alvear, Marcelo, 153
Anales de la Universidad de Chile, 27

anarchists, 115–16
Andermann, Jens, 31
Andes: Andean lakes, 7, 183–84; Andean passes, 206; Andean valleys, 82, 99, 103–9; and border debates, 37; and Mount Palique, 19; northern Patagonian Andes, 6–8, 12; topography of, 23–24; views of, 6, 10–11; watershed divide in, 37. *See also* northern Patagonian Andes
andinism, 193, 201
Andrée, Carlota, 175–76
Andriau, Ignacio, 170
anti-immigrant sentiment, 113, 116, 132–35
Antimil, Juana María, 170
Aónikenk, 1, 4, 21
Ap Iwan, Llwyd, 119–24
Araucanía, 5, 21, 23, 46–52
araucanization, of the Pampas, 5
Araucano, 166
Arauco, 46–47, 50
architectural aesthetics, 141–43; Alpine style, 170, 172; of Bariloche Civic Center, 141–42, 144, 160–65; "Bariloche style," 143; and Belle Époque, 161; in built environment, 159, 162; of Chile-Argentina Company, 146–47, 146, 152;

architectural aesthetics (*continued*)
German, 146, 163; in public buildings, 152
Ardagh, John, 34
Argentine Automobile Club (ACA), 191–92
Argentine Geographic Institute (AGI), 29–30
Argentine League against Alcoholism, 118
Argentine Patriotic League, 118–19, 138–40
Argentine Scientific Society (SCA), 29
Argentine Switzerland, 149, 151
Argentine Touring Club, 191
Argentina: Andes in, 8; anti-Chilean sentiment in, 130, 133–35, 169–70; export boom in, 11; independence of, 2; land distribution in, 70–72; learned societies in, 28–32; military coups in, 13–14, 156, 172; naval forces in, 24, 102; railroad development in, 91–92; suppression of Desert in, 5–6; voting rights in, 13
Argentum, 203–4
Arneil, Barbara, 12
Artayeta, Enrique, 166
Atacama Desert, 1, 23
Atlantic trading, 103–4, 109
automobiles, 191
Avellaneda Law of 1876, 50
Avenida Costanera, 164–65

Baden-Powell, Robert, 136
Baeza Espiñeira, Agustín, 63
Bahía Blanca, 93
Baker River, 38–39, 39
Baker River Company, 57
Balmaceda, José Manuel, 91
Baluchistan, 25
Bandieri, Susana, 122
bandits, 120–23, 130–32
Barbagelata, José Domingo, 199
Barella, Carlos, 178
Bariloche, 11–13, 82, 139, 156; Avenida Costanera, 164–65; Chilean presence in, 135; and Chile-Argentina Company, 150–52; Civic Center, 141–44, 142, 160–65, 164; railroad in, 155; Tiro Federal, 136–38; tourists in, 173–74
Barría, Bernardino, 87–88, 87
Barros Arana, Diego, 23–25, 27–28, 31–34, 36, 40–41
Barros Borgoño, Luis, 66, 68
Bayer, Osvaldo, 119
Beagle (ship), 38
Bello, Álvaro, 20
Bello, Andrés, 27
Bernal, Liborio, 76
Bess, Michael, 82
Bío-Bío River, 47, 54
blockhaus, 146–47
Bodudahue, 57
Bohoslavsky, Ernesto, 140, 155
Bolívar, Simón, 27
Bolivia, 5, 23
borderline, Chile-Argentine, 16, 20, 25
border negotiations, 11, 15–16, 19, 27; approaches to, 37, 41–42; and place-name disputes, 39–41; and scientific debate, 19, 22–25, 33–34, 42–43
Border Police, 112–13, 128–32
border regions, 10
Boy Scouts, 136–37
Branje, Cornelio, 56
Bravo River Company, 57
Brazil, 113
Brest, Romero, 136
Buenos Aires, 12, 31, 75, 113–16, 118, 160, 162
Buenuleo, Antonio, 169
Buschmann Zwanzger, Carlos, 195
Bustamante, Valentín, 111–12, 118, 140
Bustillo, Alejandro, 161–62, 167–68, 170, 172
Bustillo, Exequiel, 144, 153–58, 161, 163, 165–68, 170–73, 192, 195, 197–98

CAB (Club Andino Bariloche), 193–98, 201
CAO (Club Andino Osorno), 193, 195–96
Caimari, Lila, 113
Calén Sound, 38–39, 39
caminos de conveniencia, 80
caminos públicos, 83
caminos vecinales, 84

Campos, Amie, 54
Cañuel, 32
Canuipan, José Esteban, 67
Capraro, Primo, 103, 139, 160–61, 164, 169, 200
Cárdenas, Juan Antonio, 85–86
Carey, Mark, 193
Carlés, Manuel, 119
Carrasco Hermoza, Alberto, 191
Carrenleufu River, 34, 36
Carrera, José Miguel, 41
Carta de las comunicaciones postales y telegráficas, 75
Carter, Paul, 8
car travel, 191–92
Cassidy, Butch, 120, 122
Castillo, Ramón, 204
Castro, Juan, 46
Castro-Klarén, Sara, 22
catanga, 96
Catholicism, 160; temples, 167–68; missions, 167
Catrilef, Ceferino, 67–68
Catrilef, Sixto, 67
cattle ranching, 4, 12, 99–100, 198
Cautín, 50
Cayun, Pedro, 199
CEH (Hydrologic Studies Commission), 106–8, 108
Central Valley, 181, 190
cestoball, 136
Chaco region, 46
Chaneton, Abel, 127
Charter of Athens (1942), 161
Chasteen, John Charles, 22
Chile: and AGI map, 31; centralization, 14; civil war in, 11; European immigrants in, 5, 11, 17; immigration policy of, 49–52; independence of, 2; landed property legislation in, 47, 49–52; learned societies in, 25–28; mestizo Chileans in, 11, 50, 65, 122; military operations in, 26; naval forces of, 24, 38; north–south integration in, 103, 109; railroad development in, 90–91;

scientist-explorers in, 21; social reform agitation in, 13; suppression of Desert in, 5; US-Chilean relations, 24–25
Chilean Immigration Agency, 60
Chilean State Railway Company. *See* Empresa de Ferrocarriles del Estado (EFE)
Chilean Switzerland, 149, 151
Chile-Argentina Trading and Cattle-Breeding Company, 82, 145–48, 149, 173, 200; architectural aesthetics of, 146–47, 152; and Bariloche, 150–52; development of, 101–4; and trans-Andean trade, 143
Chiloé, 47, 49, 54–56, 59, 63
"Chos Malal a Pichachen," 96–97, 97
Christie, Roberto, 84
Chubut, 7, 34, 70, 105–6
City of Caesars, 3–4
Club Andino Bariloche (CAB), 193–98, 201
Club Andino Osorno (CAO), 193, 195–96
clubs, outdoors, 193–94
coastal boulevards, 164
Cock, Hieronymus, 1
Coihueco Island, 16, 64–70
colonization, 54; and border negotiation, 41–42; and place-names, 38; settlers, 12, 16, 46. *See also* settlement
Colonization Law of 1845, 49
Colony 16 de Octubre, 34–36, 42
Colson, Carlos, 59, 64–65
Comezaña, Julio, 165
Comisión de Estudios Hidrológicos. *See* Hydrologic Studies Commission (CEH)
Comisión Radicadora de Indígenas, 50–51, 53
commemoration, national, 138
Compañía Comercial y Ganadera Chile-Argentina. *See* Chile-Argentina Trading and Cattle-Breeding Company
Compañía Esplotadora del Baker, 57
Concepción, 99
Confino, Alon, 143
Confluencia, 93

Congrès Internationaux d'Architecture Moderne. *See* International Congresses of Modern Architecture (CIAM)
Conquest of the Desert, 28, 48, 214n45
Consejo Nacional de Educación, 136
conservation, 142–43, 154, 169
Conservative Restoration, 14, 161
Contardi, Juan, 57
Cook, James, 8
cordillera, 3, 8, 15, 23. *See also* Andes
cordillera libre, 99
Coronel Barcalá, 73
cow trails, 198
Craib, Raymond, 38
Crespo, Benito, 127, 138
criminality, 113, 120, 123, 125, 133, 139. *See also* policing; public disorder
criollismo, 178
cultural nationalism, 160, 178
Cumacure, José Antonio, 66
Cumelén, 156, 170
Currieco, Ceferino, 66, 69
Currieco, José Domingo, 65–66, 69
Currieco, Juan, 65

Darwin, Charles, 26
Dávila Larrain, Benjamín, 63
Defenders of Order, 118
Deffarges, Mauricio, 61
degeneration, theory of, 115, 127, 130
del Busto, Adrián, 129, 138
Delrio, Walter, 70
democracy, 191
Denis, Francisco, 134
de Roa, Lino, 72
Desert: as concept, 1–6, 204; and Indigenous groups, 4; and land distribution, 70; and nation-making, 18; and settlement, 45–46; suppression of, 5, 11, 16; symbolism of, 4; views of, 6
Devoto, Fernando, 119
Díaz, Manuel, 66
Dirección de Parques Nacionales. *See* National Parks Bureau (DPN)
Domeyko, Ignacy, 27

DPN. *See* National Parks Bureau (DPN)
drunkenness, 115–18, 130
Durand, Luis, 178

Edwards, Ryan, 116
EFE. *See* Empresa de Ferrocarriles del Estado (EFE)
Elflein, Ada María, 151; *Paisajes cordilleranos*, 151
Elias, Nortbert, 137
Elordi, Eduardo, 124
El Volcán, 62–63
Embrace of the Strait, 25
Empresa Cochamó, 79–80
Empresa de Ferrocarriles del Estado (EFE), 91, 175–78, 181, 199
En Viaje (magazine), 176, 178, 180–81, 183–91, 190, 199–200; cover of, 180, 181, 185, 186, 187; place-names in, 189
environmental adaptation, 115
Errázuriz, Federico, 25
Espejo, Washington, 178
Espiñeira, Agustín Baeza, 53
Estancia San Ramón, 192–93
Estrada, Ernesto de, 161–62, 170
European immigrants, 5, 11–12; German, 60; government attitudes toward, 49–50, 55, 59–60, 62–63; and holding companies, 57; and land tenure, 48; in southern territories, 17; Spanish, 62–63; Welsh, 34–36, 42, 159
Evans, Robert, 120, 124
Expedition to the Negro River, 48
explorations: Baker/Calén debate, 38–40; Chilean, 28; Cochrane/Pueyrredón case, 40
expropriation of lands, 46–47, 56–59

farming. *See* agriculture
Farrell, Edelmiro, 204
Farrington, Benjamin, 127–28
Feinmann, Enrique, 117
feminization of nature, 176, 179, 181–88, 201
Fernández Palacios, Vicente, 111–12, 114, 117–18

Ferrocarril del Estado, 93
Ferrocarril del Sud, 93
fertility, 176, 183
Figueroa, Javier Ángel, 68
Finó, Fréderic, 201
Finó, José F., 193
Fischer, Oscar, 36
Fonck, Francisco, 60, 65, 101
Fontana, Luis Jorge, 29, 34, 42
food production, 100
Fort Bulnes, 5
Foyel, 32
Francisco Moreno Museum of Patagonia, 165–66
Freitas, Frederico, 153
Freudenburg, Teodoro, 57
Frey, Emilio, 3, 106, 138, 195–96, *198*, 200–201
Frutillar, 147
Futaleufú River, 34–35, *35*, 41–42

Gädicke, Fritz, 68–69
Galladro, Pedro, 88
Galluci, Lisandro, 119
García, Agustín, 60, 62–63
García, José de, 164
garden city movement, 168
Garkovich, Lorraine, 10
gas stations, 192
Gay, Claude, 26–27
Gebhard, Mateo, 129–30
geodesy, 72–73
geographical space: and nation, 2; and spatial history, 8–9
geographic societies, 30
Geographische Zeitschrift (journal), 34
geography: monsters in, 1; and national history, 26; scholarly publishing in, 33–34
Germany, 102; German architecture, 146, 163; German immigrants, 60
Giménez, Adam, 134
Gimnasia y Esgrima, 137
glacial lakes, 7
Goedecke, Otto, 194

gold discoveries, 35–36
Gómez, Indalecio, 124, 129
González, Luis, 135
González, Telémaco, 72
González-Devoto, Josefina, 167
Gonzalorena, Julián, 121
Gorelik, Adrián, 161
government: economic intervention, 13–14; and frontiers, 4, 9; and immigrants, 55, 59–60, 63; and Indigenous peoples, 21; and nationalization, 11, 17; political repression by, 12; and tourism, 18, 183
governors of national territories, 104, 113, 124
grazing permits, 199
Greenberg, Dolores, 107
Greider, Thomas, 10
Guía del Veraneante (magazine), 176, 178, 180–82, 184, 199–200; cover of, *182, 182, 183*
Gutiérrez, Diego, 1
Gymnastics and Fencing, 137

Hansen, Justus, 59
Harambour, Alberto, 77
Harlow, Jean, 184
Harrods Gath y Chaves, 172
health, national, 113–16, 135–39
Hechenleitner, Federico, 66
Hechenleitner, Francisco, 66
Heirenmans, Amadeo, 57, 65, 69
Hess, Fernando, 101
hiking, 137, 195–96
Historical Cypress, 138–39
Hoffmann, Pablo, 57
Holdich, Thomas, 25, 34, 42
Homestead Act of 1884, 70
hotels, 188
Huaiquipan, Francisco, 66
Hube, Federico, 101–3, 150–51
Hube, Jorge, 100
Huber, Otto, 136
Huenchuman, Francisco, 88–90
Humboldt, Alexander von, 20
Hunt, John Dixon, 3

Index

Hydrologic Studies Commission (CEH), 106–8, *108*
hygienic views, 113–16, 136, 139–40
hygienists, 114–15

Ibáñez del Campo, Carlos, 13–15, 177
immigrants: and anarchism, 116; anti-immigrant sentiment, 113, 116, 132–35; and banditry, 122–23; and cities, 103–4, 116; government preferences about, 49–50, 55, 59–60, 63; and Indigenous groups, 51; and land tenure, 48–52; and nationalism, 153. *See also* European immigrants
Inacayal, 32
Indigenous Chileans, 67, 122
Indigenous groups, 4–6; attitudes to, 122; and Desert, 4; genocide of, 28, 206; images of, 157, *157*; and immigrants, 51; land titles of, 50–51, 70, 76–77; and military operations, 21, 23, 26, 28–29, 46–48, 172–73; as museum artifacts, 32; and nationalization, 162, 166–67; native place-names, 38; redistribution of, 48–49; removal of, 5–6, 16, 64–70, 169–70; and settler colonialism, 12, 46. *See also* Mapuche
Indigenous Settlement Commission. *See* Comisión Radicadora de Indígenas
indios chilenos. *See* Indigenous Chileans
industrial settlement, 107–8
inebriation, 115–18, 130
Infamous Decade, 14
Instituto Nacional, 21, 26
Instituto Pedagógico, 21
International Boundary Committee, 25
International Congresses of Modern Architecture (CIAM), 161
Iribarne, Guillermo, 72
Itzigsohn, José, 152

Jesuits, 3–4
Jockey Club, 156
Jones, Jarred, 100
Jönsson, Juan, 54

Jordan, Dorothy, 184
Julio A. Roca statue, 163, 166, 172–73, 205
Justo, Agustín P., 14, 153, 163

Knapp, Reynaldo, 195
knowledge creation, 15–16, 20
Koessler, Rodolfo, 168–69
Kootz, Luther, 172
Kramer, Pablo, 36
Krüger, Pablo, 36

La Alemana, 100–101, 146, *148*
labor strikes, 115, 118–19, 140
La conquista del desierto. *See* Conquest of the Desert
Lake Buenos Aires, 41
Lake Cochrane, 40
Lake General Carrera, 41
Lake Lacar, 37
Lake Llanquihue, 148–49
Lake Nahuel Huapi, 11, 17–18, 48, 199; Chilean influence on, 143–45, 148; Chile-Argentina Company, 82; exploration and development of, 100–101; projected city on, 107; Switzerland trope, 148–49
Lake Pueyrredón, 40
lakes, Andean, 7, 183–84
Lake San Martín/Chacabuco, 31, 41
Lamarckism, 115
Landaeta Sepúlveda, Wenceslao, 178
land concessions, 57–61
Landi, Elisa, 184
land legislation: effectiveness of, 77; and immigrants, 48–52; and Indigenous groups, 50–51, 70, 76–77; and land distribution, 70; Law of 1866, 50–51; and road use, 83, 86, 88–89; and settlers, 47, 49–52, 65, 109
landscapes, 10; and city life, 194; and female imagery, 182–84; national, 144–45, 152; value of, 143; and women's bodies, 184–87, 201
La Nueva Era (newspaper), 117, 121–22, 129–30, 131, 135, 139

La Pampa, 21, 29
La Plata Museum, 29, 31–32
Larrain, Carlos, 87
Las Heras, 38, 40
Latorre, Mariano, 178
Lavinas Picq, Manuela, 46
law enforcement, 113, 119, 121, 127. *See also* policing
Law of 1866, 50–51
Law of Residency (1902), 116
Law of Social Defense (1910), 116
Lekan, Thomas, 142
Lenton, Diana, 5
Lerdo Law, 46
Lesser, Jeffrey, 5
Lezama, Julio, 133
Liberal Reformists, 104, 116, 158
Liga Argentina Contra el Alcoholismo, 118
Liga Nacional de Templanza, 118
Liga Patriótica Argentina, 118–19, 138–40
Lima, 23
Linares, Ildefonso, 72
Linde, Juan, 85–86
Lindh, Jerardo, 61
Lindholm, Víctor, 60
Lista, Ramón, 29
Llanquihue, 5, 7, 17; expropriations in, 57; foreigners in, 49; Indigenous communities in, 46–47; population of, 56, 95; settlement in, 51–52, 54; as tourist destination, 190; wheat production in, 100
Llao Llao Grand Hotel, 171–72, *171*, 192
Locke, John, 12
Lois, Carla, 73, 75
Lolich, Liliana, 161
Longabaugh, Harry Alonzo, 120
lonko, 52
López y Planes, Vicente, 193
Los Lagos, 105, 125–27, *125*

Mackenna, Benjamín Vicuña, 52
Mackenzie, Alister, 172
Magallanes (territory), 46, 51
Magellan, Ferdinand, 1

Maldonado, Daniel, 92
male tourist gaze, 175–76, 178–81, 201
Mange, Pablo, 164
"Mapa geográfico-comercial con la red completa de ferrocarriles de las repúblicas Argentina, Chile, Uruguay y Paraguay," 126, *126*
maps, 15; AGI map, 30–31; *Carta de las comunicaciones postales y telegráficas*, 75; Chile and Argentina, *xviii*; "Chos Malal a Pichachen," 96–97, *97*; grids, 73; Gutiérrez and Cock's map detail, *xxiv*; Lake Llanquihue, *xxii*; Lake Nahuel Huapi, *xxiii*; land concessions, *58, 59*; "Mapa geográfico-comercial con la red completa de ferrocarriles de las repúblicas Argentina, Chile, Uruguay y Paraguay," 126, *126*; map of 1873, 27; and native place-names, 38; northern Patagonian Andes, *xxi*; Pérez Rosales Pass, 146, *147*, 200, *200*; *Plano del territorio de La Pampa y Río Negro*, 28–29; *Plano demostrativo del estado de la ierra pública en los territorios nacionales del Sud*, 73, *74*, 75–76; Provinces and National Territories, *xix*; rivers and streams, 106; road maps, 192; and settlement, 73–77; southern Chile and northern Patagonia, *xx*; and surveying, 76
Mapuche, 4–6, 11, 67, 149; and land legislation, 49–53; military campaigns against, 21, 23, 26; and nation-making, 166–67; removal of, 46–47
Mapuche-Huilliche, 47, 67
Martínez, Carlos Walker, 68
Mascardi, Nicolás, 167–68
masculinity, 113–14, 136
Massey, Doreen, 8, 206
McGee Deutsch, Sandra, 119
Meiling, Otto, 194–97
Menéndez, Francisco, 100–101
Menge, Ana, 85–86
mestizo Chileans, 11, 50, 65, 122
Mexican Revolution, 13

Mexico, 46, 113
military coups, 13–14, 156, 172
military nationalism, 152–53
military operations, 13–14, 70, 153, 166; and cartography, 29; and expropriation laws, 46–47; and Indigenous people, 21, 23, 26, 28–29, 46–48, 172–73
Military Prizes Act, 70–71
mining districts, 190
mining permits, 35–36
Ministry of Agriculture, 198
Minte, Augusto, 79–80, 89, 103
missions, Catholic, 167
Mitre, Bartolomé, 26
modernism, architectural, 161–62
Mödinger, Guillermo and Lorenzo, 89
monsters, 1
Montero, Maximiliano, 129
Montt, Jorge, 64
Morel, Benedict, 115
Moreno, Francisco, 3, 31, 41–43, 98, 106, 138, 158, 166, 173, 193, 204–5; and the Argentine Switzerland, 149; and border negotiations, 20–21, 23–25, 28, 32; and Boys Scouts, 136; burial of, 204; explorations of, 19, 29, 193; and museums, 31–32, 48, 165; and national parks, 107, 154; and railroads, 93–94, 106; and Theodore Roosevelt, 138, 154; and scientific societies, 29–30; and Hans Steffen, 33–34, 36–41
mountaineering clubs, 193–94, 201
Mount Palique, 19, 37
movement of settlers, 80. *See also* railroads; roads
Moyano, Carlos, 41
Muñoz, Froilán, 130, 134
Muñoz Sougarret, Jorge, 57
Münstermann, Gaspar, 85
Museo Nacional de Historia Natural (Chile), 27
Museum of Natural History (Argentina), 31
Musters, George, 29

Nahuel Huapi National Park, 144, 153–54, 159, 177; Alpine style, 162–63; de-Chilenization of, 162, 168–69; and DPN development, 192
Namuncurá, Beatus Ceferino, 167
nation: and state, 22; views of, 2. *See also* nation-making
National Agriculture Society, 56
National Association of Labor, 118
National Education Council, 136
National Geographic Society, 30
national guard, 134
nationalism, 152–53, 155, 160–61, 165, 173. *See also* nation-making
national landscapes, 144–45
National League for Temperance, 118
national museums, role of, 31
National Museum of Natural History (Chile), 27
national parks, 152–59; Iguazú, 144, 153, 162–63, 168. *See also* Nahuel Huapi National Park
National Parks Bureau (DPN), 142–44, 154–73, 177, 192, 196–99
National Snow Queen, 196
national territories, 7, 21, 34; public health of, 115; voting rights in, 116
nation-making: and aesthetics, 142–43, 174; and anti-immigrant sentiment, 113, 133; and border delineation, 22; and border regions, 10, 206; and built environment, 162; and culture, 160, 178; and the Desert, 18; and Indigenous groups, 162, 166–67; and Indigenous place-names, 38; and movement of settlers, 80–81; and national politics, 9; and natural resources, 26; and physical fitness, 113–14; and scientific exploration, 20–21, 43; and spatial history, 8–9, 77; and sports clubs, 194, 197; teleological history of, 166; and tourism, 199; views of, 206–7
nature: feminization of, 176, 179, 181–88, 201; and male tourist gaze, 179–81; and

sports clubs, 194–95; virgin, 180–81; and women's bodies, 184–85
naval arms race, 24–25, 102
Navarro Floria, Pedro, 110
Negro River, 48
Neumeyer, Juan Javier, 195–97
Neuquén, 7, 12, 23, 82, 105; Coronel Barcalá colony in, 73; Indigenous-Chilean migrants in, 122; policing in, 124–27
Newkirk, Pamela, 32
Niklitschek, Walter, 195
Nöbl, Hans, 171, 197
Northern Ice Field, 7
northern Patagonian Andes, *xxi*, 6–9, 10–11, 16–17, 110, 145, 192, 204, 206; aesthetics of, 143; agrarian character of, 11–12, 18, 174, 199; and Chilean/Argentine Switzerland, 143, 149–50; danger in, 119, 121; immigrants and settlers in, 11–12, 17, 80, 131–33, 140; movement across, 2–3, 104; as national landscape, 159, 173, 178; policing in, 124, 128; recreation in, 159, 194–95; roads in, 109; significations of, 18, 110, 152, 206–7; as transnational region, 10, 205
Norwegian immigrants, 59–60
Ñuble-Rupanco Company, 67–69
Núñez, Andrés, 83, 103

Obelisk, 162
Office of Colonization, 66, 73
Office of Hydrography, 27–28
Office of Lands, Colonies, and Agriculture, 73. *See also* Office of Lands and Colonies
Office of Lands and Colonies, 54, 57, 76, 168
Office of Land Surveys, 54, 59
Office of Topography, 28
O'Higgins, Bernardo, 41
O'Higgins Lake, 41
Öhlander, John, 64
oil drilling, 192
Olascoaga, Manuel, 28, 94, 106

organicist views, 114
orography, 34, 37
Ortiz, Roberto, 109, 153
Ortiz Basualdo, Luis, 156
Osorno, 47, 65
outdoors clubs, 193–94
Oyarzun, Luis, 127–28
Oyarzún, Manuel, 85–86

Paichil (Paisil), José María, 72, 170
Paisajes cordilleranos (Elflein), 151
Palena River, 34–36, 35, 41–42
Pamir Mountains, 25
Pampa de Buenuleo, 169
Panama Canal, 92, 102
Paraguay, 124
Parker, Robert LeRoy, 120
parks, national. *See* national parks
Pastor, Luis, 139
Patagonia: challenges to progress, 158–59; early map of, 1; hygienic views of, 113–14; images of, 203–4; Indigenous groups in, 4–5; meanings of, 6–7; and nation-making, 4; postcolonial colonization of, 6; Spanish jurisdiction in, 22; territorial division of, 105–6, *105*; as tourist destination, 17–18; views of, 2–3, 6, 17–18, 112–13
Patagonian Andes, northern. *See* northern Patagonian Andes
Patagonia y sus problemas, La (Sarobe), 158
Patroni, Adrián, 199, 201
peak-water-forest triad, 180–81, 201
Pedersen, Cristian, 55
penal colony, 113
Pérez Colma, Juan Carlos, 136
Pérez Rosales, Vicente, 65–66
Pérez Rosales Pass, 100–101, 145–47, 150, 200
Perito Moreno glacier, 204
Perito Moreno Hotel, 137
Perón, Juan Domingo, 14, 153
Peru, 5, 23
Pescado River, 88

Philippi, Rodulfo, 27
physical education, 136
physical fitness, 113–14, 136
Pichachen, 96
Pinochet, Ricardo del Río, 61
Pinto, Jorge, 48
Pissis, Amadeo, 27
Pius XI, 167
Pizarro, Rafael B., 65–66
Plano del territorio de La Pampa y Río Negro, 28–29
Plano demostrativo del estado de la tierra pública en los territorios nacionales del Sud, 73, 74, 75–76
plant species, imported, 169
Plaza Expedicionarios del Desierto, 163
Policía Fronteriza. *See* Border Patrol
policing, 112–13, 115–16, 123–32, 140
politics, national, 9
polygamy, 52–53
Popular Front, 15, 177
postcolonial colonization, 6
prison abuses, 127–28
privatization of lands, 46, 56–59
Production Development Corporation, 177
profiling, 130
public disorder, 112–14, 116–19
public health, 113–16, 135–39
Pucón, 182
pueblos originarios, 12
Puelche, 21
Puerto Montt, 5–6, 12, 47, 49–50, 57, 147–48
Puerto Varas, 5, 47, 147
Puntiagudo volcano, 180, 182
Puyehue Pass, 100, 199

Radical Civic Union, 13–14
railroads, 12, 17, 80–81, 90–95, 155; and Andean valleys, 82, 94, 106, 108–9, *108*; development of, 90, 110; and settlers, 92; trans-Andean, 94–95, 98–99, 102; and vacation travel, 188
Ramos, Manuel, 128
Ramos Mexía, Ezequiel, 82, 106–7, 116, 158–59

ranching, cattle, 4, 12, 99–100, 198
rastrilladas, 80
Rawson, Franklin, 94–96, 98–99
recreational activities, 194
Rehren, Otto, 55, 63
Reiche, Carlos, 36
Revolution of May 25, 1810, 160
Riesco, Germán, 68
Río Negro, 7, 12, 23, 29, 70, 82, 104–6
Ríos, Arturo, 135
roads, 16–17, 80–90, 109–10; and car tourism, 191–92; communal/private, 84–86; condition of, 83–85; conflicts on, 86–90; costs of, 97–98; kinds of, 81; and land legislation, 83, 86, 88–89; rights of use of, 85–86; and settlement, 80–81, 85, 109; state-built, 83–84; trans-Andean, 95–98
Roca, Julio A., 25, 48, 100, 142, 158, 163, 165–67, 172–73, 205
Roca, Julio A., Jr., 14, 153
Romero, Luis Alberto, 13
Roosevelt, Theodore, 136–38, 145, 150, 154, 204, *204*
Roth, Ricardo, 103, 200
Royal Geographical Society, 30, 34
Royal University of San Felipe, 27
Ruggiero, Kristin, 112
Rupanco Colonizing, Farming, and Cattle-Breeding Company, 16, 65, 67–69
Rural Code of the National Territories (1894), 118, 120
rustlers, 131

Saavedra, Juan María, 72
Sáenz Peña Law (1912), 116
Saldivia, Candelaria and Genaro, 66–67
Sales, Manuel, 111–12
Salessi, Jorge, 114
San Antonio, 109
San Carlos de Bariloche, 100
San Martín Cattle-Breeding Company, 104
San Martín de los Andes, 37
San Ramón, 147
Santiago, 26, 190

Santos Bascuñar, José, 128
Sarmiento, Domingo, 26
Sarobe, José María, 158; *La Patagonia y sus problemas*, 158
Sayhueque, Valentín, 70, 138
Schley, Winfield, 24
Schulten, Susan, 10
scientific institutions: in Argentina, 28–32; in Chile, 25–28
scientific knowledge, 20; and border negotiations, 23–25, 42–43; and conflicting observations, 25; and liberal order, 20; and military operations, 22; and nation-making, 20–21, 43
scientist-explorers, 20–21, 25
Sentinel Island, 204–5
Serrano, Pedro, 136–37
Serrano Montaner, Ramón, 38
servidumbre, 85
settlement, 2, 45–77; and Christian morality, 52–53; costs of, 60–61; and the Desert, 45–46; economic policy of, 104; and environmental knowledge, 61–64; estate expropriation, 56–59; and foreign colonists, 49; and health challenges, 56; images of, 157, 157; and immigration policy, 49–52; and inspection, 54–56; and land concessions, 57–61; landed property legislation of, 47, 49–52, 65, 109; land tenure disputes over, 51–52, 65–66; mapping of, 73–77; and Mapuche communities, 46–47, 49–53, 67; populations in, 56, 109; privatization of, 49; records of, 54–55; relocation of, 55; removals of, 64–70; requirements for, 59–60; and roads, 80–81, 85, 109; and surveying, 71–72, 76
settler colonialism, 12, 16, 46. *See also* colonization; settlement
sheep farming, 198
shipping times, 75
shooting clubs, 113–14, 136, 138
Silvestri, Graciela, 161
skiing, 195–97
Snead, James, 81

Sobral, José María, 192
social unrest, 119, 140
Sociedad Científica Argentina (SCA), 29
Sociedad de Fomento Fabril, 56
Sociedad Esplotadora de Río Bravo, 57
Sociedad Nacional de Agricultura, 56
Society for Industrial Development, 56
Southern Ice Field, 7
Southern Land Company, 104
Southern Railway Company, 93
space: geographical, 2, 8–9; and landscape value, 143; and male tourist gaze, 179, 201; nationalist views of, 165, 173; ordering of, 86; re-signification of, 197–202; as subjective trajectories, 206; Switzerland trope of, 148–49; trans-Andean, 150–51, 199–200. *See also* spatial discourses
Spain, 3–4
Spanish immigrants, 62–63
spatial discourses, 8–9, 204–6; hygienic views as, 114; and nation-making, 8–9, 77; and social control, 112, 114
spatial history, 8–9, 77, 203–7
sports clubs, 137, 193–95
Stange, Pablo, 36
state: and nation, 22; and science, 27. *See also* government
State Oilfields Company (YPF), 191–92
State Railway, 93
steamboat proposal, 101–2
Steeger, Francisco, 85–89, 92, 109
Steffen, Hans, 20–21, 33–41, 43
Steffen Sound, 38
Stegmaier, Alberto, 68
Stepan, Nancy, 115
Strait of Magellan, 3, 24
Sundance Kid, 120, 122
surveying, 15, 20; border, 23, 27; funding of, 29, 33; and land grants, 71–72, 76; maritime, 27
Swedish immigrants, 64
Switzerland trope, 148–52, 173

Tauschek, José, 100–101
Tehuelche, 1, 4, 21

Tellez, Ramón, 85–86
temperance movements, 118
Temuco, 54, 199
Teski Club, 195
Tierra del Fuego, 113
Tiro Federal, 135–38, 140
Toledo, Natalia, 127
Toltén River, 47
topography, 23–24, 34–37, 42
toponymy, 38–41
Tornero, Juan, 57, 59
Torrontegui, Juan, 132
tourism, 143–44, 163–72, 175; and democracy, 191; and government, 18, 183; and nation-making, 199; and sports clubs, 194–95
tourist gaze, 17, 144, 175–76, 201–2; male, 175–76, 178–81, 201
touristscapes, 182–84, 199–200
trade routes, 17, 104
Tragic Week, 118
trails, hiking, 195–96
trans-Andean trade, 81–82, 95–103, 143, 200
trans-national region, 9
transnational space, 150–51, 199
transportation, 83, 185, 200
travel, 188–92, 201
travel publications, 179, 182, 188, 201
Treaty of 1881, 23, 33, 37
Treaty of Friendship, Peace, and Navigation (1856), 22
Tres Montes Peninsula, 84
Turismo Austral (magazine), 200–201
Turismo en la provincia de Llanquihue a través de la Suiza chilena y argentina (Wiederhold), 151
Turismo en las provincias australes de Chile (Gerike, Manriquez, and Thies), 150
Tutzauer, Herbert, 196
Tuza, Pablo, 60–61

Unión Cívica Radical, 13–14
United States, 154
University of Chile, 21, 26–27

Uriburu, José Félix, 14, 153
Urmeneta, Errázuriz, 63
Urry, John, 17, 144, 175
USS *Baltimore*, 24
uti possidetis iuris, 22

Valdivia, 46–47, 49, 51, 54, 57, 190
Valdivian Forest, 8
Vallmitjana, Ricardo, 3
Vega, Nicolás, 55
Velázquez, Pedro, 127
Vereertbrugghen, José, 131–35
veterinary offices, 198–99
Vicuña, Francisco Rivas, 68
Vidal Gormaz, Francisco, 27
Villa Catedral, 171
Villa La Angostura, 72, 170
Villarrica, 47
villas, tourist, 168–72
virgin forests, 179
vom Hau, Matthias, 152
voting rights, 139

Wakild, Emily, 153
War of Occupation (1881–83), 47–48
War of the Pacific (1879–84), 5, 24, 177
War of the Triple Alliance, 24
watershed divide, 37
Weber, Alfredo, 63, 126, 126
Wecker, Alberto, 36
Welsh immigrants, 34–36, 42, 159
wheat production, 100
Whiteside Toro, Arturo, 55
Wiederhold, Carlos, 92, 100–101, 103, 146, 151–52, 161–62, 164; store of, 148
Wiederhold, Germán, 151; *Turismo en la provincia de Llanquihue a través de la Suiza chilena y argentina*, 151
Willis, Bailey, 106–8
Wilson, William, 120, 124
women: feminization of nature, 176, 179, 181–88, 201; fertility, 183; social roles, 187–88; and touristscapes, 182–84
Woodhouse, Jorje, 59

World War I, 12–13, 102, 104, 118–19, 177, 200
World War II, 187, 191

Yacimientos Petrolíferos Fiscales (YPF), 191–92
Yelcho River, 34
yellow fever, 114

Yrigoyen, Hipólito, 13–14, 118–19, 138, 153, 156

Zañartu, Sady, 178
Zeballos, Estanislao, 24–25, 30
Zorrilla, Benjamín, 76

www.ingramcontent.com/pod-product-compliance
Lightning Source LLC
Chambersburg PA
CBHW032217230426
43672CB00011B/2587